VIRUSES

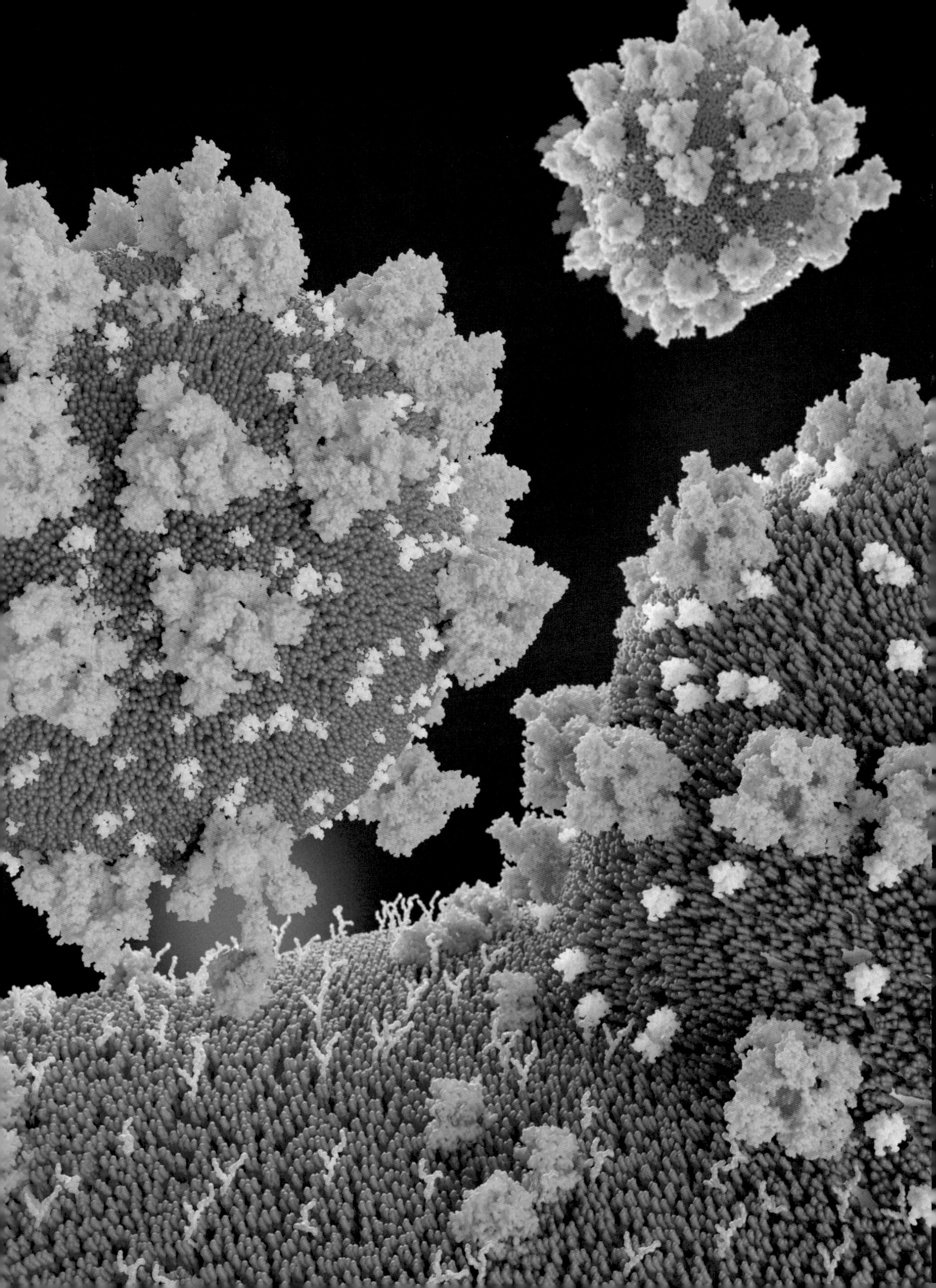

VIRUSES

A NATURAL HISTORY

Marilyn J. Roossinck

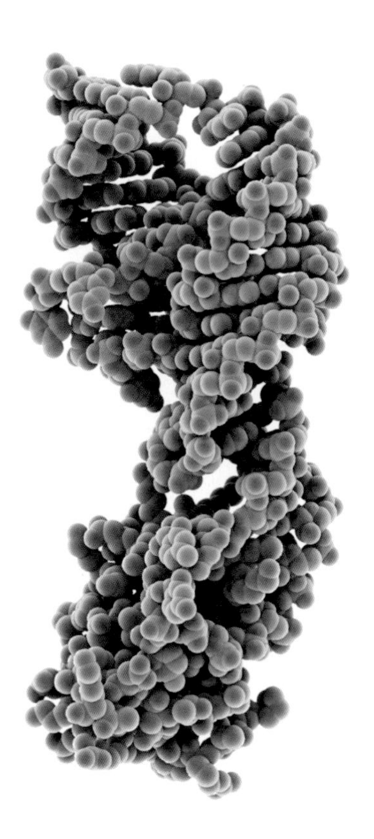

PRINCETON UNIVERSITY PRESS
PRINCETON AND OXFORD

Published in 2023 by Princeton University Press
41 William Street, Princeton, New Jersey 08540
99 Banbury Road, Oxford OX2 6JX
press.princeton.edu

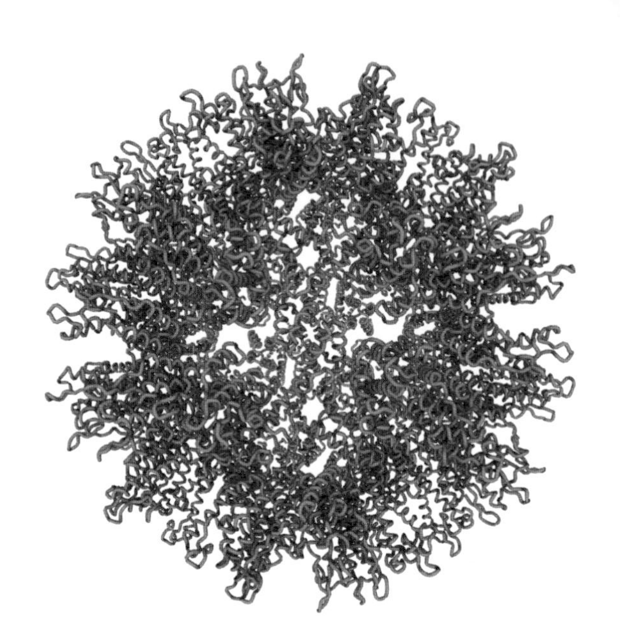

Library of Congress Control Number 2022945508
ISBN 978-0-691-23759-6
Ebook ISBN 978-0-691-24080-0

Typeset in Bembo and Futura

Printed and bound in China
10 9 8 7 6 5 4 3 2 1

British Library Cataloging-in-Publication Data is available

This book was conceived, designed, and produced by
UniPress Books Limited
Publisher: Nigel Browning
Commissioning editor: Kate Shanahan
Project manager: David Price-Goodfellow
Art direction: Wayne Blades
Design: Lindsey Johns
Cover designer: Wanda España
Copy editor: Susi Bailey
Picture researcher: Sharon Dortenzio
Illustrators: Caitlin Monney (Monney Medical Media), Tejeswini Padma, John Woodcock, Martin Brown, and Lindsey Johns
Maps: Les Hunt

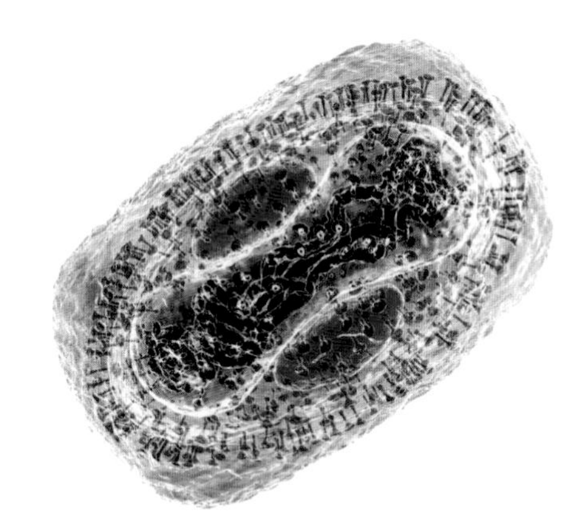

Cover images: (Front cover) False-colour TEM of virions of influenza virus by NIBSC; (back cover and spine) Chikungunya virus particle, illustration by Ramon Andrade 3Dciencia. Both images courtesy of Science Photo Library.

Previous page, main image: Artist's impression of SARS-CoV-2 viruses binding to receptors on the cell surface.

Title page: Hepatitis deltavirus ribozyme complex.

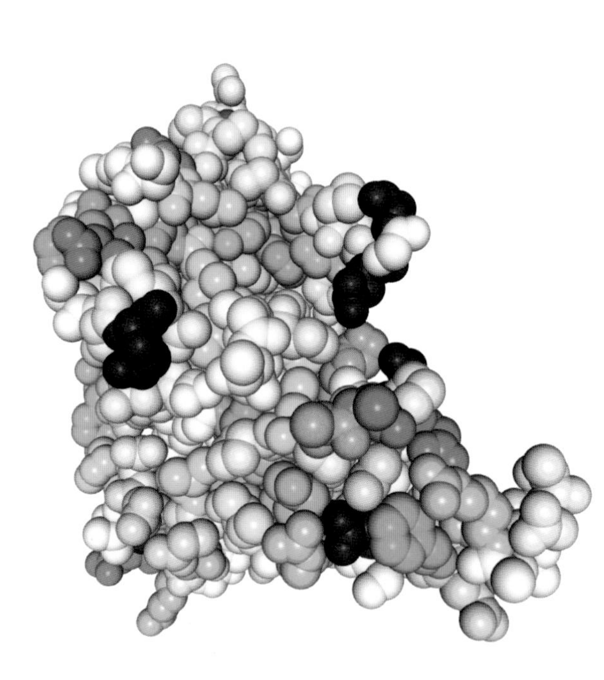

Author Acknowledgments
Writing this book has been a wonderful opportunity for me to present my favorite things in the world: viruses! I would like to thank my colleagues at the Center for Infectious Disease Dynamics at Penn State University for numerous helpful discussions on key points in the book, and the team at UniPress for their efforts on the book's organization, layout, and illustrations, which together make it so attractive.

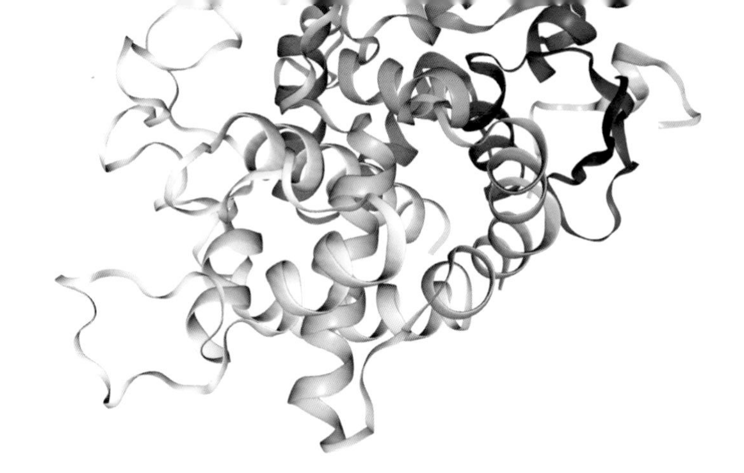

CONTENTS

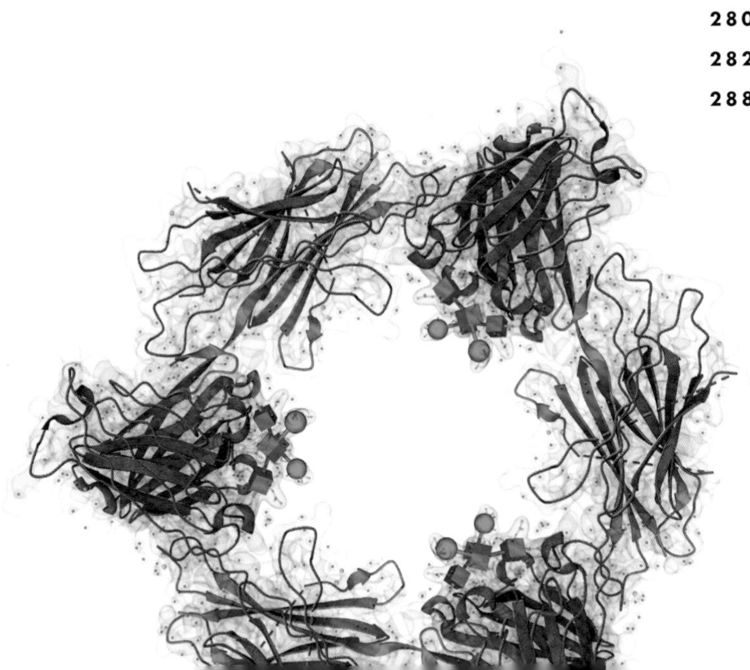

INTRODUCTION

What are viruses?

Since the emergence of severe acute respiratory syndrome coronavirus 2 (SARS-CoV-2) in late 2019 and early 2020, the world has become very aware of viruses and how much they can impact our lives. The news, and even the field of virology, became overwhelmed by details of this virus, which causes coronavirus disease 2019 (COVID-19). While the effects of SARS-CoV-2 have been felt worldwide, this is just a tiny part of the story of viruses. This book will take you on a fascinating journey beyond COVID-19 and into the realm of the most diverse entities on Earth.

Finding a definition that fits all of the different types of viruses is difficult. The *Oxford Learner's Dictionary* defines a virus as "a living thing, too small to be seen without a microscope, that causes disease in people, animals and plants." However, even the very first phrase here—"a living thing"—is controversial (see below). Viruses infect more than just people, animals, and plants; in fact, they infect every life-form known, and most of them probably don't cause disease. Finally, this definition doesn't distinguish viruses from bacteria.

The *Oxford English Dictionary* has a slightly different definition: "An infectious, often pathogenic agent or biological entity which is typically smaller than a bacterium, which is able to function only within the living cells of a host animal, plant, or microorganism, and which consists of a nucleic acid molecule (either DNA [deoxyribonucleic acid] or RNA [ribonucleic acid]) surrounded by a protein coat, often with an outer lipid membrane." We are making some progress here, although many giant viruses are larger than some bacteria, and not all viruses have a protein coat.

There are a few features that all viruses do have in common: they have genomes of RNA or DNA, they require a host for all of their functions, they may carry the genetic material for many sophisticated functions, and they cannot generate their own energy. An ongoing discussion concerns whether viruses are alive or not. When they were first discovered it was assumed that they were alive, but when tobacco mosaic virus was made into a crystal in 1935, some thought viruses were more like a chemical than a life-form. Some have suggested that viruses are alive when they are infecting a host cell, and that they are more like seeds or spores when they are outside of a cell.

In short, there is no simple answer to the question "Are viruses alive?" There have been many arguments on both sides, but rarely by virologists. In general, virologists find their favorite entities fascinating, and whether they are alive or not has little relevance because they certainly impact the lives of everything on Earth.

→ High-resolution, cryo-EM structure of Zika Virus.

What are cells?

A cell is the basic unit of life. There are two types of cells: prokaryotic and eukaryotic (see diagram below). Prokaryotic life includes bacteria and archaea, which mostly comprise single cells, although some can form multicellular structures. Eukaryotic cells include everything else.

The basics of cellular life

All life is made up of cells, which are either prokaryotic or eukaryotic. Bacterial and archaeal cells are prokaryotic, meaning they lack a nucleus and generally have a wall surrounding them. Animal and plant cells are eukaryotic, meaning they have a nucleus that houses the organism's genome. Animal cells do not have cell walls, but most other eukaryotic cells do. Structures within eukaryotic cells are called organelles and are surrounded by their own membranes. Mitochondria in most eukaryotic cells and chloroplasts in plant cells generate the cells' energy and are derived from ancient bacteria. They have their own DNA genomes, but they cannot survive independently. Cells are shown as average sizes for the type of cell, but the size actually varies enormously. The largest known cell by volume is the ostrich egg.

↓ Artist's visualization of a short piece of DNA double helix.

BACTERIAL CELL (PROKARYOTIC)

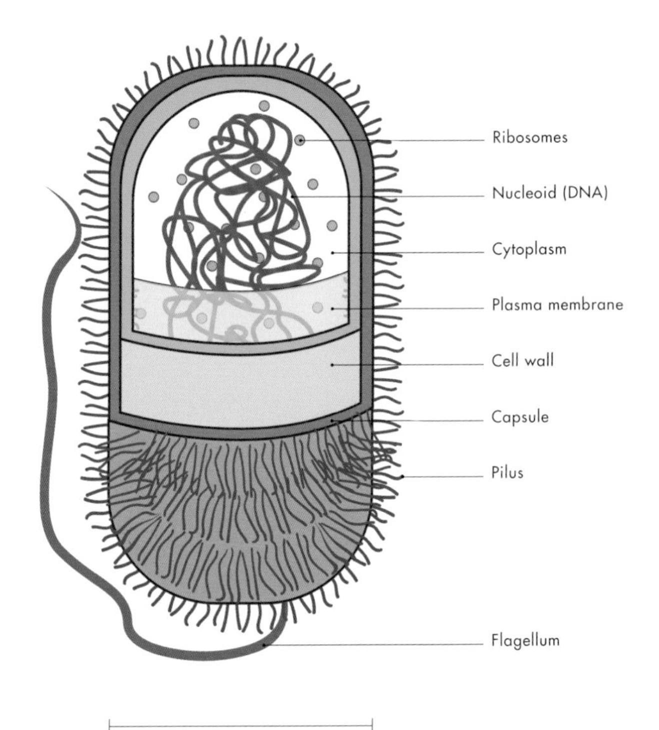

ANIMAL CELL (EUKARYOTIC)

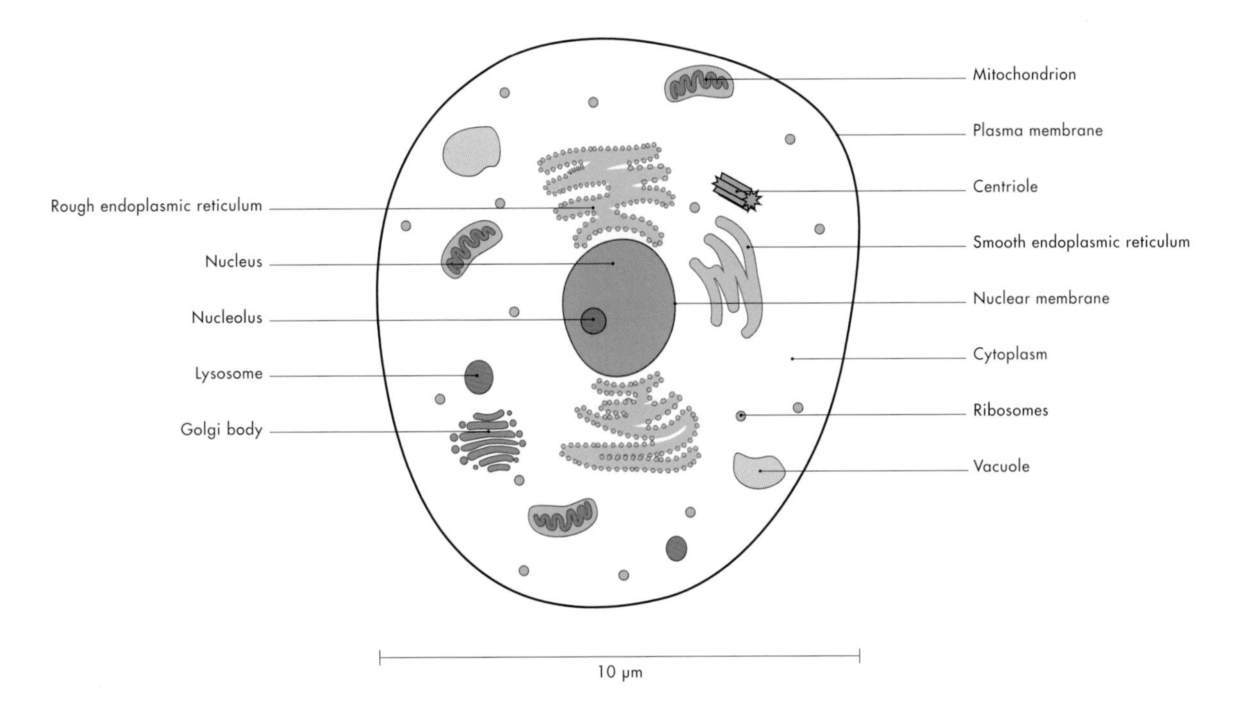

PLANT CELL (EUKARYOTIC)

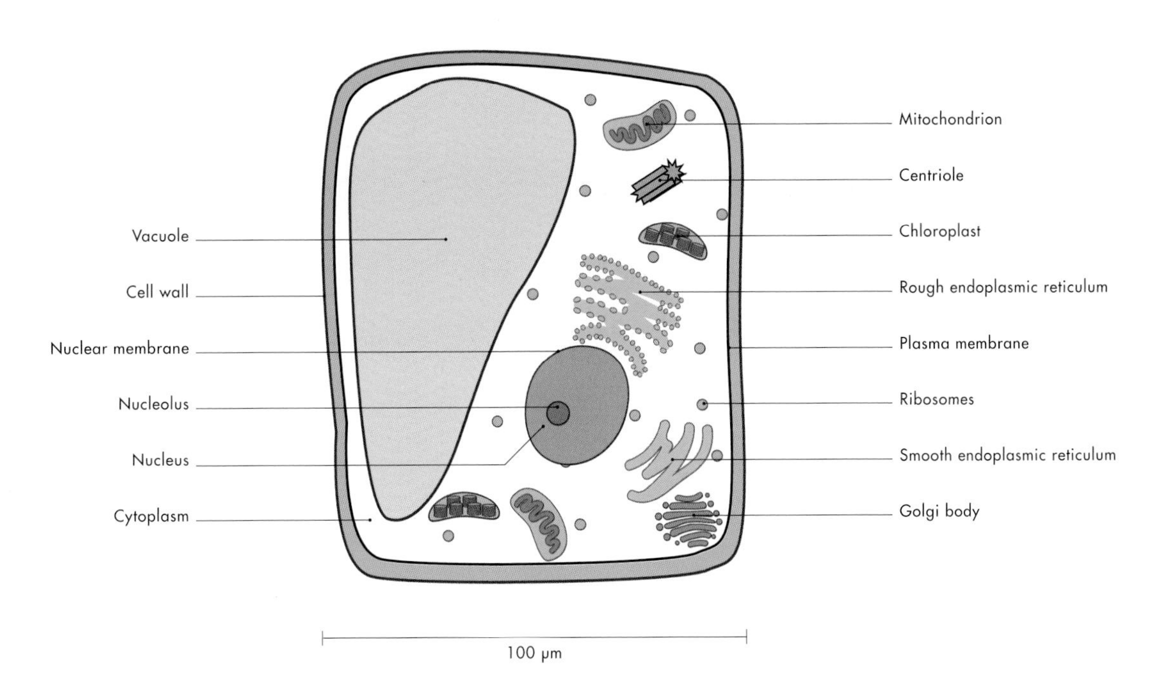

DNA and RNA

Our genomes, and the genomes of all life that has cells, are made of DNA, long chains of deoxyribonucleotide bases, of which there are four (see page 66). The genome is a type of code containing all the information that is needed to direct the cell to make proteins. Proteins are chains of amino acids, and each amino acid uses three nucleotides called a codon for its code (see table). The parts of the genome that have the codes for proteins are called the coding regions.

Since there are four nucleotides, and 22 amino acids that need to be coded for, plus a code for stopping translation, there is usually more than one codon for each amino acid as there are 48 codon combinations. The table shows the first, second, and third nucleotide for each codon (abbreviated as U, C, A, and G), and which amino acid this code tells the translation machinery to insert when the protein is being made.

THE GENETIC CODE

Each amino acid in the table is represented by a three-letter abbreviation. For example, Ser stands for serine, Leu stands for leucine, and His stands for histidine.

	SECOND BASE				
FIRST BASE	U	C	A	G	**THIRD BASE**
U	UUU Phe UUC Phe UUA Leu UUG Leu	UCU Ser UCC Ser UCA Ser UCG Ser	UAU Tyr UAC Tyr UAA STOP UAG STOP	UGU Cys UGC Cys UGA STOP UGG Trp	U C A G
C	CUU Leu CUC Leu CUA Leu CUG Leu	CCU Pro CCC Pro CCA Pro CCG Pro	CAU His CAC His CAA Gln CAG Gln	CGU Arg CUC Arg CGA Arg CGG Arg	U C A G
A	AUU Ile AUC Ile AUA Ile AUG Met	ACU Thr ACC Thr ACA Thr ACG Thr	AAU Asn AAC Asn AAA Lys AAG Lys	AGU Ser AGC Ser AGA Arg AGG Arg	U C A G
G	GUU Val GUC Val GUA Val GUG Val	GCU Ala GCC Ala GCA Ala GCG Ala	GAU Asp GAC Asp GAA Glu GAG Glu	GGU Glu GGC Gly GGA Gly GGG Gly	U C A G

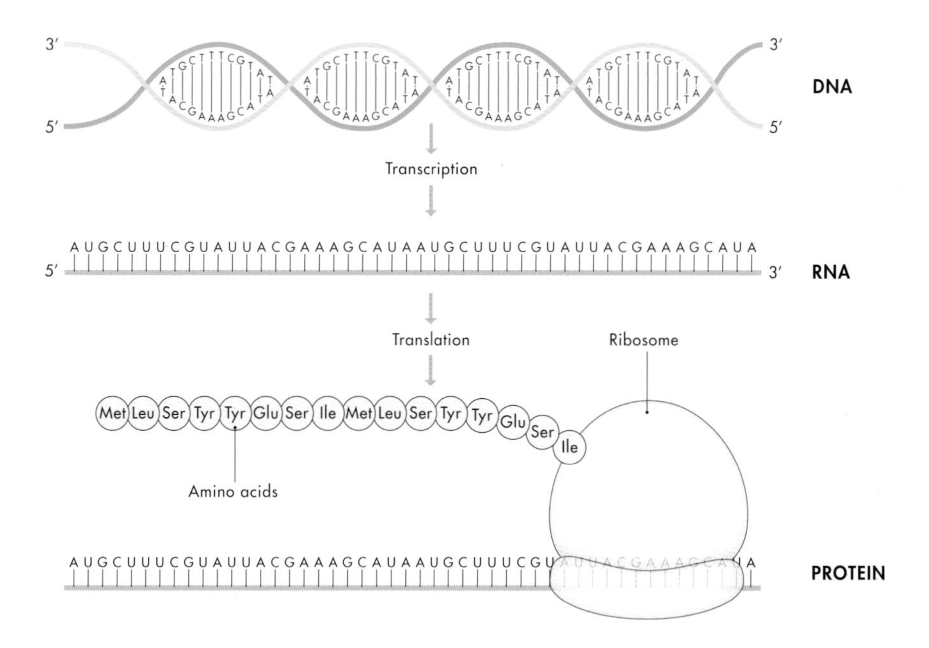

DNA to RNA to protein

DNA is the genetic material of all life that has cells. In a cellular genome it comprises two long strings of complex sugar molecules, each with a nucleotide base attached. There are only four bases in DNA: adenine (A), cytosine (C), guanine (G), and thymine (T). Each base pairs with a complementary base—A with T, and C with G—so that the two strands are also complementary and together form a double helix. DNA is transcribed into RNA, and the RNA in turn carries the code for proteins. RNA has a very similar structure to DNA, but the thymine base is substituted for uracil (U). This central dogma of molecular biology—DNA to RNA to protein—held up until 1970, when two independent American scientists, David Baltimore and Howard Temin, discovered a new enzyme made by viruses that can convert RNA into DNA. In viruses DNA genomes can be single-stranded.

There are many other elements in DNA that do not code for proteins but are important in regulating when and how proteins are made. In fact, there is a lot more of the non-coding DNA in the genomes of most cells than there are coding portions—for example, the coding part of the human genome is only about 1.5 percent of the total genome. The purpose of much of this non-coding DNA is currently still unclear.

In contrast, virus genomes can be made of either DNA or RNA. What's the difference? Chemically, a DNA base contains one less oxygen atom (hence "deoxy-") than an RNA base. Biologically, this small change can make a big difference: different enzymes are used to copy the different bases, they have different structures, and RNA has a lot more biological activity beyond just coding for genes. RNA can act as an enzyme itself, and it is part of a lot of the complex machinery inside cells, such as the ribosomes that translate genes into proteins. A major difference between the genomes of viruses and cellular life is that most viruses contain very little RNA or DNA that is non-coding.

The genomes of all cellular life are double-stranded DNA. In eukaryotes they are linear, but in bacteria and archaea the genomes are often circular. In viruses the genomes can be DNA or RNA, single-stranded or double-stranded, and either linear or circular. While all cells use a lot of single-stranded RNA to carry out various functions, double-stranded RNA is unique to viruses, other than very small molecules. Most cells recognize double-stranded RNA as something foreign, and this can trigger an immune response (see chapter starting on page 160). Viruses with double-stranded RNA genomes have evolved ways to hide their genomes from the cells they infect.

How viruses are named

The first level of virus classification is often called the Baltimore scheme, named after American biologist David Baltimore. In this scheme viruses are put into seven categories based on their genome type (see below and table on pages 34–35). Different classes of viruses infect different hosts.

Virus names are usually first assigned by the discoverer, and later elaborated or approved by the International Committee on Taxonomy ofViruses (ICTV).

The names of plant viruses usually include the name of the first host where the virus was isolated and the symptoms it produces, an example being banana streak virus, which induces yellow streaks in the leaves of banana (see page 54). In contrast, the names of human viruses often include the organ where the virus is found, such as the hepatitis viruses found in the liver and the rhinoviruses that infect the upper respiratory tract. The names of fungal viruses include the Latin genus and species names of the host, such as Saccharomyces cerevisiae virus L-A,

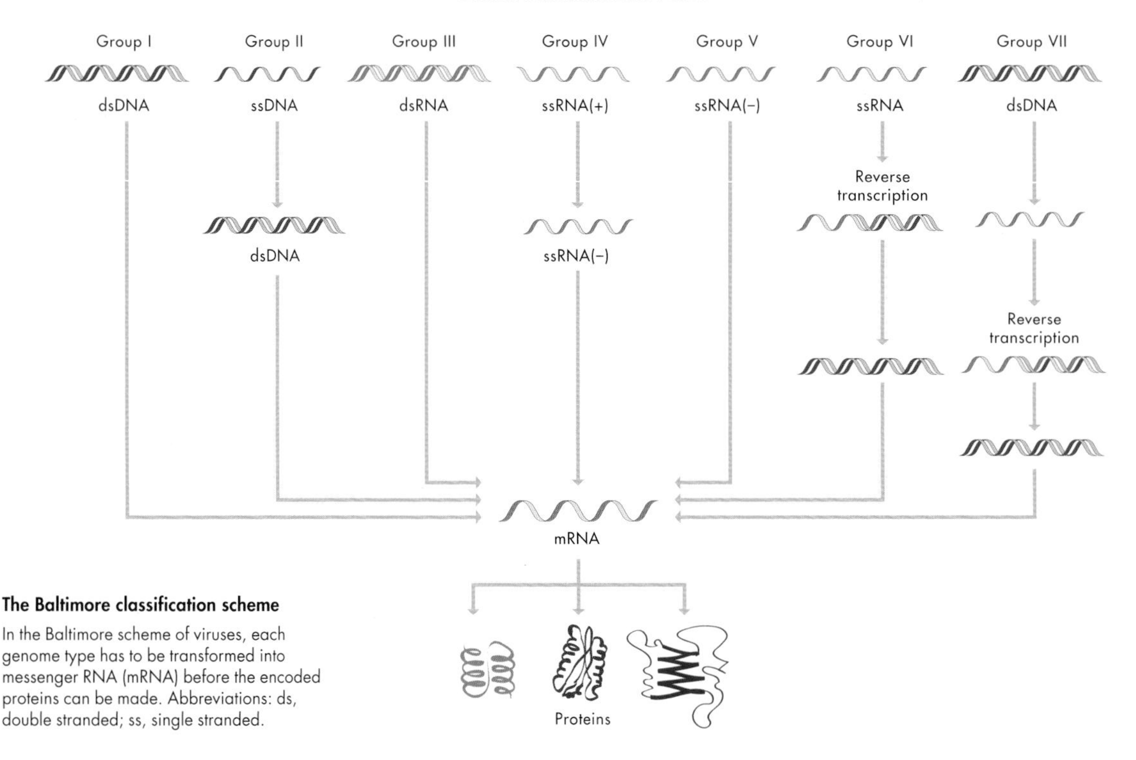

The Baltimore classification scheme

In the Baltimore scheme of viruses, each genome type has to be transformed into messenger RNA (mRNA) before the encoded proteins can be made. Abbreviations: ds, double stranded; ss, single stranded.

which infects yeast (see page 246). Virus names can be confusing because viruses are not always discovered in their natural hosts, and they may infect many other hosts. For example, cucumber mosaic virus infects about 1,200 plant species, but not most modern cucumber cultivars, which are resistant to it. The virus profiles in this book use the names from the 2020 ICTV report. The abbreviations for viruses are given in many instances, but it should be noted that these are not always used in the same way by different virologists, and more than one virus may have the same abbreviation. For example, rous sarcoma virus and respiratory syncytial virus, both profiled in the book, use the abbreviation RSV.

Taxonomic classification of viruses differs from that of cellular life-forms in a couple of ways. First, the highest level of classification for viruses is realm, as opposed to domain in cellular life. The remaining levels are the same. And second, in viral taxonomy all the levels of classification are written in italics, whereas in other taxonomy only the genus and species names are written in italics. Although the use of latinized names for viruses is now generally accepted, the rules around when italic type should be used and when it should not vary. To avoid confusion, all virus names in this book are therefore written in roman type. Common names are also included in the descriptions where readers may be more familiar with these.

REPLICATION OF VIRUSES

The most important function of viruses is to replicate, to make more copies of themselves to infect other host cells and other hosts. The details of this process vary depending on the type of genome the virus has, and the type of host it infects. Details of this process will be described in "Viruses making more viruses" starting on page 62. This chapter is the most technical chapter in the book, and is provided for those who want to take a deeper dive into how viruses work.

→ Tobacco mosaic virus causes a pattern of light and dark green on the leaves of infected tobacco plants. The virus is concentrated in the light green areas of the leaves.

Do viruses have colors?

No virus found to date makes pigments. Pigments are biologically costly to make, and they always have a specific purpose in biology, such as attracting mates or deterring predators. Viruses don't have any need for color, so they are colorless, with the exception of the iridoviruses. The latter are large by virus standards, and they have thousands of facets in their capsid structure that reflect light, creating iridescent colors that can sometimes be seen in the infected hosts (see the image below).

Although most viruses are colorless, they may have a dramatic effect on the color of their host by affecting the pigments their host makes. For example, many stripes or mottles in flowers and leaves are caused by viruses disrupting the genes that make pigments, and viruses also affect pigment production in fungi.

Most of the pictures of viruses in this book are generated by computers using complex data, with the color added to make it clearer to see certain features. The latest methods use cryo-EM, which is a type of electron microscopy (EM) where the samples are flash-frozen and imaged in the frozen state. This is a big advance over older methods, where the samples had to be chemically fixed, often causing their structure to change. With cryo-EM, thousands of individual images are merged to produce a very detailed structure.

Another way to look at virus structure is through x-ray crystallography. Viruses are quite easy to make into crystals because they usually have very regular shapes. When a beam of x-rays passes through a crystal, the rays are diffracted in different directions based on the molecular structure within the crystal. The diffractions are then translated into a structure by computer programs. Some of the images in this book were generated in this way.

Iridovirus structure

Iridoviruses have so many capsids that they reflect different colors of light, much like the way a butterfly can look iridescent thanks to the tiny scales that cover its wings.

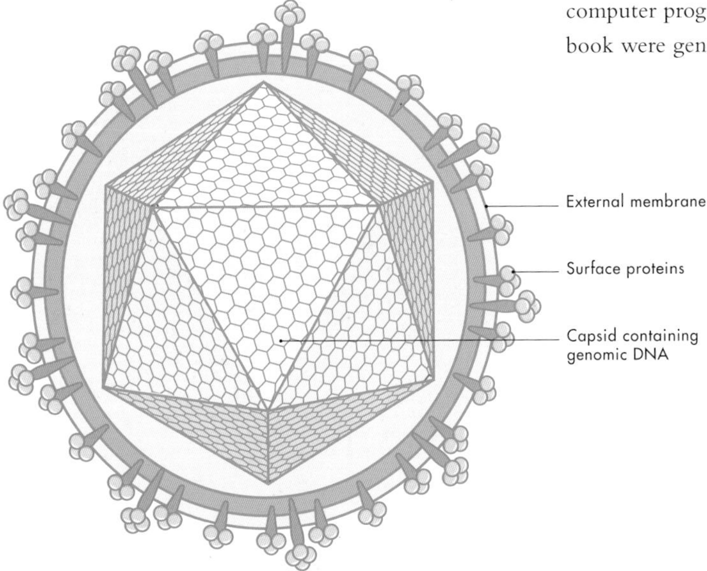

→ Tulips infected with tulip breaking virus have stripes. In the seventeenth century the Dutch were so enamored with these beautiful flowers that infected tulip bulbs resulted in "tulipomania" in Holland. However, because the virus was sometimes lost when the tulips were propagated, the colors were unstable.

The history, and future, of virology

The first hint that an agent other than bacteria or fungi was causing infections came in 1892, when the Russian biologist Dmitri Ivanovsky (1864–1920) demonstrated that a mosaic disease in tobacco plants could be transmitted by the sap of the plant. He concluded that there was a poison in the sap. In 1898, Dutch microbiologist Martinus Beijerinck (1851–1931) passed the sap of mosaic tobacco plants through a fine porcelain filter that could exclude bacteria, and found that the filtered sap was still infectious. He concluded that there was an infectious agent smaller than bacteria in the plant sap, and called it a living contagious fluid. Beijerinck later used the word virus for the agent, from the Latin word meaning "poison."

→ A large model of tobacco mosaic virus, designed by English chemist Rosalind Franklin, was displayed at the 1958 Brussels World's Fair. Here it is seen under construction.

Later that same year, German bacteriologists Friedrich Loeffler (1852–1915) and Paul Frosch (1860–1928) showed that the infectious agent for foot and mouth disease was also a filterable virus, and the field of virology was born. By 1901, US Army doctor Walter Reed (1851–1902) had demonstrated that the agent for yellow fever was also a virus, and in the next decade leukemia and solid tumors were shown to be transmissible by viruses in chickens. In 1915 bacterial viruses were discovered by two independent scientists.

Viruses were pivotal in many major advances in biology. The basic components of tobacco mosaic virus were shown to be RNA and protein, and its structure was seen in an electron microscope in the 1930s. The ability of plant viruses to mutate was also discovered in the 1930s, and this was demonstrated for bacterial viruses in the 1940s. In the 1950s, English chemist Rosalind Franklin (1920–1958) made a detailed structural model of tobacco mosaic virus using x-ray crystallography, a technique she later used to show the structure of DNA. This led to the discovery of RNA as a genetic material. Viruses were also used to decipher the genetic code.

Viruses contributed many fundamental tools for the study of molecular biology throughout the twentieth century. The first enzymes for determining the sequence of DNA were isolated from viruses and many of the tools for DNA cloning came from viruses.

THE FUTURE OF VIROLOGY

The beneficial roles of viruses for life on Earth (see chapters starting on pages 194 and 220) are only just becoming clear, and this is an area that should receive a lot of attention in the next decades. With the increases in technology that allow the discovery of more and more viruses (see chapter starting on page 26), scientists will uncover many examples of viruses that do not cause disease.

The overwhelming global effects of the COVID-19 pandemic have made it clear that more energy needs to be placed on understanding how viruses emerge to cause severe disease (see chapters starting on pages 160 and 248). Better surveillance methods are also critical, so that potential pandemics can be stopped in their tracks. COVID-19 has also stimulated the development of new technology for vaccine research, and highlighted how much more is still needed for understanding the immune response and for developing durable vaccines. Development of treatments for virus diseases is also very important (see "The battle between viruses and hosts," page 161).

In the coming decades we can look forward to more virus-based technology, mitigating the issues of antibiotic resistance, providing gene delivery to treat genetic diseases, and giving us better tools for understanding the planet and our relationships with the environment.

← Martinus Beijerinck (1851–1931) was a Dutch microbiologist who is best known for his early work on viruses. His experiments showed that the mosaic disease of tobacco was caused by an infectious agent smaller than any known bacteria. He coined the term "virus" to describe this agent, which he considered to be a "contagious poison." Beijerinck had another critical role in agricultural microbiology: he discovered that bacteria colonizing the roots of legumes (beans, lentils, peas, etc.) could "fix" nitrogen. Nitrogen is abundant in the air, but it cannot be used by plants in this form. The bacteria convert nitrogen to a form that plants can use. Native American farmers already knew this indirectly because they grew their corn and beans together. The beans provided excess nitrogen through the bacteria in their roots, and the corn stalks provided a support for the vining bean plants. Nodules in a legume root, where the nitrogen-fixing bacteria reside, are shown in the image above. In some cases, infection of the host plant by a virus can reduce the size and abundance of these nodules.

→ Rosalind Elsie Franklin (1920–1958) was a British scientist who studied chemistry and x-ray crystallography. She is best known for her work on the structure of DNA, for which she received little credit during her lifetime. One of her discoveries in her DNA work was the A and B forms of the DNA double helix (see pages 34–35). X-ray crystallography is a very powerful tool to determine the structure of large molecules such as nucleic acids and proteins. The molecules are crystallized and an x-ray is passed through the crystal, producing a diffraction pattern that can be interpreted to reveal the structure. Franklin applied this technique to determine the structure of viruses, and since her time it has been used to establish many virus structures, some of which are shown in this book. You can see the diffraction pattern for tobacco mosaic virus (TMV) from Franklin's work below; it may not look like much to the untrained eye, but it allowed the scientist to build a model of the virus that was displayed at the Brussels World's Fair in 1958 (see page 19). Franklin died very young, and it was only after her death that the critical role she played in determining the DNA structure was recognized.

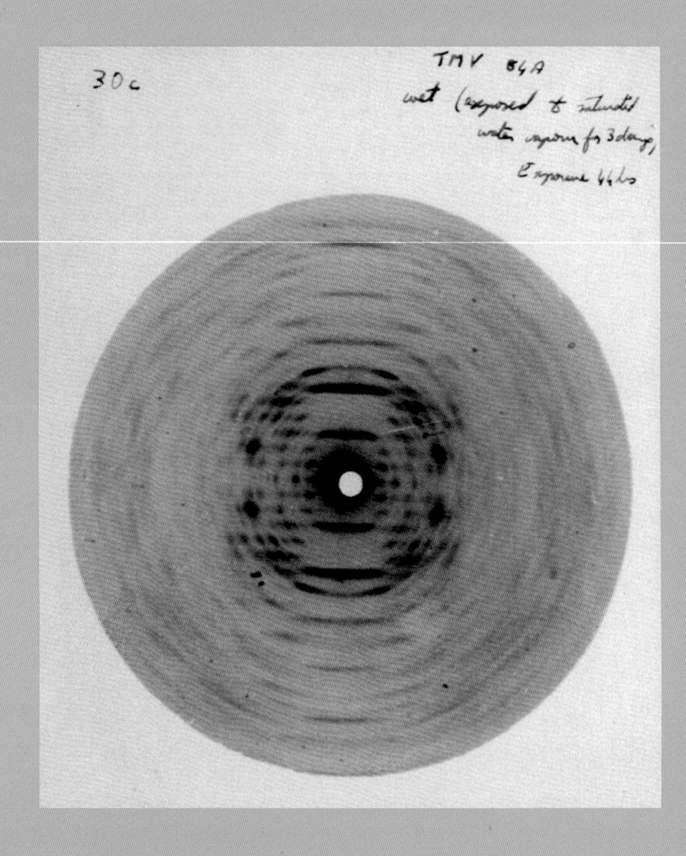

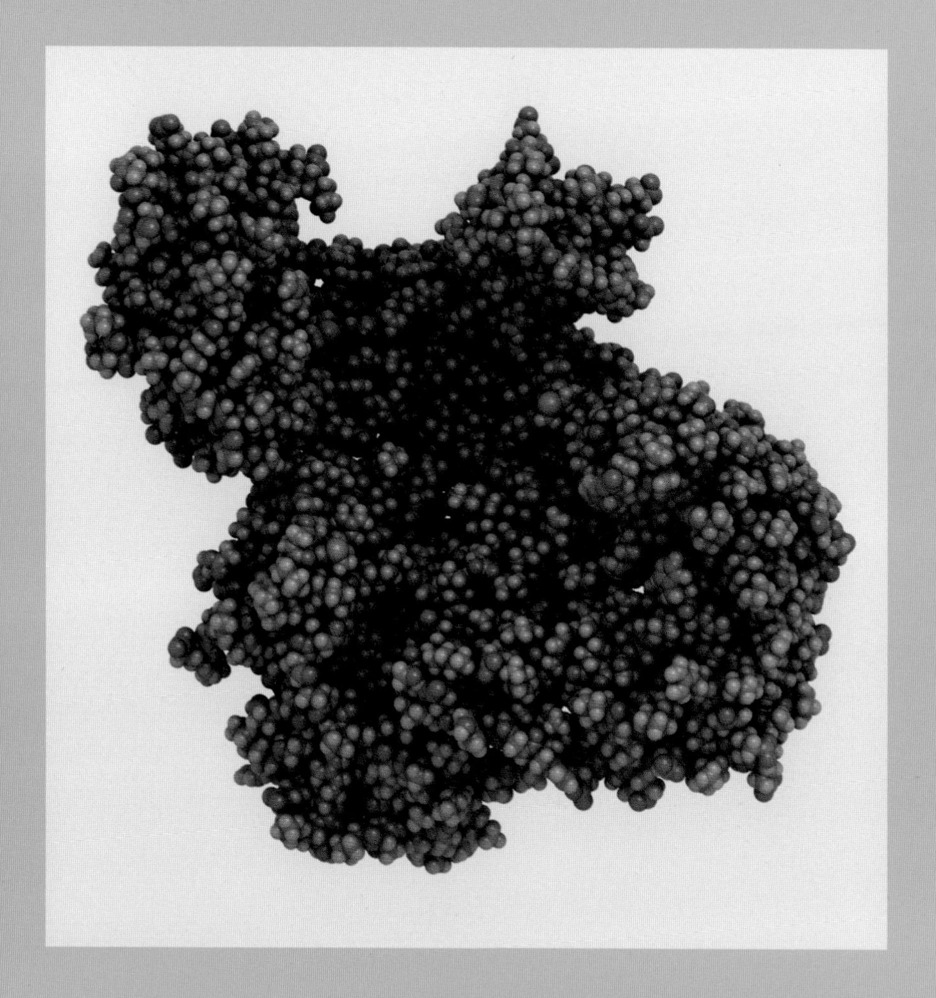

← Howard Martin Temin (1934–1994) was an American virologist. He studied Rous sarcoma virus (see page 100) during his graduate and postdoctoral years, and was recruited by the University of Wisconsin–Madison in 1960. He discovered that the genome sequences of this RNA virus could be found in the DNA of the infected host cell. He concluded that the virus had a way to convert its RNA into DNA. He had discovered the enzyme reverse transcriptase, shown here as a model derived from x-ray crystallography. American virologist David Baltimore (b. 1938) made a similar discovery at the same time, using a different virus, and the two shared a Nobel Prize in 1975. The world of molecular biology was thrown into chaos by these findings, because they violated the central dogma of biology (see page 13). Since its discovery, reverse transcriptase has become an essential component of the molecular biology toolbox. Among other things, it allowed scientists to determine the sequence of RNA molecules for the first time, and to clone genes from their messenger RNA.

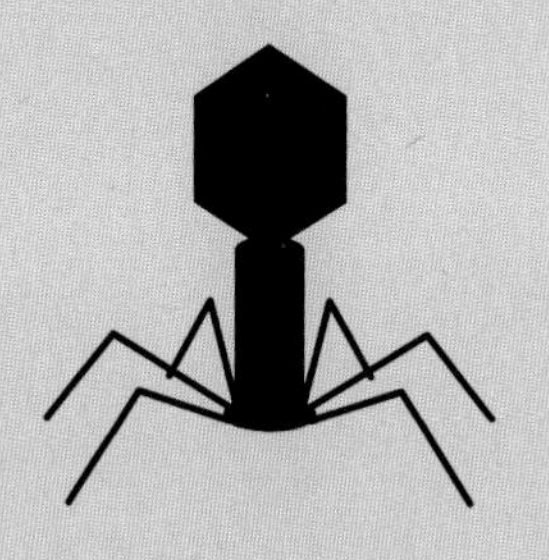

THE DEPTH AND BREADTH OF VIRUSES

Introduction

Virology is in an age of discovery, with hundreds of new viruses being described every day. For a majority of new viruses we know only the sequence of their DNA or RNA genomes, but we can guess what they look like, which organisms they infect, and how they function based on their genes and their well-studied relatives.

A BRIEF HISTORY OF THE VIRUS HUNTERS

Before the mid-2000s, the discovery of new viruses was a painstaking process involving electron microscopy and cell culture, and attempts to determine the genome sequence of the virus once evidence of it was found. Establishing the entire genetic code of a single small virus could take more than a year. During the first severe acute respiratory syndrome coronavirus (SARS-CoV) epidemic in 2002–2004, the determination of the virus's genetic sequence was completed in just a few months, which was considered astounding. Discovery of other microbes was much easier than for viruses, because all living cells have genes in common, and very short regions of these common genes are identical. This allowed researchers to obtain the genetic sequence for these common genes. In contrast, viruses don't have any shared common genes, so large studies were limited to looking for viruses related to others that were already known. However, once researchers automated the process for determining DNA sequences, a huge amount of genetic code could be determined very rapidly. The SARS-CoV-2 virus sequence was determined in just a few days, for example.

It wasn't long before methods were developed to carry out genetic sequencing randomly, without the need for any previous knowledge. This process is called metagenomics. Computer programs that analyze all this genetic information are continuously being improved, but we have now described an enormous number of viruses. Even so, we still understand very little about them, and we are very far from knowing all of them.

← In 2005, the Connecticut-based 454 Life Sciences company made the first high-throughput DNA sequencing machine. This allowed researchers to determine the sequence of tens of thousands of nucleotides in one experiment and opened the way for the discovery of new viruses.

↗ Mosquitoes carry a diverse range of both insect viruses and mammalian viruses, and as such can be a starting source for virus discovery.

→ Many emerging disease-causing mammalian viruses are carried by bats, including Ebola, SARS-CoV-2, MERS, and rabies viruses. Fieldworkers must therefore take care when collecting wild bats to avoid becoming infected.

The diversity of life

To understand how diverse viruses are, we have to start with the diversity of life. We divide all life into three domains: Bacteria, Archaea, and Eukarya. Bacteria and archaea are structurally simple and usually single-celled, and the cell lacks a nucleus. They mainly reproduce asexually. The domain of Archaea was not identified until 1977. Before that, these cells were thought to be bacterial, but beyond some structural similarities they are actually very different. The first archaea were identified in extreme environments such as deep-sea vents, acid hot springs, and salt flats, but we now know that they occur everywhere, including inside the human gut.

Eukarya are structurally more complex than bacteria and archaea, although they are often simpler biochemically. This is because they depend on bacteria to produce many of the basic chemicals they require. For example, a human body cannot synthesize many critical nutrients and instead depends on food and on the bacteria that live inside it to make these. Bacterial synthesis of vitamins B_{12} and K are especially important in human nutrition. Eukaryotic cells have a nucleus that houses the genome and the machinery for copying this and transcribing it into RNA. They also have other internal structures called organelles. Some of these—including the mitochondria, and chloroplasts in plants and algae—are derived from bacteria that moved from being independent to living inside the cells of ancient eukarya (see illustration on page 11).

Life-forms are further divided into kingdoms. The domains Bacteria and Archaea each contain a single kingdom, while the domain Eukarya is divided into four kingdoms: Plantae (plants), Animalia (animals), Fungi, and Protista (protists). Some organisms don't fall very clearly into any of these kingdoms. The majority of people are most familiar with the animal kingdom, because we are part of it.

Viruses don't form an independent domain or kingdom, but rather are associated with hosts from all kingdoms of life. In general, viruses don't cross domains, but they do cross kingdoms. For example, a number of viruses can infect both plants and fungi, or both plants and insects.

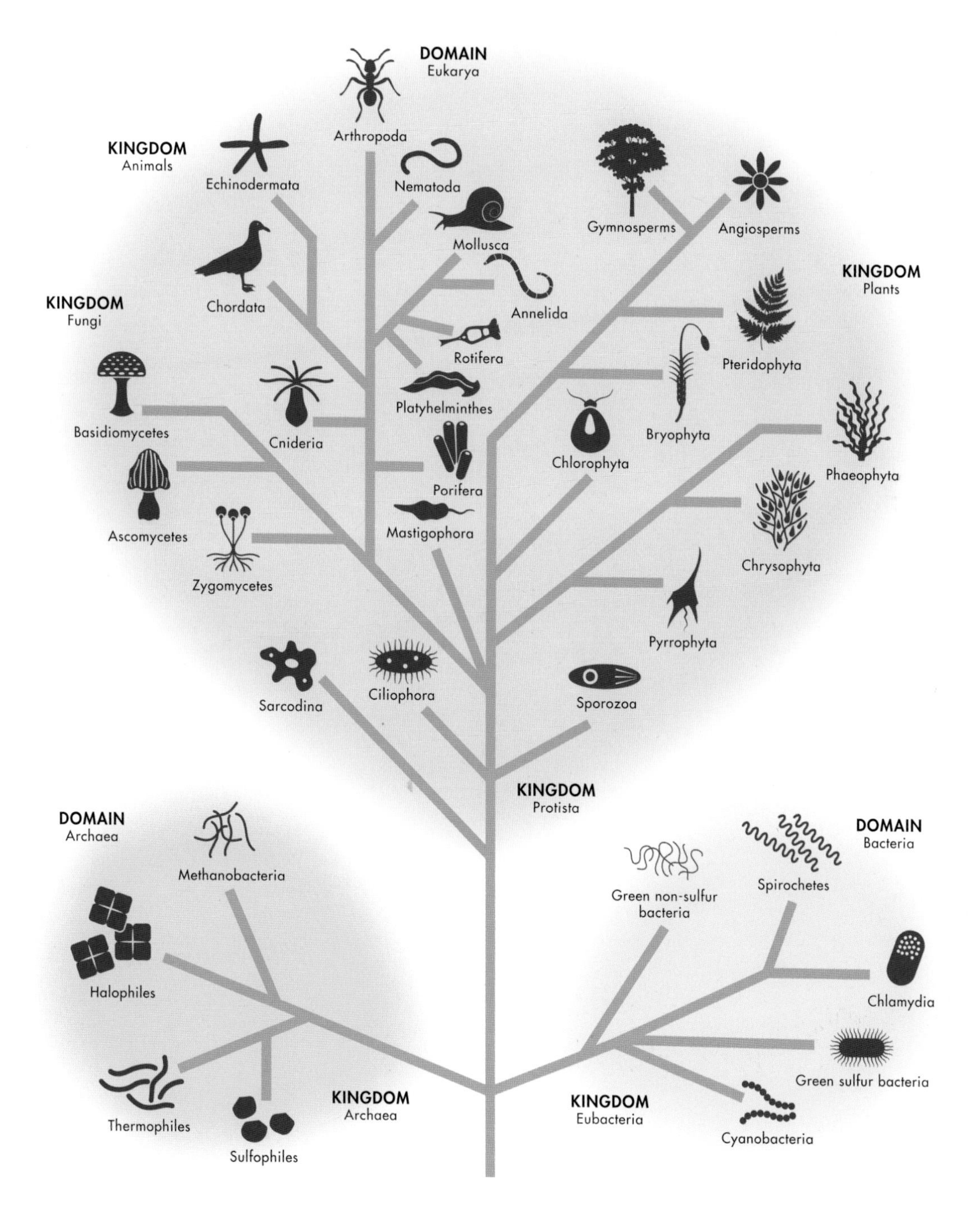

The domains and kingdoms of life

Life is divided into three domains and six kingdoms. Here, the domains—Bacteria, Archaea, and Eukarya—are shaded in the background, with the kingdoms indicated in each. Bacteria is the oldest domain of life. Bacteria and Archaea are prokaryotes and are usually single-celled; the cell doesn't have a nucleus. In contrast, Eukarya cells do have a nucleus where the DNA genome is stored, and often other organelles such as chloroplasts and mitochondria, which are derived from ancient bacteria.

SIZE OF VIRUSES

Viruses vary enormously in their physical size and the size of their genomes. The smallest viruses are only 17 nm (nanometers) across. To get an idea of the size of a nanometer, imagine slicing the thickness of your fingernail (which is about a millimeter) into 1,000 slices. Each slice would be about a micrometer (μm) thick. Then take a single slice and cut this into 1,000 slices—each of these is about a nanometer thick. The largest known virus is 1.5 μm across, 90 times larger than the smallest virus and larger than many bacteria. Genomes vary in size too; the smallest viral genome comprises just over 1,700 nucleotides and the largest comprises almost 2.5 million, a difference of about 1,500-fold.

Along with the enormous variation in genome size comes the enormous variation in the number and types of proteins that are encoded by viruses. The simplest viruses can make only two proteins: an enzyme to copy their genome, and a coat protein to cover up and protect their genome. There are even a few viruses that skip the coat protein and exist simply as naked RNA. The virus with the largest genome can make more than 2,500 proteins that do many things a cell can do, but not everything. So far, no virus has been found that can completely direct the manufacture of its own proteins. While they have the genes to make the proteins, viruses need the machinery of the cell they infect to turn those genes into proteins. Viruses are also incapable of generating their own energy, and rely on the host cell for this too.

SHAPES OF VIRUSES

Viruses come in many different shapes, these often resembling geometric structures. The classic virus shape is an icosahedron, a geometric structure with 20 equivalent faces. These are often drawn in two dimensions as six- or eight-sided polygons. The faces of a virus particle are often subdivided into many subfaces, called capsomeres. The largest icosahedral viruses have more than 2,000 capsomeres, while the smallest have only 12.

Another common shape for viruses is the helix. Tobacco mosaic virus, the first virus ever discovered (see page 18), has a helical shape. Some helical viruses are rigid, while others are flexuous. Viruses that have an envelope (an outer membrane layer) have a more relaxed outer shape, but often have a rigid internal shape. Many bacterial viruses (also called bacteriophages) have complex "landing gear" structures. However, the most diverse shapes of all are seen in archaeal viruses. Many of these are unique, such as the bottle-shaped ampullaviruses. When these viruses were first discovered, no one could figure out the functions of their genes, because they were so different to those found in other viruses.

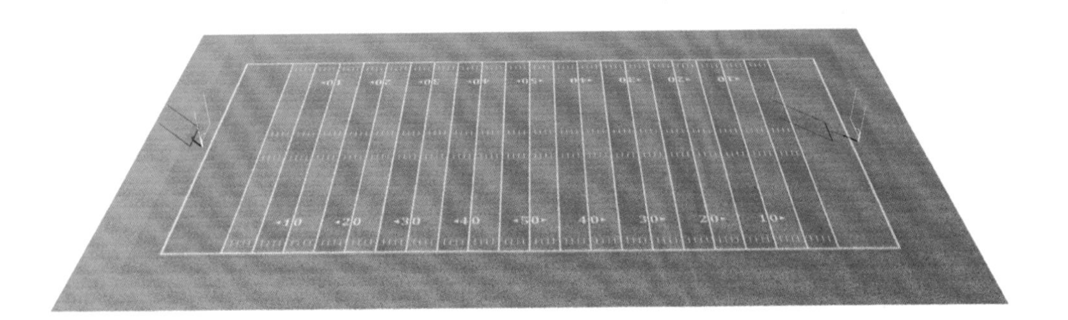

↙ If we imagine a plant cell blown up to the size of a football field, a virus would be about the size of a baseball.

Size of viruses

The smallest known virus is porcine circovirus 1 (PCV-1) and the largest by physical size is Pithovirus sibericum, although Megavirus chilensis is the largest virus with an icosahedral structure (viruses are not drawn to a comparative scale).

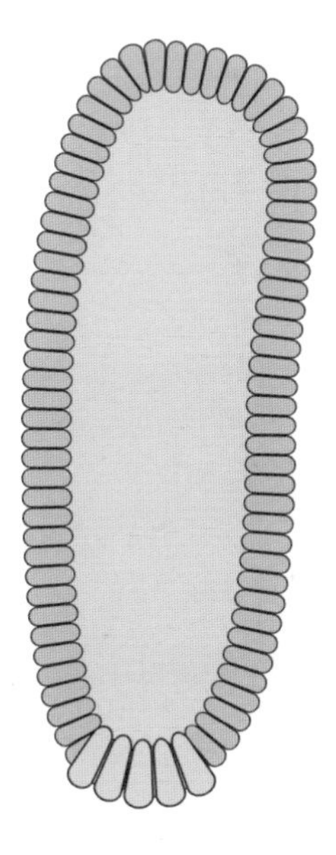

Pithovirus sibericum
1.5 μm long

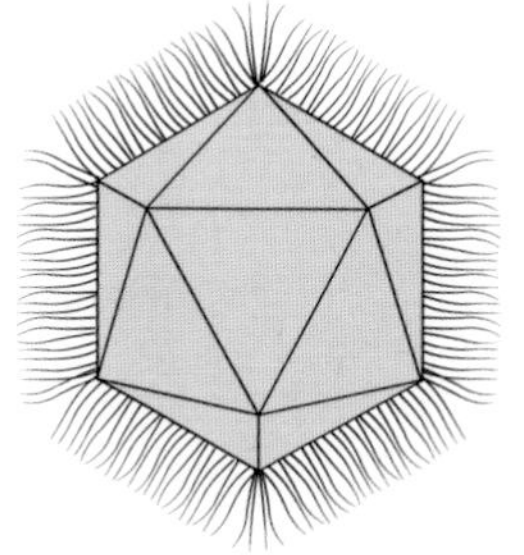

Megavirus chilensis
440 nm

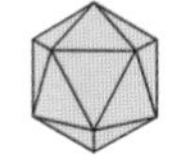

Porcine circovirus
17 nm

0.25 μm

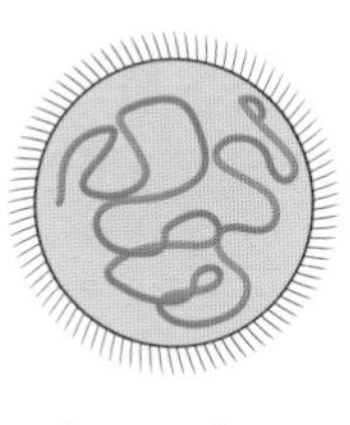

Paramyxovirus (mumps)

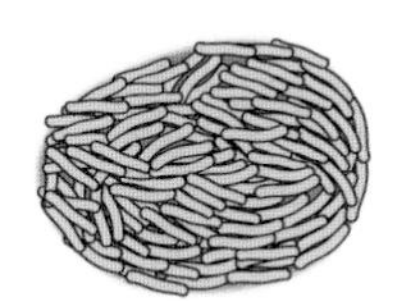

Vaccinia virus

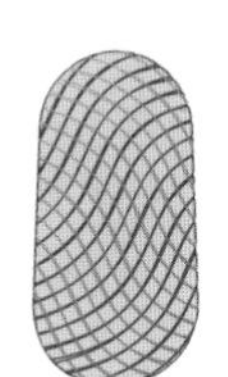

Orf virus

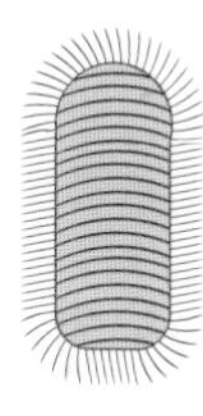

Rhabdovirus

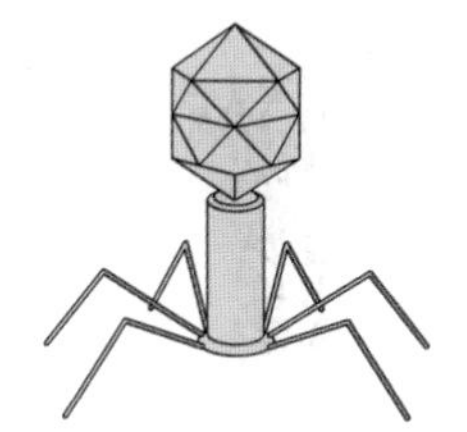

T-even coliphage

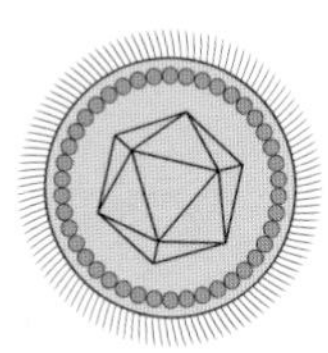

Herpesvirus

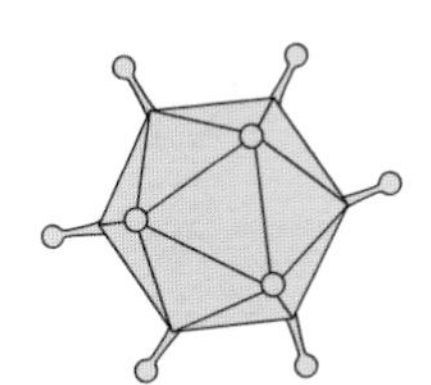

Adenovirus

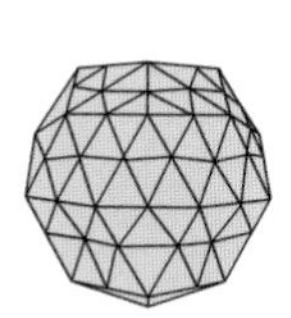

Polyomavirus

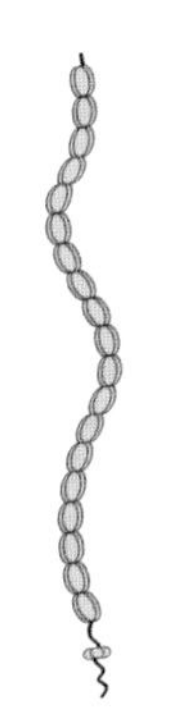

Tubulovirus

Flexuous-tailed phage

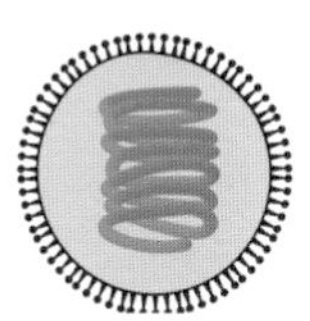

Influenza virus

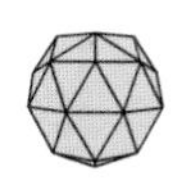

Picornavirus

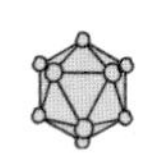

ϕX174

Shapes of viruses

Viruses come in a variety of shapes and structures beyond the iconic icosahedron or helix, and some have a membrane on the outside known as an envelope.

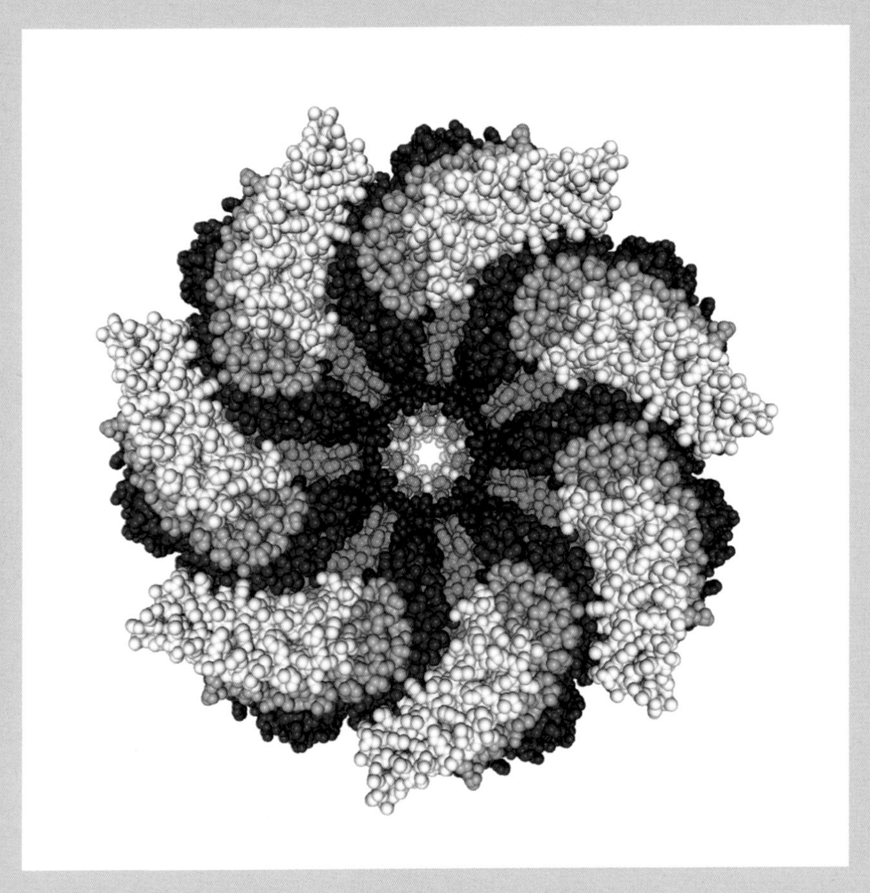

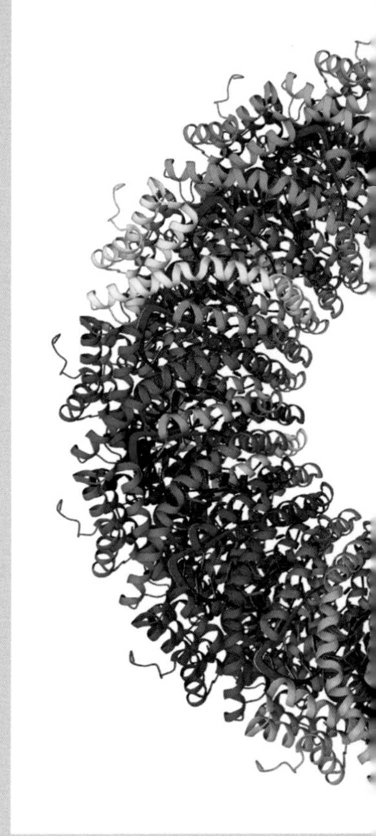

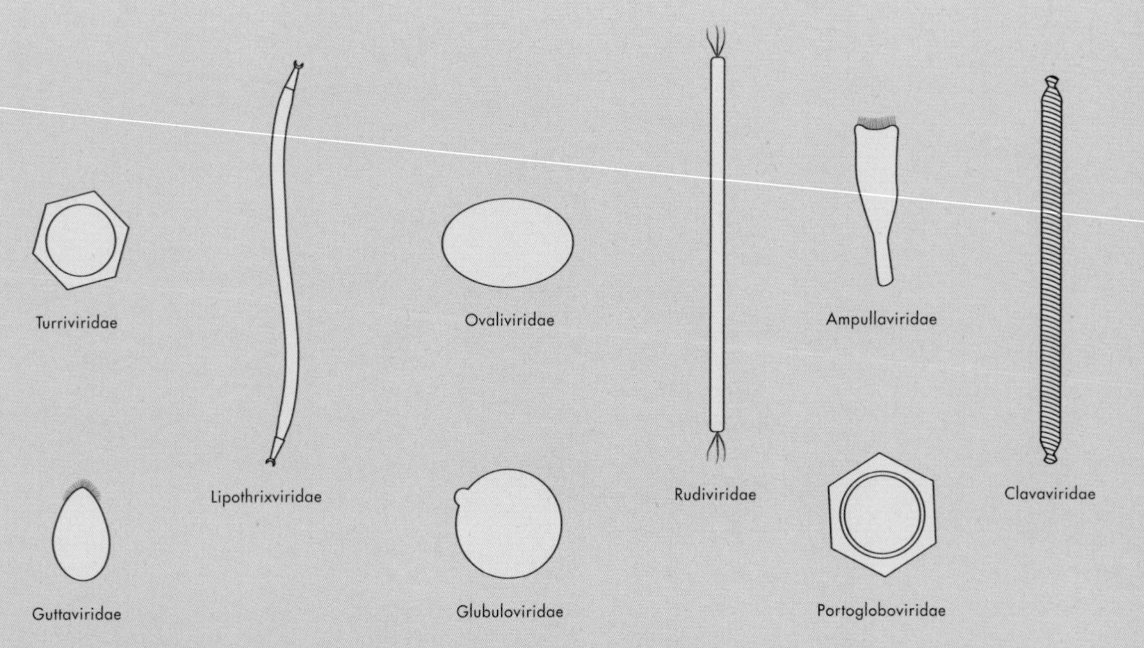
Turriviridae
Lipothrixviridae
Ovaliviridae
Rudiviridae
Ampullaviridae
Clavaviridae
Guttaviridae
Glubuloviridae
Portogloboviridae

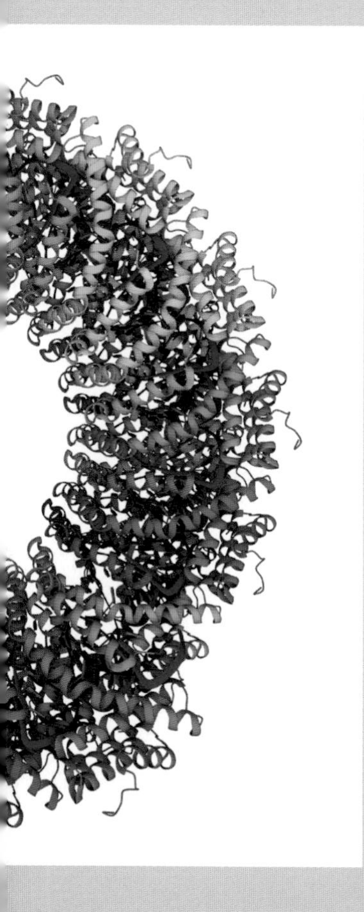

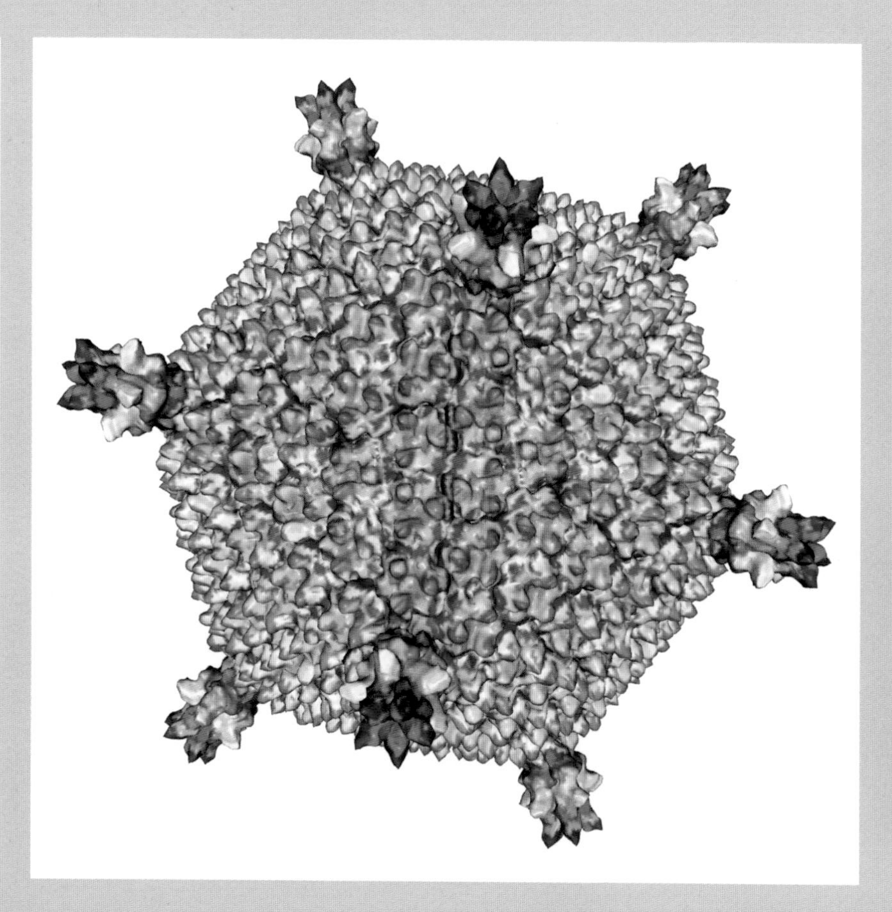

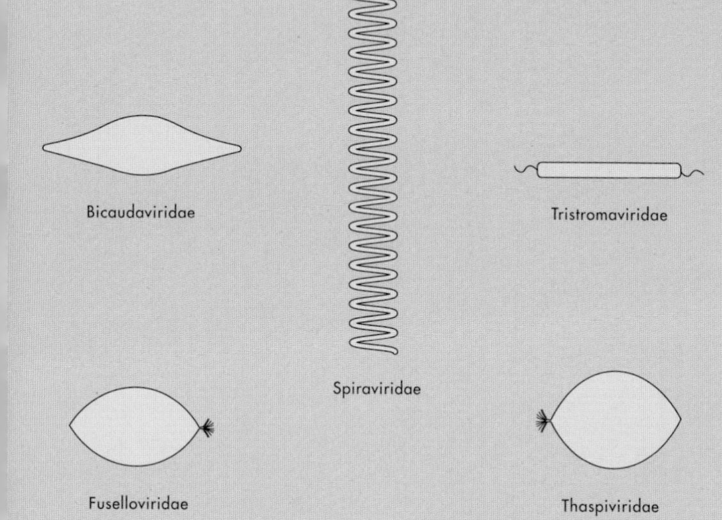

↖↖ Model of a cross section of the tail structure of Sulfolobus spindle-shaped virus 19 as deduced by cryo-EM data.

↖ Ribbon diagram of the interior of Sulfolobus islandicus rod-shaped virus from cryo-EM data. The host of this virus lives in very acidic water at 80°C (176°F). The DNA can be seen within the structure as a helix, but this is a different form of DNA known as the A-form. DNA in this form is more stable in extreme environments.

↑ Structural model of Sulfolobus turreted icosahedral virus derived from cryo-EM data.

← The huge variety of shapes that are found among viruses that infect Archaea.

Classification of viruses

Virus genomes are made from either DNA or RNA, which may be either double-stranded (ds) or single-stranded (ss) (see table overleaf). Viral genomes may be linear or circular, and they may be in one or more segments—something like our own genomes, which are in several segments called chromosomes (we have two each of 23).

The genome types of viruses are restricted to infecting specific kingdoms of life, with Type II ssDNA being the only type found in all kingdoms. There are no RNA viruses found in the Archaea, and there are no Type I dsDNA viruses found in plants other than algae. The reasons for these differences are not always clear, but some may be related to aspects of the host. For example, most dsDNA viruses are large, and plant cells have very restricted openings between them that are too small for large viruses to fit through. In general, plants have the largest cells among all the kingdoms of life, and most of the smallest viruses.

So, how many viruses are there? The first big studies of virus biodiversity, in the early 1990s, looked at viruses in the sea. These simply counted virus particles in seawater, visualized by electron microscopy or fluorescent methods. Using these data, scientists estimate that there are about 10^{30} (a 1 with 30 zeros after it) viruses in the sea, or 10 million times the number of stars in the universe. Even though viruses are individually very small, this still represents a huge total biomass—equivalent to about 15 times the total biomass of all the whales in our oceans in pre-whaling days. If the average size of a virus in the sea is 100 nm and you could string them all together, they would stretch far beyond the Milky Way.

We know much less about terrestrial viruses. Although many terrestrial systems have also been sampled for viruses, there are technical challenges to studying them. The numbers given above for marine viruses refer to the number of estimated individual virus particles, but how many different species of viruses are there? The answer to this is that we simply don't know. More viruses are being identified all the time, but the International Committee on Taxonomy of Viruses recognizes only a little over 9,000 species. This is almost certainly just a small fraction of the number that really exist on our planet.

→ Images of viruses from various sources including artists' renditions and electron microscopy showing a range of shapes: (A) influenza virus; (B) cytomegalovirus; (C) variola virus; (D) rabies virus.

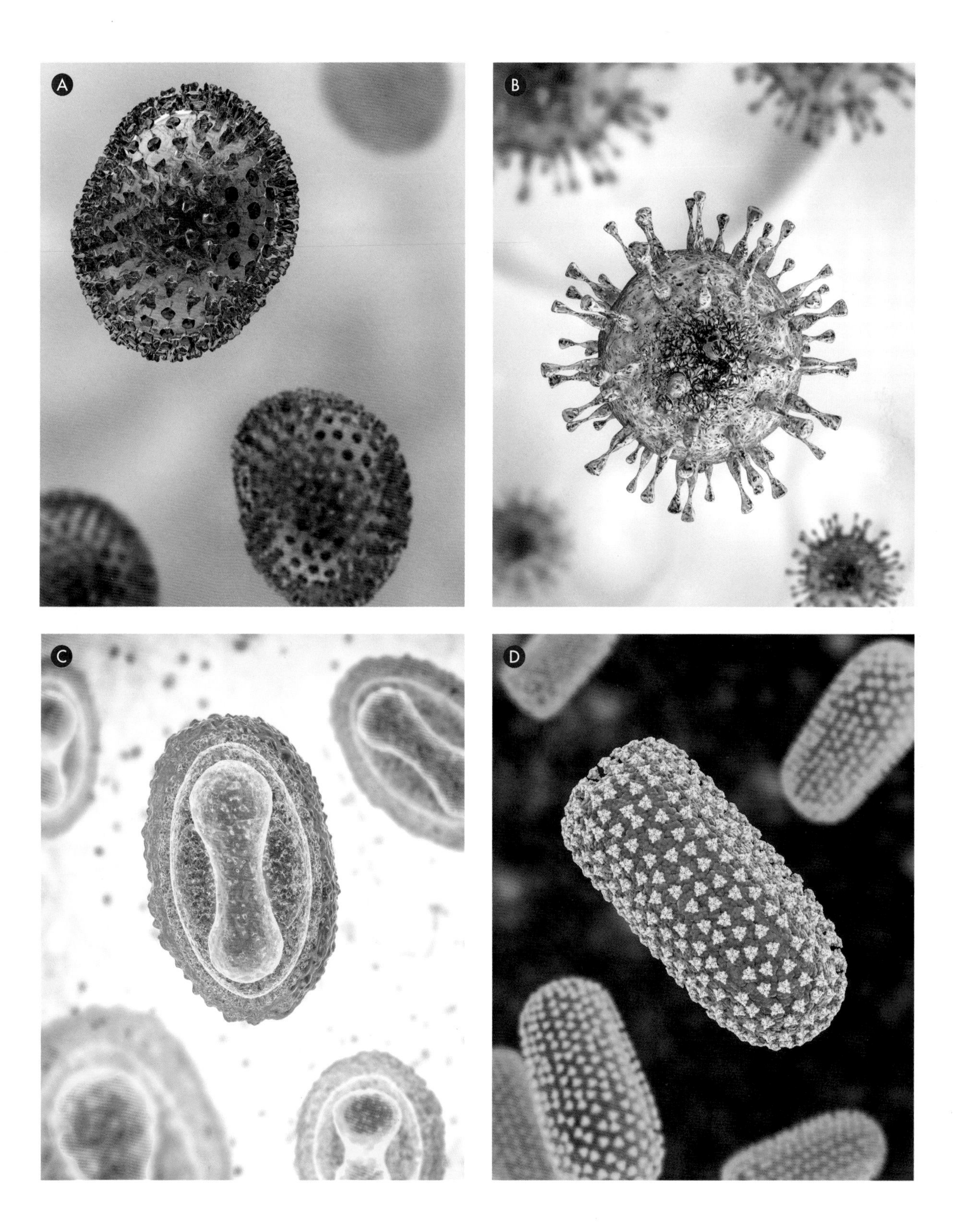
A
B
C
D

BALTIMORE SCHEME AND HOST TYPE

Viruses in different Baltimore classification groups infect different kinds of host. The top half of this figure shows the distribution of various classes of virus in the three domains of life, while the lower half shows the distribution of the different classes in three large groups of Eukarya.

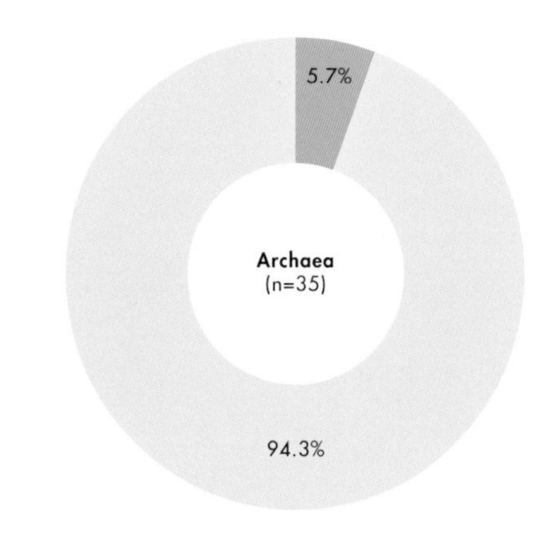

↓ Different categories of eukaryotic cells

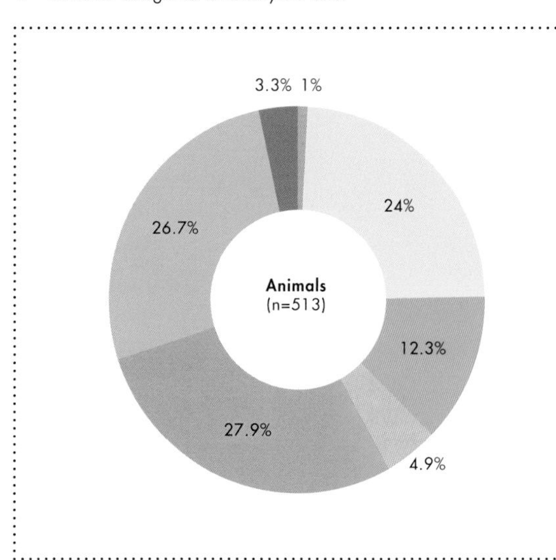

dsDNA = double-stranded DNA

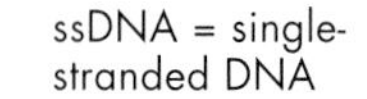
ssDNA = single-stranded DNA

dsRNA = double-stranded RNA

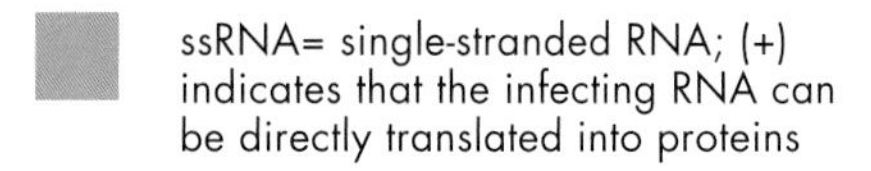
ssRNA= single-stranded RNA; (+) indicates that the infecting RNA can be directly translated into proteins

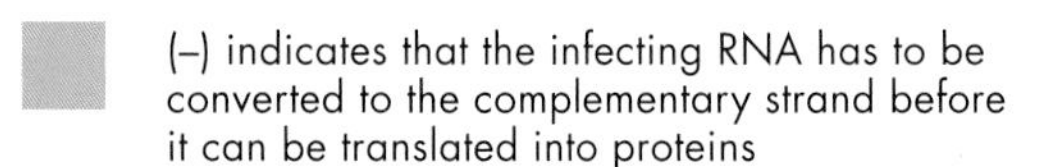
(–) indicates that the infecting RNA has to be converted to the complementary strand before it can be translated into proteins

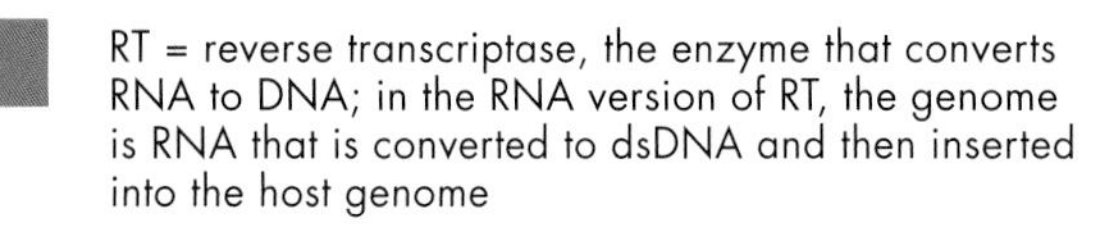
RT = reverse transcriptase, the enzyme that converts RNA to DNA; in the RNA version of RT, the genome is RNA that is converted to dsDNA and then inserted into the host genome

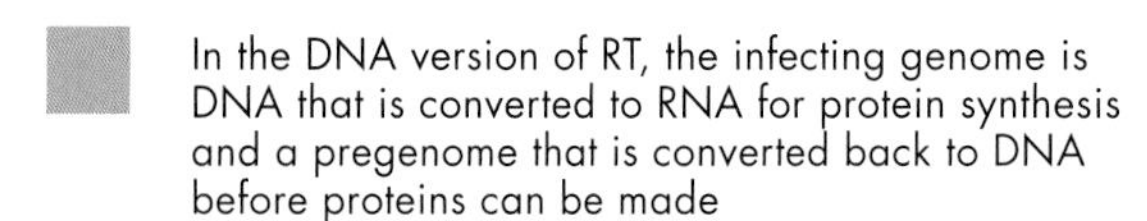
In the DNA version of RT, the infecting genome is DNA that is converted to RNA for protein synthesis and a pregenome that is converted back to DNA before proteins can be made

↓ Types of hosts infected by different types of viruses by genome

GENOME TYPE
Animal
Plant
Fungus
Protist
Bacteria
Archaea

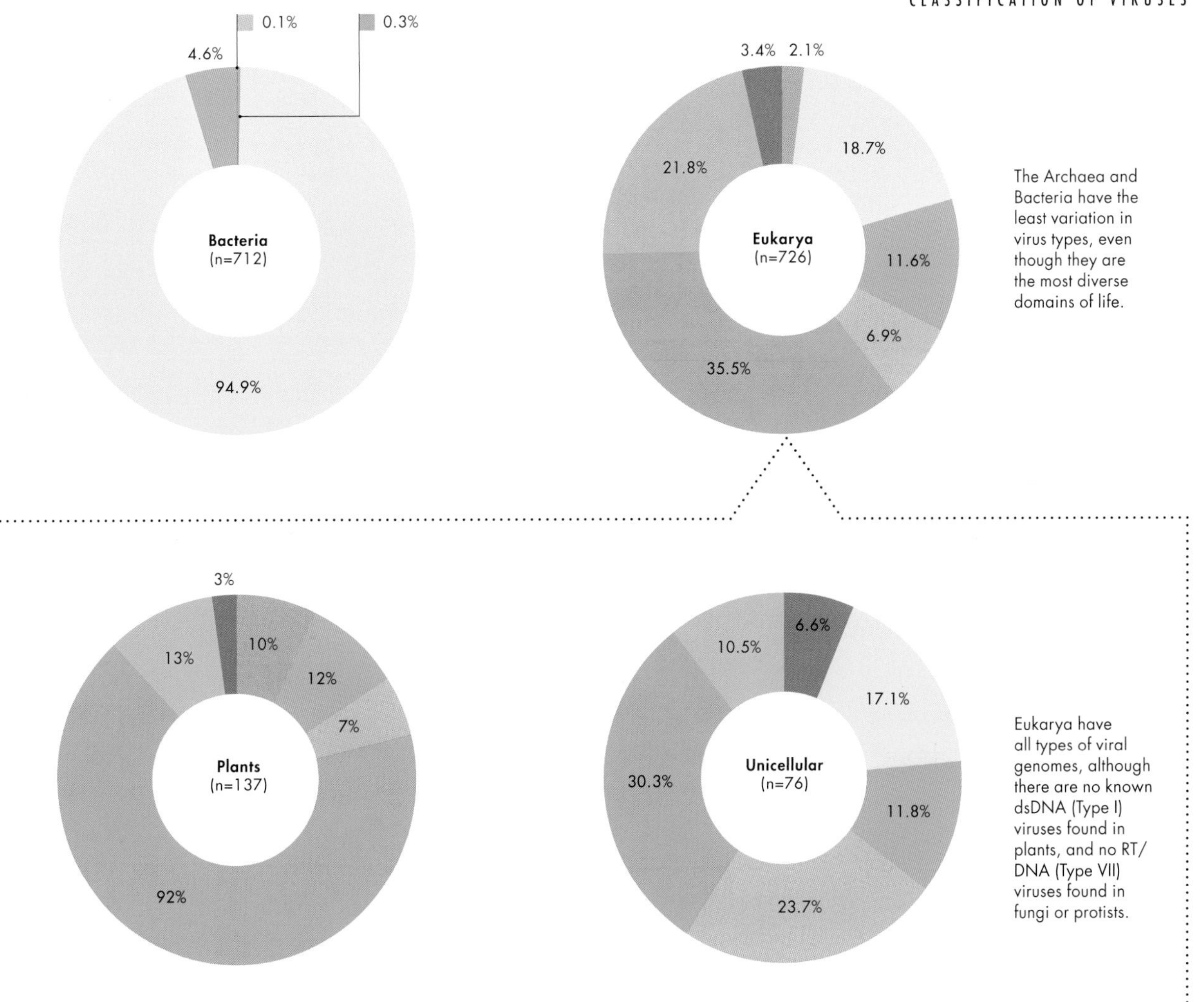

The Archaea and Bacteria have the least variation in virus types, even though they are the most diverse domains of life.

Eukarya have all types of viral genomes, although there are no known dsDNA (Type I) viruses found in plants, and no RT/DNA (Type VII) viruses found in fungi or protists.

I dsDNA[1]	II ssDNA[2]	III dsRNA[3]	IV ssRNA (+)[4]	V ssRNA (–)[5]	VI RT RNA[6]	VII RT DNA[7]
yes	yes	yes	yes	yes	yes	yes
no	yes	yes	yes	yes	yes	yes
yes	yes	yes	yes	yes	yes	no
yes	yes	no	yes	no	no	no
yes	yes	no	yes	no	no	no
yes	yes	no	no	no	no	no

Viruses in the sea

Most of the viruses in the sea infect bacteria or other microbes, which together make up the greatest biomass of the oceans. These viruses are important for many cycles of life and energy on the planet, and are discussed in more detail in "Viruses in ecosystem balance" (page 195). To date, more than 150 different marine samples have been collected and analyzed for viruses, from Arctic, temperate, tropical, and Antarctic oceans. Samples from temperate and tropical oceans have been taken at multiple water depths.

↙ First noticed in 2013, starfish have been suffering from a wasting disease whose cause is unknown. A densovirus has been found in echinoderms, but it is present in both diseased and healthy animals, so is probably not the cause.

↓ Seals and other marine mammals can be infected with influenza viruses (although not the strains that infect humans). They are also often infected with a virus related to measles. It is not clear if these infections cause any problems for the animals.

↘ Researchers from San Diego State University collecting marine samples to search for viruses.

Viruses of other marine life-forms, including fish, crustaceans, plants, and mammals, have not been very well studied, but diseases in ocean life often prompt a search for viruses as a potential cause. For example, when starfish began to die off along the West Coast of the United States in 2013, scientists looked for viruses and found a type of densovirus common in starfish. This virus was blamed for the starfish decline, but there was never any actual evidence for this, and in fact the virus is common in both diseased and healthy starfish.

Researchers have looked for viruses in mollusk and crustacean species important for human consumption. They have been found in shrimp, oysters, crab, lobsters, and crayfish, and in many cases are most problematic in farmed species, including farmed shrimp and oysters.

Aquaculture is a very old practice, but it became widespread for growing fish only relatively recently. Fishermen rarely observed diseases among wild-caught fish, but once fish viruses were identified in fish farms, virologists carried out more studies to look for them in wild populations. Viruses have been studied in both freshwater and marine fish, and a large number have now been described. Interestingly, viruses that cause diseases in domestic or farmed fish are often present in wild fish without causing disease.

Marine mammal viruses have been very poorly studied aside from those belonging to two major groups: those related to influenza viruses, which are commonly found in seals and Walrus (*Odobenus rosmarus*); and those in a large group of viruses that includes the measles virus (see page 156). The few studies that have more generally looked for viruses in marine mammals have found many other viruses in addition to these groups.

Terrestrial viruses

There are many diverse terrestrial and freshwater environments on Earth, all supporting a huge range of life-forms, including plants, animals, fungi, bacteria, and archaea. Viruses are also found in all these environments, and many recent studies have focused on trying to find all the viruses associated with a particular environment or host, called the virome. Efforts to describe the human virome have revealed that our bodies are full of many different viruses, some of which infect our cells and many that infect our microbes.

→ Tomato plant infected with tomato yellow leaf curl virus.

↓ Banana streak virus causes yellow streaks between the major veins of banana leaves.

VIRUSES IN THE SOIL

Viruses are abundant in the soil. Most of these are viruses of bacteria or archaea, but some viruses of eukarya are very stable and can survive as dormant particles in the soil for long periods of time. A very interesting soil virus discovery project has been conducted since 2014, involving undergraduate students from nearly 200 colleges and universities, mostly in the United States. The students collect samples, process them for sequence analysis of their genomes, and use sophisticated computer tools to find out what viruses are present. This project has led to the description of about 20,000 bacterial viruses.

Other soil virus studies have been carried out across many different ecosystems, including deserts, salt flats, pristine Antarctic soils, agricultural soils, forest soils, river sediments, and wetlands. The numbers of viruses found in these studies vary tremendously. In the richest environments, such as forest or wetland soils, more than a billion virus particles may be found in a single gram of soil, whereas in a desert the figure may be as low as 1,000.

VIRUSES OF PLANTS AND FUNGI

Plants were the focus of much early terrestrial sampling for viruses. Many studies have looked at viruses in crop plants, but very few have focused on wild plants. Most wild plants have multiple viruses yet rarely show any evidence of disease. In some cases viruses found in wild plants are also found in crop plants. However, it is not clear whether the wild plants are sources for crop plant viruses, or the viruses are moving the other way, from crops to wild plants.

Fungi represent one of the most poorly studied host kingdoms for viruses, but in recent years new metagenomics technology has allowed researchers to describe an enormous variety of apparent viral infections in fungi. Most fungal viruses don't cause disease. Some are beneficial, and others are being used to control fungi that cause plant diseases.

VIRUSES OF INSECTS

Insects are in decline around the world, with an estimated loss in total numbers of about 9 percent every decade since 1990. These losses are due to a variety of factors, which are unlikely to include viruses. A few viruses can cause serious diseases in insects, and large die-offs have been attributed to baculoviruses, large Type I viruses. These die-offs were well studied before the cause was known and are thought to be part of a natural cycle that controls insect populations. When any population of virus hosts becomes too dense, virus spread can be very rapid.

Insects host a huge variety of viruses, reflecting the incredible diversity of insect life on the planet. The insect viruses that have been studied in most detail also infect plants and mammals, as the insects act as the agents of transmission (see page 117). The Western Honey Bee (*Apis mellifera*) has been analyzed for viruses, and a few pathogenic viruses are known to be responsible in part for the general decline of domestic honey bees.

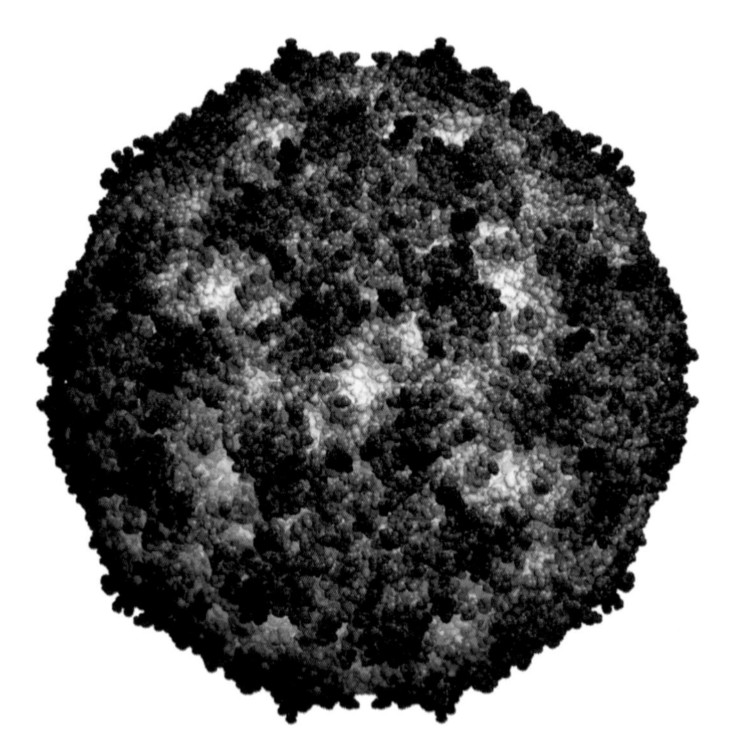

↑ The Greater Wax Moth (*Galleria mellonella*) is a predator of honeybees and now has a global distribution. Its larvae feed on wax and pollen in beehives, and can cause a lot of destruction and economic loss.

← Galleria mellonella densovirus infects the Greater Wax Moth. Many densoviruses of insects are lethal to the host, and are being explored as biocontrol agents of insect pests.

↗ Frogs have suffered serious declines in many parts of the world. One contributing factor is frog virus 3, a ranavirus found all over the world.

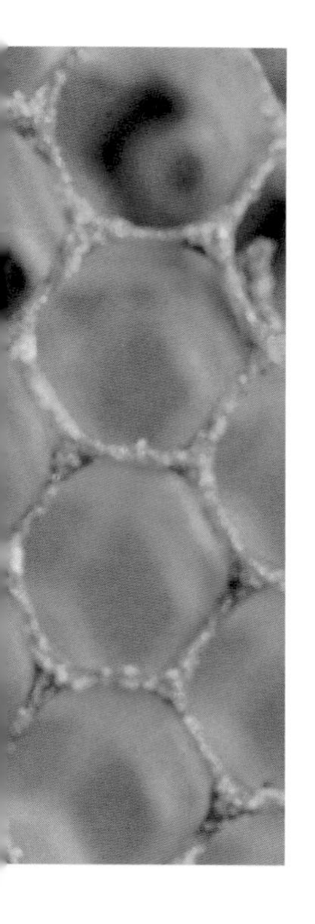

VIRUSES OF VERTEBRATES

Viruses of amphibians and reptiles have been poorly studied, although, as with many hosts, the numbers discovered are increasing and viruses from all major families have been found. Some have been surveyed in a search for disease agents, and a virus has been implicated in a neurological disease of captive boas and pythons. Frogs in particular have suffered from serious declines in recent decades. One major agent of frog disease is a fungus, but viruses have also been implicated. Ranaviruses are found in frogs all over the world.

Most studies of bird viruses have focused on domestic birds. Most research has been carried out with chickens, but ducks, turkeys, and geese have all been studied too. Wild bird studies have focused mainly on specific viruses, such as influenza in wild waterfowl and West Nile virus in crows and related birds. Because many birds migrate, they are good candidates for moving viruses over long distances, so there has been some focus on looking for human and domestic animal pathogenic viruses in migratory species. Nearly 300 bird species have been found to be infected with the influenza A virus (see page 252). As with other host groups, there are many viruses in birds, and most wild birds often don't show any symptoms of viral diseases—although there are a few exceptions, especially among viruses that affect fledglings.

Bats are full of viruses, many of which infect other mammals, including humans. Many viruses causing emerging diseases in humans—including Ebola, Middle East respiratory syndrome (MERS), SARS-CoV, and SARS-CoV-2—most likely originated in bats. The majority of bat viruses don't seem to cause disease in the bats, although rabies is an exception. Bats are long-lived (about 40 years for the North American Little Brown Bat, *Myotis lucifugus*, for example) and move hundreds of kilometers throughout the year, so, like birds, they are good candidates for the movement of viruses. On the other hand, bat–human contacts are rare. Viruses often pass to humans via an intermediate host—for example, MERS seems to pass from bats to camels to humans.

VIRUSES AND HUMANS

Of all the mammal viruses, those affecting humans have been studied in most detail. The human virome has been explored in a large number of studies and is estimated to comprise about 10 trillion viruses. These not only infect human cells, but also the microbes that reside here (including bacteria and archaea), while some are just transients passing through on our food. The profile of viruses in humans can change a lot during various disease states—for example, both severe malnutrition and type 1 diabetes result in a reduction in virus diversity, while colorectal cancer leads to increased diversity. It seems likely that most other mammals, and indeed most life-forms, have similar numbers of viruses to those in humans.

While much more research has been carried out into the viruses of domestic animals than wild animals, surveys of the latter have also been undertaken. This is because disease-causing viruses can often be transmitted from wild or domestic animals to humans, a process called zoonosis. Humans share a number of viruses or virus groups with other mammals, and viruses of other primates are probably more similar to human viruses than those of any other animals.

← The North American Little Brown Bat (*Myotis lucifugus*) is in severe decline due to a fungal disease called white nose syndrome. In the northeastern United States about 90 percent of these bats have died from the disease. The fungus is infected with a virus that may be contributing to the disease.

→ Colorized transmission EM of Nipah henipavirus in infected tissue. Nipah henipavirus is a serious human pathogen that is carried by fruit bats. To infect humans, the virus may pass from bats to horses first.

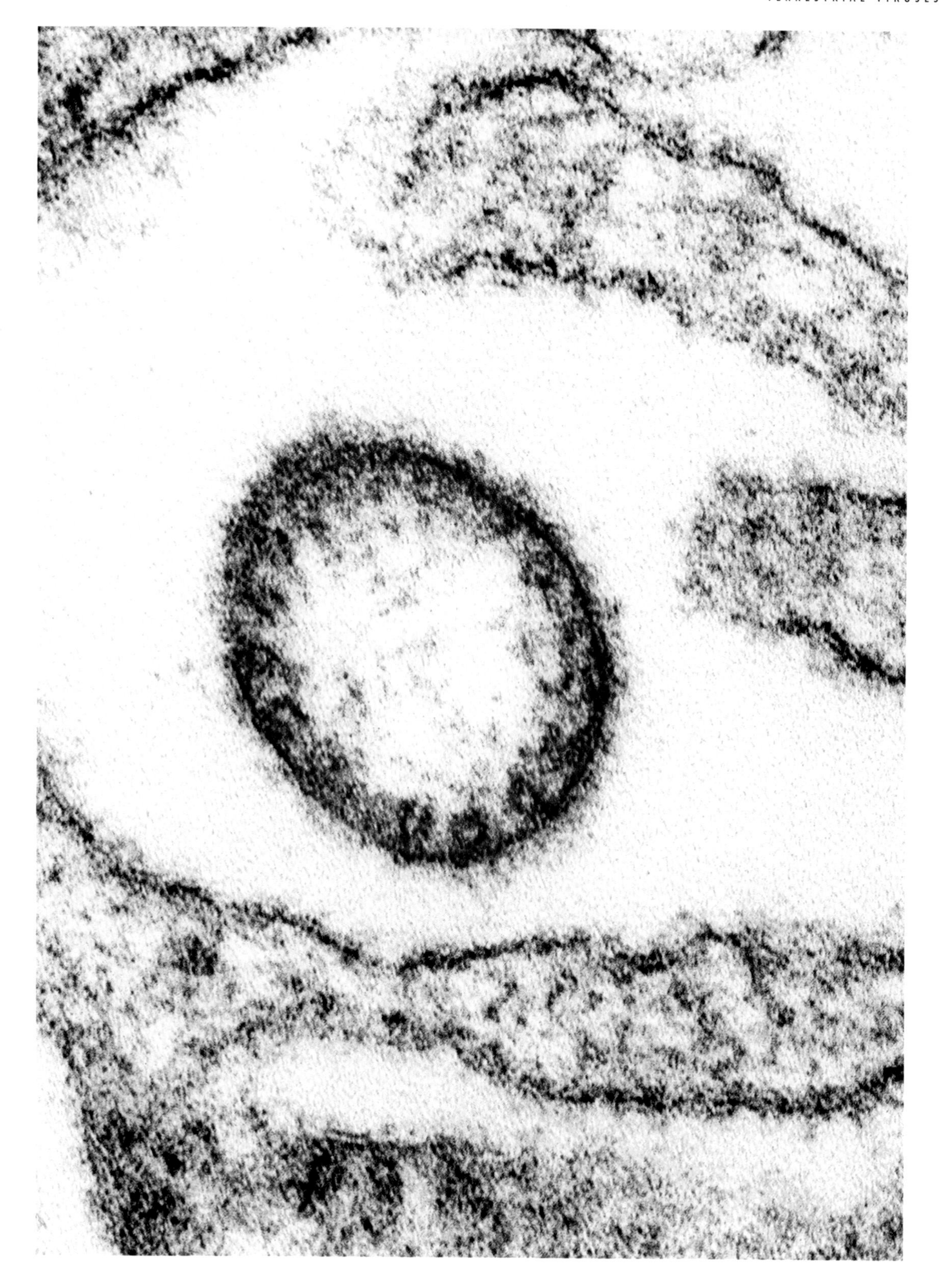

Viruses inside our genomes

Many pieces of viruses are found as part of our genome and of the genomes of all life. These are called endogenous viruses, meaning "inside the genome." Of these, the endogenous retroviruses have been studied most closely.

RETROVIRUSES

Retroviruses (Type VI viruses) have RNA genomes that are converted into DNA once they infect a cell, and they then insert this DNA into the genome of the host cell. All retroviruses do this in every cell they infect. Most of the time this has little consequence, but occasionally the DNA is inserted in a location that changes how a gene is used. Very rarely, these viruses infect germ-line cells—eggs or sperm. When that happens, the virus can be passed on to the next generation inside the genome of the host. This has occurred numerous times throughout evolutionary history—about 8 percent of the human genome comprises retroviruses. Many other viruses or virus genes are also found in genomes, and are like a fossil record of past virus infections. Studying these endogenous virus genes has led to a new field, called paleovirology.

We are very far from knowing all the viruses on our planet, and a lot of modern virology is revealing just how much we don't know. What are all these viruses doing? The view of viruses as simply agents of disease is changing, and a deeper look at what they do is covered in the following chapters.

↓ Artist's rendition of a double-stranded DNA molecule, the genetic material for all cellular life—and for many viruses.

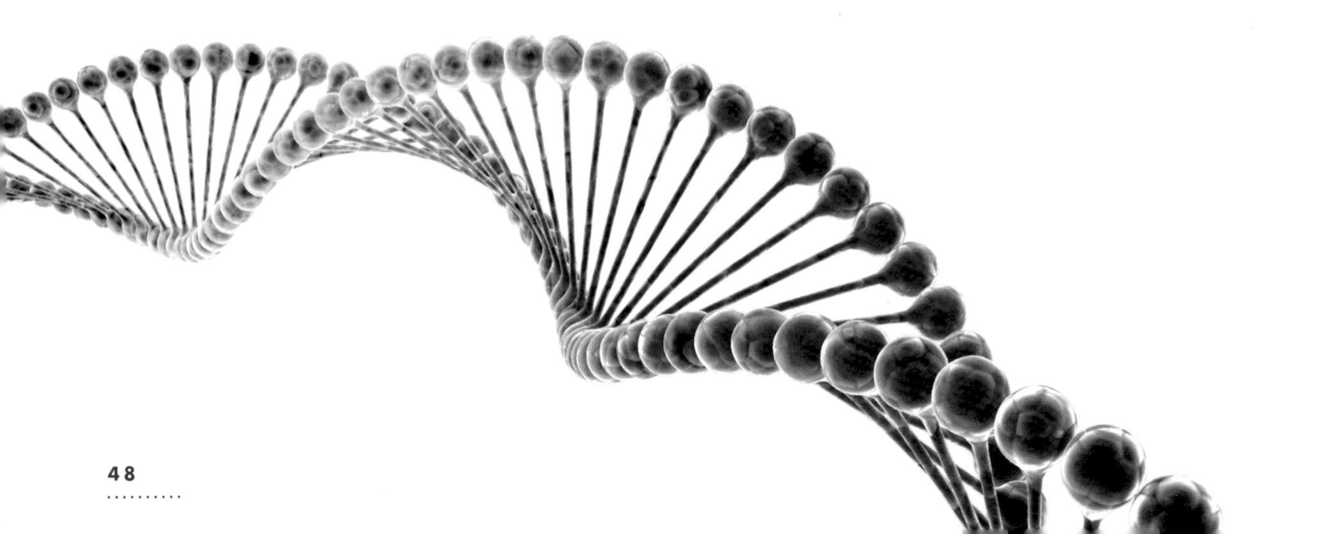

SUBVIRAL ENTITIES

Viruses are not the smallest things around—viroids, RNA molecules that infect plants, are even smaller. They are usually less than 400 nucleotides in length, and they don't code for any proteins. Instead, all of their biological activity comes from the RNA molecule and they use the host enzymes to copy themselves.

Viroids are best known for the diseases they cause, including potato spindle tuber (see page 60), avocado sunblotch, citrus exocortis, and coconut cadang-cadang. They are transmitted by contact between infected and uninfected plant material, and some may hitchhike with viruses or be transmitted by insects.

Some viruses have viruses of their own, called satellite viruses. These code for a coat protein but they use the host virus (called a helper virus) for everything else. A related entity is called a satellite RNA; this doesn't encode a coat protein and sometimes doesn't encode any protein at all. Some satellite RNAs have a very dramatic effect on the symptoms the helper virus causes, either making them better or worse. For example, tomato plants (*Solanum lycopersicum*) are infected by a satellite RNA that uses the cucumber mosaic virus as its helper, causing a lethal disease that kills the plants in about 10 days.

↓ Potato spindle tuber viroid causes potato tubers to become elongated and spindle-shaped, and may result in stunted plants. The viroid also infects other garden plants, including tomatoes.

PCV-1

Porcine circovirus

The smallest known virus

GROUP	II
FAMILY	Circoviridae
GENUS	Circovirus
GENOME	Circular, single-component, single-stranded DNA of about 1,760 nucleotides, encoding two proteins
VIRUS PARTICLE	Icosahedral
HOSTS	Wild and domestic pigs (*Sus* spp.)
ASSOCIATED DISEASES	None, but the related porcine circovirus 2 (PCV-2) causes wasting and diarrhea in piglets
TRANSMISSION	Contact
VACCINE	Engineered virus or heat-killed virus, used to treat PCV-2

With a tiny genome and a virus particle measuring just 17 nm, porcine circovirus 1 (PCV-1) is a benign virus. However, scientists now recognize four different types of related porcine circoviruses (types 1–4).

Porcine circovirus 2 (PCV-2) causes wasting diseases in pigs, especially piglets, and has become a serious problem worldwide for the hog industry. PCV-1 is very similar genetically, but it has a very different impact on its hosts. The reason for this is not known.

PCV-1 replicates in the nucleus of the host cells, using the same the enzyme (the DNA-dependent DNA polymerase) that the host uses to copy its own DNA. The genome is copied by a rolling circle method, where the polymerase continuously copies the DNA to make a long string of genomes that later get cut to size. Although the virus encodes only two proteins, it can make two different versions of one of these, the Rep protein that controls replication of its genome. This strategy of using the same genetic sequences for multiple purposes is common among small viruses.

Porcine circoviruses are part of a large group of viruses called CRESS (circular, Rep-encoding, single-stranded) viruses. Recent studies have found these viruses integrated into the genomes of hosts across the whole Eukarya domain. Most have never been studied, with the exception of those that cause diseases, such as the geminiviruses in plants. CRESS viruses always encode a Rep protein, which directs their unique pattern of rolling circle replication.

→ Ribbon diagram of the PCV-1 capsid derived from cryo-EM data.

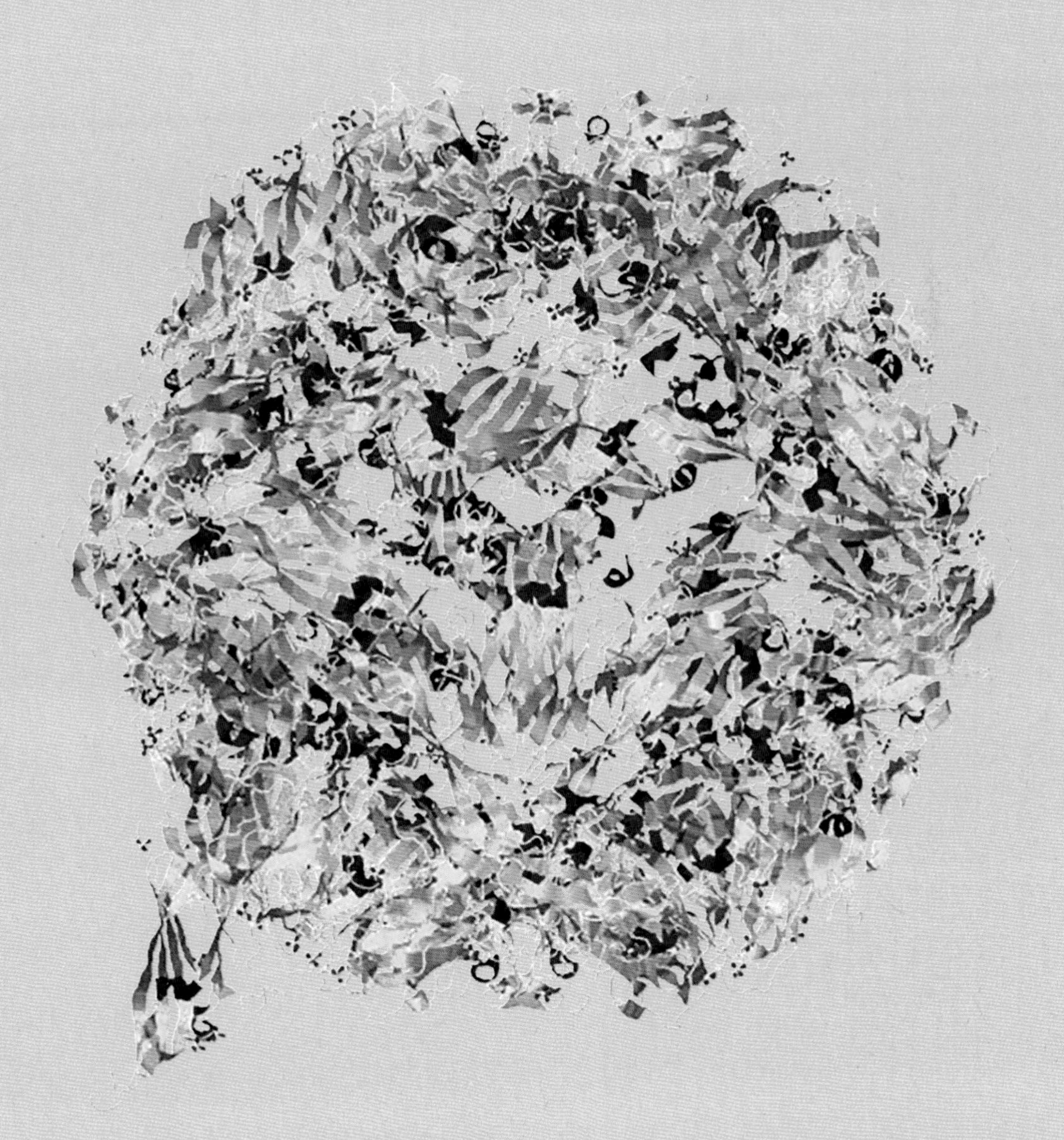

Pandoravirus salinus

The largest known virus genome, bigger than some bacteria

GROUP	I
FAMILY	Unassigned
GENUS	Pandoravirus
GENOME	Linear, single-component, double-stranded DNA of about 2.5 million nucleotides, encoding about 2,500 proteins
VIRUS PARTICLE	Elongated oval with a pore on one end
HOSTS	Amoebas
ASSOCIATED DISEASES	Nucleus degradation
TRANSMISSION	Diffusion in water

Pandoravirus salinus has the largest genome of any virus, although it is not the virus with the largest measurement; that distinction belongs to pithovirus sibericum, which is about 50 percent larger and measures 1.5 μm in length.

In the last 20 years a number of new giant viruses have been discovered that challenge the definition of the group as a whole. These are large enough to be easily seen through a simple light microscope; they encode thousands of proteins, including some that are used for making proteins; and interestingly, many infect single-celled protists such as amoebas. When Pandoravirus was discovered, its unusual shape led to its name, and the researchers who found it felt that it would challenge our understanding of what a virus is, opening a pandora's box of knowledge.

Other large viruses that infect algae have been known since the 1970s, but these are larger and more complex.

Pandoravirus salinus is so different from any other virus that it has not been classified further than its genus and species name. It was found in a search of sediment in the coastal waters off Chile. A related virus was found in a similar search in fresh water in Australia and named Pandoravirus dulcis. The discovery of these related viruses so geographically distant from each other and in different environments hints at their ancient history. They both infect the amoeba *Acanthamoeba castellanii*. Both the host and the environment are very poorly studied, so it is possible that there are many other related viruses that have yet to be discovered.

→ Electron micrograph of Pandoravirus dulcis, a close relative of Pandoravirus salinus. The pandoravirus family was discovered in 2013 by scientists from the Laboratoire Information Génomique et Structurale, associated with the Laboratoire Biologie à Grande Échelle.

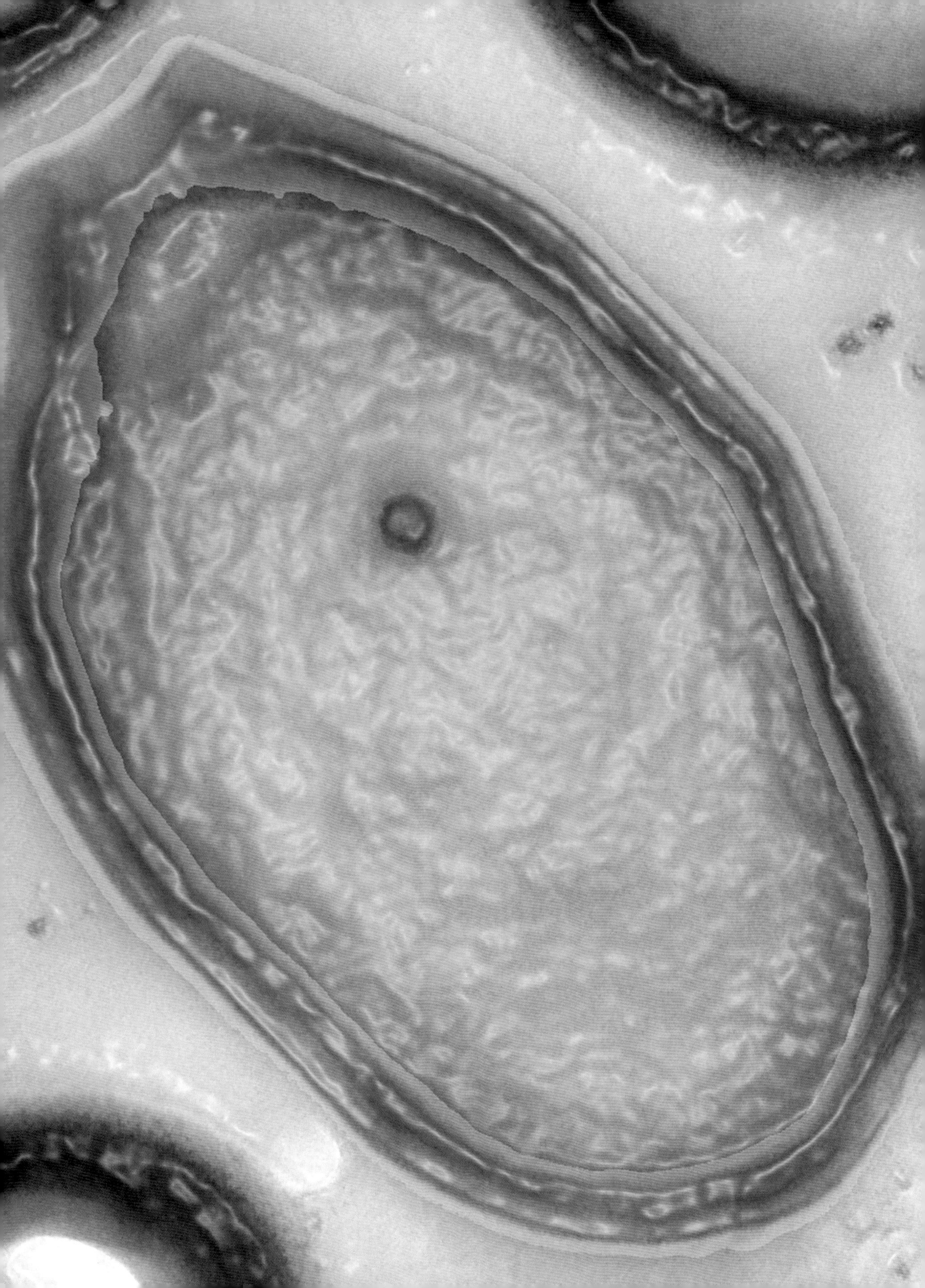

BSV

Banana streak virus

A remarkable virus that jumps in and out of the host genome

GROUP	VII
FAMILY	Caulimoviridae
GENUS	Badnavirus
GENOME	Discontinuous circular, double-stranded DNA of about 7,400 nucleotides, encoding six proteins, some via a polyprotein
VIRUS PARTICLE	Non-enveloped elongated faceted particles, 150 nm long by 30 nm wide
HOSTS	Banana (*Musa* spp.)
ASSOCIATED DISEASES	Banana streak
TRANSMISSION	Mealybugs, stress

Banana streak virus (BSV) is transmitted by mealybugs and causes a serious disease in parts of Africa. Once a rare problem, it has become more common in recent years following the adoption of new methods of propagating bananas known as micropropagation.

There are two species of banana, abbreviated as AA or BB. Domestic bananas are tetraploids of AAAA (meaning they have a double set of chromosomes) or hybrids of AAB (with a full double set of AA chromosomes and one set of B chromosomes). AAAA are common dessert bananas, while AAB are plantains.

The wild banana ancestor BB is immune to BSV, and it turns out that this is because it has an endogenous version of the virus integrated into its genome. As with most endogenous viruses, this virus stays in the genome of BB plants and is passed on from generation to generation. However, when the AAB hybrids are stressed, as happens during micropropagation where new plants are grown from small amounts of plant tissue, the virus comes out of the AAB genome, a process known as exogenization. This causes the virus to become infectious and spread. The AAB plants still have a copy of the endogenous virus in one of their chromosomes but it doesn't protect them from the virus, and mealybugs spread the virus among AAAA and AAB genotypes.

Plants infected with the exogenous form of the virus have streaks on their leaves (see page 42). The virus interferes with chlorophyll production in the areas between the major veins of the plant.

→ Transmission electron microscope image of banana streak virus particles. Their length is uniform, although a few shorter, broken particles can be seen—this usually occurs during the processing of the sample.

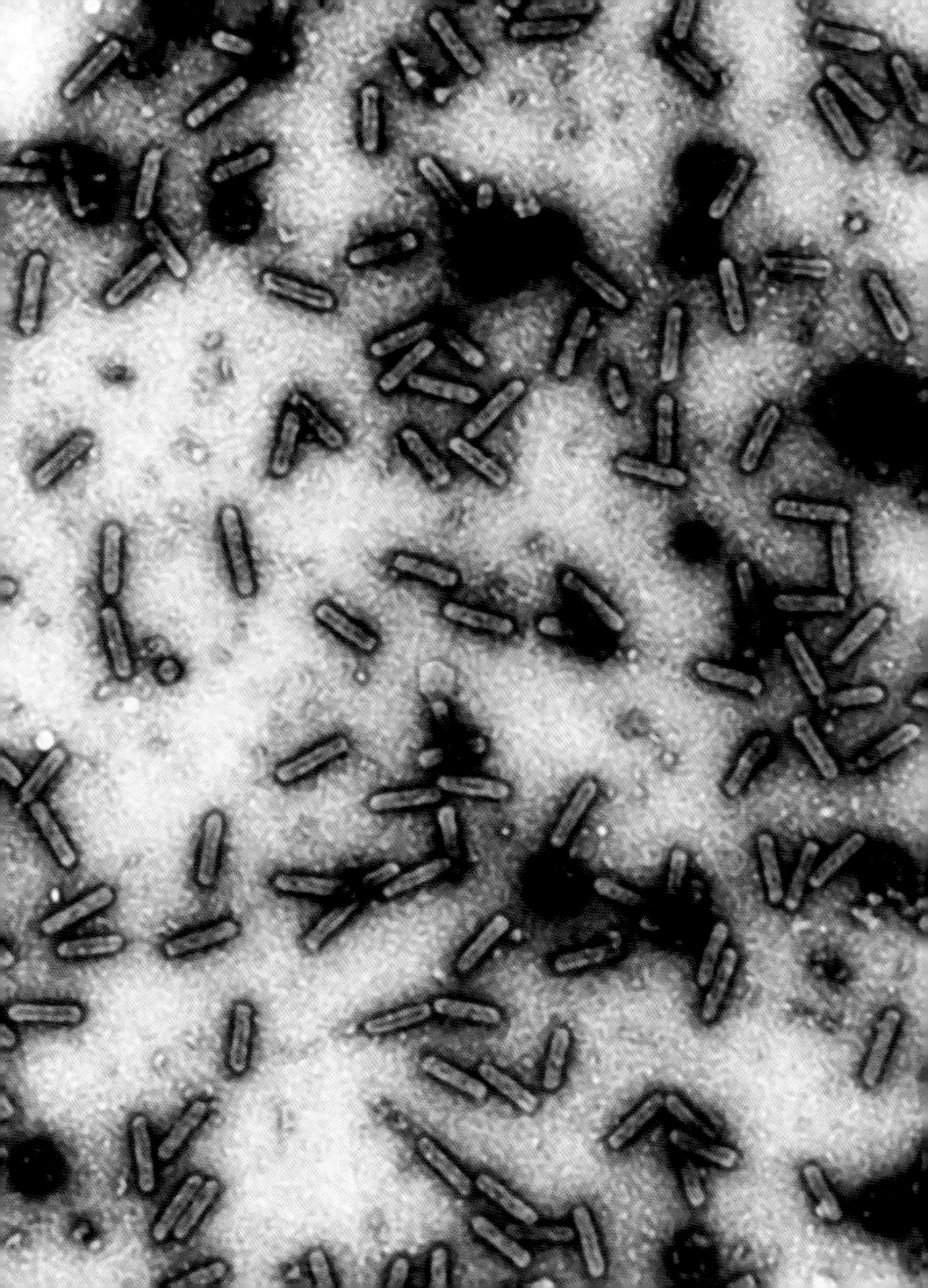

HERV-K

Human endogenous retrovirus K

The youngest of the human endogenous retroviruses

GROUP	VI
FAMILY	Retroviridae
GENUS	Betaretrovirus
GENOME	Proviral
VIRUS PARTICLE	None
HOSTS	Humans; related viruses in other great apes
ASSOCIATED DISEASES	Possibly cancer
TRANSMISSION	Vertical through the genome

Human endogenous retrovirus K (HERV-K) is not a single virus, but in fact a rather large group of partial or complete retrovirus sequences found throughout the human genome. Of these, the most well studied has about 90 copies in the human genome.

HERV-K is "active" in that the RNA and proteins from the proviruses can be found in various human tissues, although they are most common in embryos and testicles. Scientists have found a link between expression of viral genes and cancer, but the details are not yet definitively understood. HERV-K is one of the few truly human versions of human endogenous retroviruses, as it is not found in other primates. This means that it first infected humans after the split between humans and other great apes. The places in the human genome where these virus sequences are found are not the same in every individual, meaning that it has been actively moving around in the genome fairly recently in evolutionary timescales.

So, are these endogenous retroviruses doing anything? Some certainly are. Members of the human endogenous retrovirus W (HERV-W) group produce a protein called syncytin, which is critical for the establishment of the placenta—without HERV-W, there would be no placental mammals. In other cases, the placement of the virus seems to affect how nearby genes are turned off or on.

→ HERV-W is integrated across many sites in the human genome. This image was created using a fluorescent probe, which recognizes the HERV-W sequence, applied to a spread of human chromosomes and viewed under a microscope.

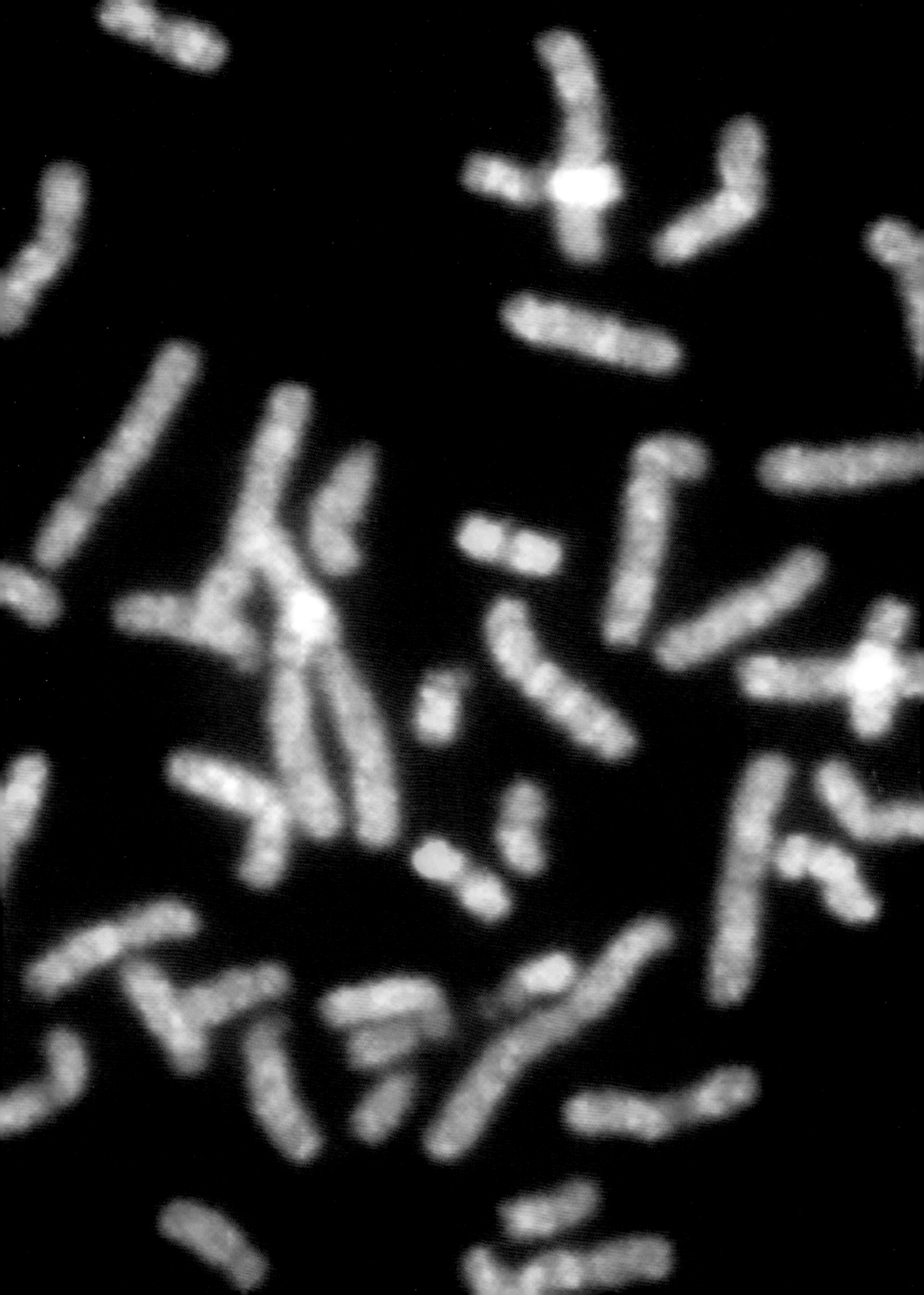

HDV

Hepatitis deltavirus

A viral hitchhiker

GROUP	V
FAMILY	Kolmioviridae
GENUS	Deltavirus
GENOME	Circular, single-stranded RNA of about 1,700 nucleotides, encoding one protein
VIRUS PARTICLE	Enveloped, spherical, about 22 nm, lacking an inner core
HOSTS	Humans
ASSOCIATED DISEASES	Acute hepatitis
TRANSMISSION	Sexual contact, bodily fluids, vertical
VACCINE	Hepatitis B virus vaccine

Hepatitis deltavirus (HDV) is a satellite virus sometimes found with hepatitis B virus (HBV) infections. It requires HBV, the helper virus, for packaging, and uses helper virus proteins in its envelope.

HDV has been found in human infections in different parts of the world, and it has been divided into eight distinct species in the genus Deltavirus; however, HDV is rarely referred to by these species names, and hence they are not defined here. HDV replicates in a rolling circle process, much like a viroid, using the RNA polymerases of the host. Unlike a viroid, however, it codes for a protein, the delta antigen. The virus makes two versions of this protein, one early in infection and a later one that inhibits the replication of the helper virus. The late version is required to assemble the delta virus.

People can be infected with HBV for years and then acquire HDV, or they can acquire the two together. Infection with HDV makes the symptoms of HBV worse, and this is especially true when the two viruses are acquired together.

When HDV copies itself, the product is a long string of genomes linked together, called a concatemer. The virus RNA has an enzyme-like portion called a ribozyme, which cuts the long string into genome-sized pieces; these then re-form into circles. Molecular biologists have used the HDV ribozyme as a tool to cut RNA molecules to the correct size inside a cell.

HDV particle

The genome of HDV (left) uses the proteins of HBV (right) to encapsidate its genome.

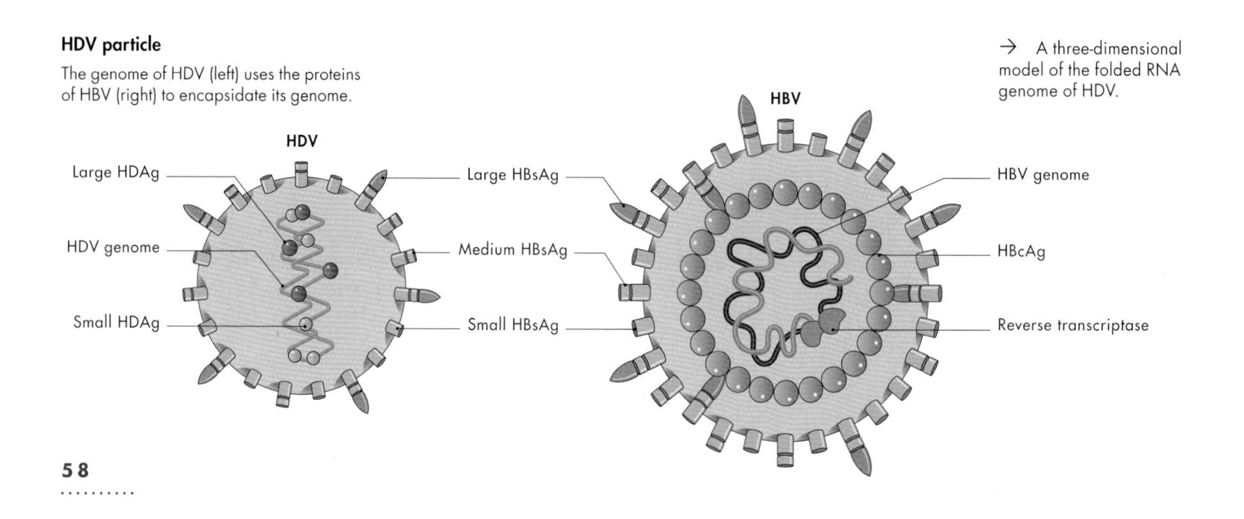

→ A three-dimensional model of the folded RNA genome of HDV.

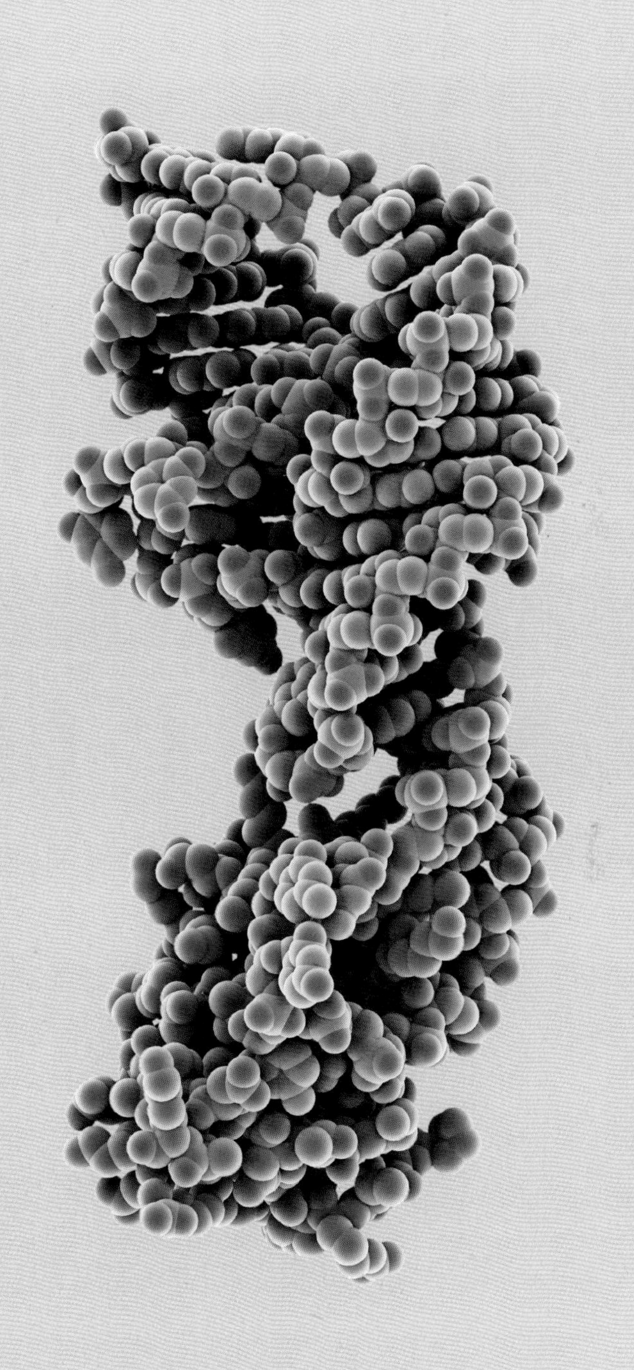

PSTVd

Potato spindle tuber viroid

A remnant of the RNA world?

GROUP	N/A
FAMILY	Pospiviroidae
GENUS	Pospiviroid
GENOME	Circular, single-stranded, single-component RNA of about 360 nucleotides, encoding no proteins
VIRUS PARTICLE	None
HOSTS	Potato (*Solanum tuberosum*), tomato (*Solanum lycopersicum*), and other members of the nightshade family
ASSOCIATED DISEASES	Potato spindle disease, tomato stunting
TRANSMISSION	Seeds, pollen, insects when in association with plant viruses

Viroids are simple circular RNA molecules that do not encode any proteins. They fold into complex structures because most of their nucleotides are complementary and can form base pairs. Different parts of the structure are responsible for the viroid's replication and the different effects it has on its hosts. Because viroids have a lot of biological activity but don't code any proteins, some researchers have speculated that they are a remnant from a precellular world where RNA ruled.

Viroids replicate in a rolling circle manner, which produces long RNA molecules with many copies of the genome strung together. Some viroids contain a ribozyme, an enzyme-like RNA molecule that is thought to be a remnant of a world before cellular life, and that cuts the long RNA into genome-sized sections. However, the potato spindle tuber viroid (PSTVd) doesn't contain a ribozyme, and instead co-opts an enzyme from the host to cut its RNA.

The most obvious symptom in potato plants infected with PSTVd is the spindly shape of the potato tubers. However, the viroid also infects tomato plants, where it can cause stunting, changes in plant pigments, and death of the plant tissue along the veins. PSTVd is related to chrysanthemum stunt viroid, tomato apical stunt viroid, and citrus exocortis viroid, which cause serious diseases worldwide in their respective hosts.

PSTVd genome

The genome of potato spindle tuber viroid (PSTVd) and the secondary structure that is formed by basepairing of nucleotides. Different regions of the genome that are involved in biological activity are shown.

→ A PSTVd-infected potato plant with very mild symptoms. Although potatoes are usually propagated via the tubers, this plant has fruits on it, and it is thought the viroid was probably spread around the world by the true seeds of potatoes.

Left terminal | Pathogenicity | Central conserved segment | Variable | Right terminal

GAACU AACU GUGGUUCC GGUU ACACCU CUCC GAGCAG AAGA AGAAGGCGG--CUCGG GAGC UUCAG UCC-CCGGG CUGGAGCGA UGGC AAAGG GGUGGGG GUG CC GCGG CG AGGAG CCCG GAAA AGGGU

CUUGG-UUGA-CGCCAAGG CCGA UGUGGG GGGG UUCGUU UUCU UUUUUCGCC GAGCC--CUCG--AAGUC AAG GGCCC GGCUUCGCU--GUCG UUUCC CCGCUCC CAC GG CGCC GC UCCUU GGGC CUUU CUCCCA

VIRUSES MAKING MORE VIRUSES

The infection cycle

If we imagine that viruses have a goal, it is simply to make more of themselves. They are not driven to cause disease or to do good; they just want to make more viruses. Sometimes, in this drive to reproduce, they benefit their hosts, and if that happens there may be strong selection to maintain the relationship. At other times they accidentally cause harm to their hosts, especially if they and their host have a new relationship that has yet to be honed through adaptation and evolution. Ultimately, a virus will adopt anything that furthers its cause to reproduce.

The whole process of viruses making more of themselves starts with infecting a host. The details of how they get into (and out of) their hosts is described in the next chapter, but for now we will assume that a virus has managed to get into a host cell. For many viruses the next step in the infection cycle is to free their genome of its packaging, sometimes referred to as uncoating. This process varies for different viruses. Many viruses stay inside their protective coats until they get to their target location in the host cell. Some types of viruses, such as those with genomes of double-stranded RNA, never uncoat at all. Retroviruses stay inside the infecting particle until they have converted their RNA genome into DNA. Other viruses, including bacteriophages and the phycodnaviruses that infect algae, inject their genome directly into the host cell, never allowing their capsid (the intact virus particle) to enter the cell.

Once the infection cycle begins, the virus has a few jobs: make messenger RNAs (mRNAs) that can direct the production of proteins; make copies of their genome; and package the genome into new particles that can go on to infect other cells or hosts. The way these steps are accomplished depends on the type of virus (genome type; see pages 14 and 38), the host type, and whether or not the virus has a membrane around it, called an envelope (see page 108).

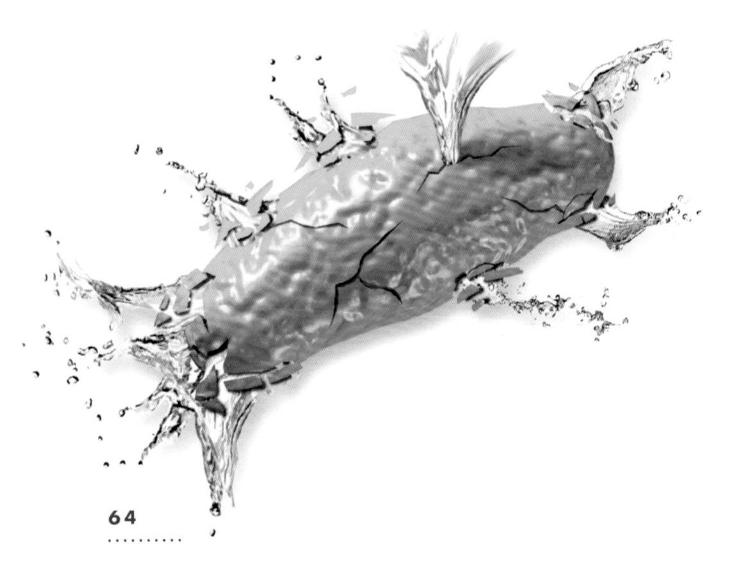

← Artist's rendition of a bacterial cell bursting open.

Complexity of a double-stranded DNA virus genome

A complex double-stranded DNA virus such as Marek's disease virus makes about 70 different messenger RNAs (mRNAs), shown in this map of the genome by the colored arrows going in both directions. Although the genome is linear, it is drawn here as a circle so that it is easier to see all the genes. Before the start of each messenger RNA there are signals in the sequence that tell the RNA polymerase where to start making the RNA. There are other signals that tell the polymerase when to stop.

Base pairing of nucleotides

The chemical structure of nucleotides, the building blocks of DNA or RNA, has a phosphate in the 5′ position of the ribose sugar, and a hydroxyl group in the 3′ position. This is where the individual nucleotides are joined to make the long string of molecules that make up DNA, shown here. The bases of each molecule pair with their complementary bases through hydrogen bonding, shown as dotted lines, to create double-stranded DNA. The cytosine–guanine (C–G) pair is stronger than the thymine–adenine (T–A) pair because there are three bonds in the C–G and only two in the T–A. By convention, DNA and RNA molecules are written in the 5′-to-3′ orientation, with the phosphate group at the start and the hydroxyl group at the end. RNA has a very similar chemical structure, but thymine is replaced with uracil (U) and it has one more hydroxyl (OH) group.

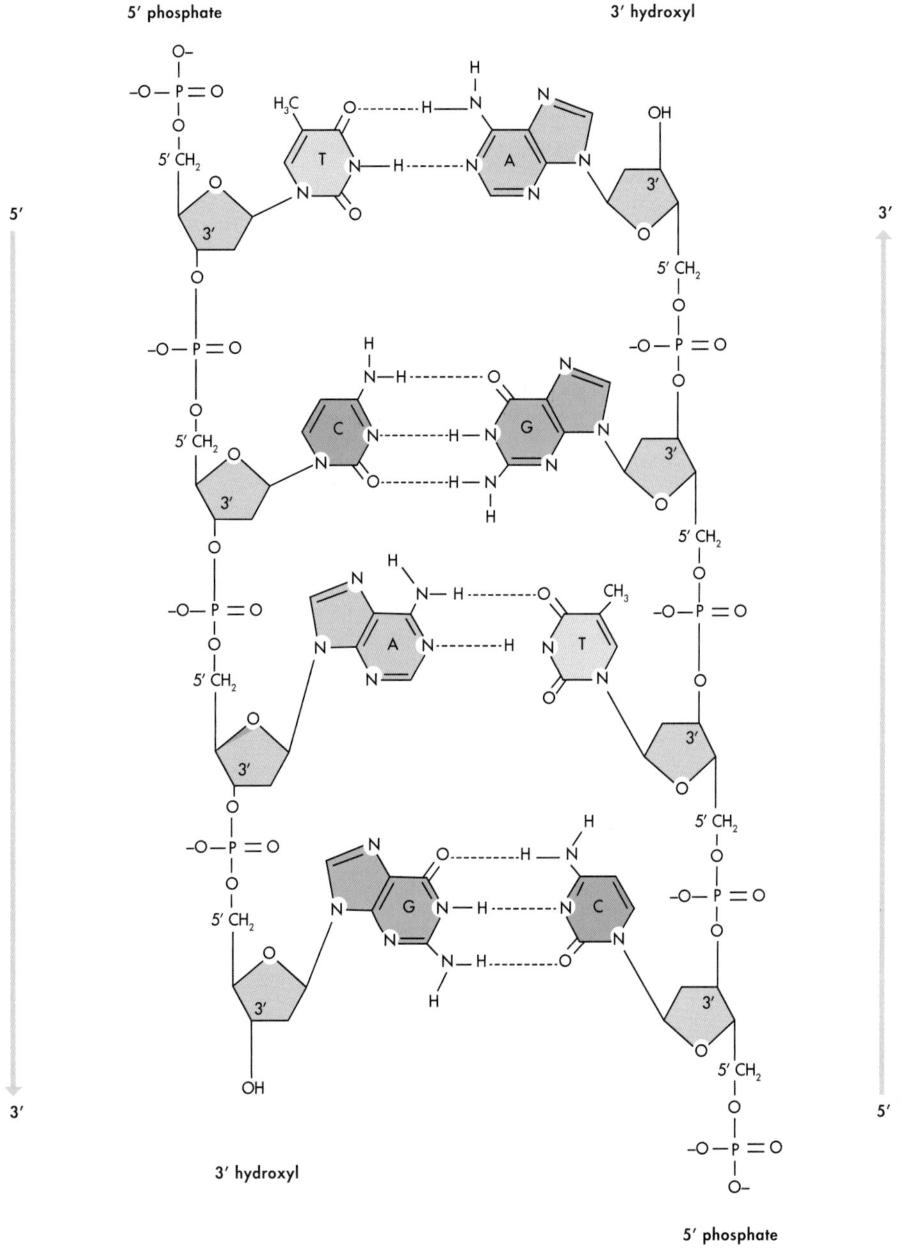

When RNA or DNA is copied, the correct nucleotide sequence is determined through base-pairing. This is a fairly accurate process—adenine (A) always pairs with thymine (T) (or uracil, U, in RNA), and guanine (G) always pairs with cytosine (C)—but rarely mistakes can be made, leading to a mutation (see page 136). When cellular genomes are copied, the enzymes involved have methods to check that they have not made any mistakes, called proofreading. Some viruses that use the host's enzymes for copying their genomes benefit from this proofreading. However, other viruses—especially RNA viruses—don't have the same kind of proofreading, and hence are prone to making more mistakes that result in mutations.

↓ The spike protein of SARS-CoV-2 showing three copies of the protein with the mutation at D614G highlighted in red. This mutation emerged in Europe in 2020.

DNA viruses

Since all host cells copy their genome using double-stranded DNA replication, DNA (dsDNA) viruses often use the host cell's enzymes to make messenger RNA and copy their genome. In eukaryotic host cells these enzymes are in the nucleus, so most of the virus life cycle takes place in that organelle. Host cells usually make these enzymes only during the replication phase of their life cycle, so the virus has to time its replication to match. However, many viruses have ways to alter the host cell cycle so that they can have access to the enzymes they need when they need them.

DOUBLE-STRANDED DNA VIRUSES

These viruses are often very large and complex, and may have hundreds of genes that encode their proteins (see diagram on page 70). By contrast a few RNA viruses have only one gene, and many small RNA viruses have only two genes.

RNA and DNA are generally made in only one direction, which by convention is written as 5' to 3' (see diagram on page 65). Double-stranded DNA is found as a double helix, which must unwind into separate strands before it can be copied. This is done with the help of enzymes called helicases and topoisomerases. If the DNA is being used to make RNA, then the 3'-to-5' strand of the DNA is copied in the 5'-to-3' direction in the RNA. If the DNA is being replicated, however, both strands have to be copied. This is a more complicated process, because only one strand can be copied in a continuous string of nucleotides all going in the 5'-to-3' direction. The other strand has to be copied in short lengths that are then spliced together (see diagram on page 85). The enzymes that copy DNA need to bind to something before they can act, and this is usually an RNA molecule called a primer. Other enzymes (primases) make these RNA primers to kickstart the process, then these have to be removed before the short lengths are linked together by the enzyme DNA ligase.

Most large DNA viruses use a different strategy to make the first copy of their genome and then a modified process to make additional copies. Poxviruses and many small DNA viruses use a strand-displacement method of replication, while adenoviruses are able to use one end of the genome that is complementary to the other end as a primer.

↗ A Laysan Albatross (*Phoebastria immutabilis*) chick infected with avian pox virus. Birds usually recover from this form of the disease within a few weeks.

→ Artist's rendition of a cross section of a pox virus; the inner core where the genome resides is shown in red.

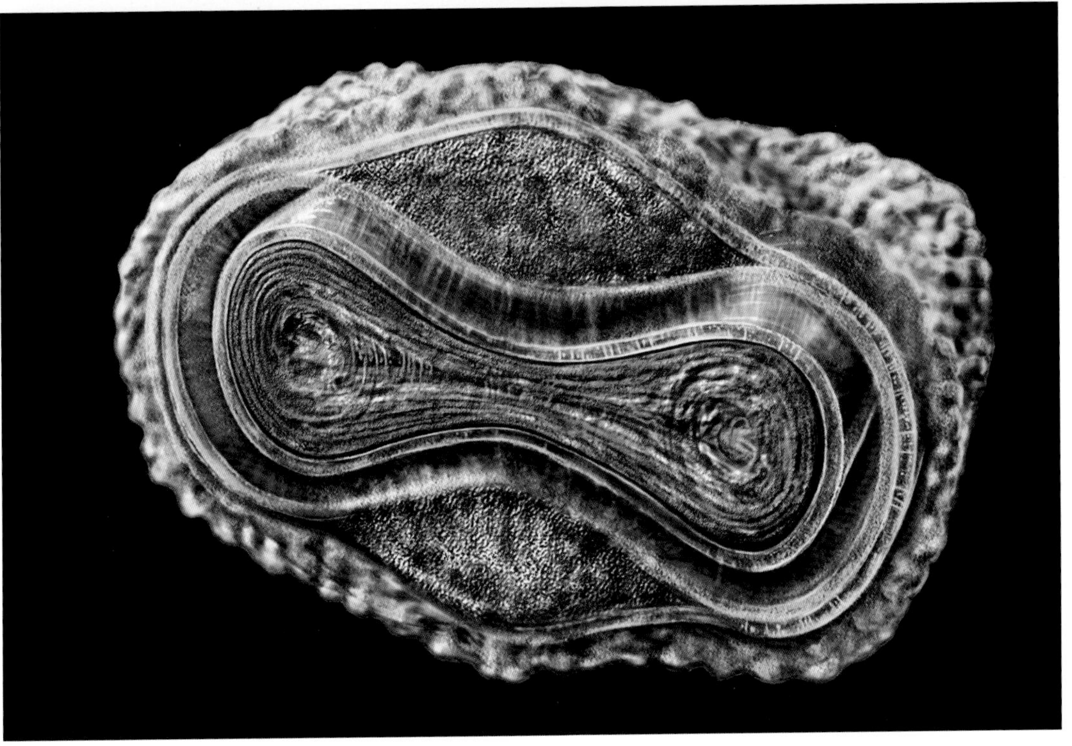

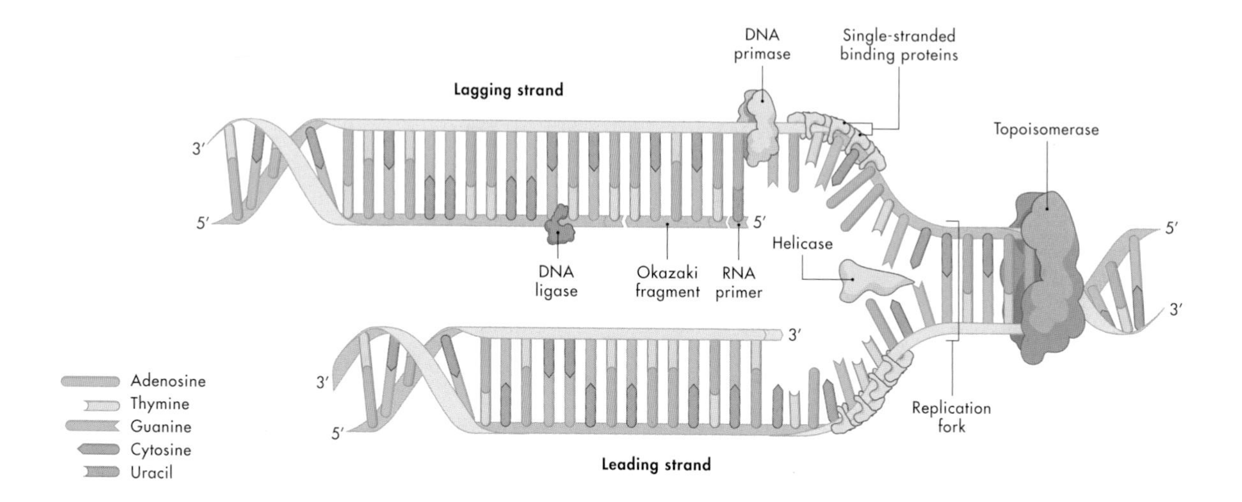

Details of DNA replication

Double-stranded DNA is made in the 5′ to 3′ direction in both strands. For one strand (the leading strand), this is a continuous copying as the DNA is unwound. For the other strand (the lagging strand), this has to be done in short pieces, called Okazaki fragments after the pair of Japanese scientists who first described them. The fragments are then spliced together by an enzyme called DNA ligase. Enzymes called helicases and topoisomerases are needed to unwind the double-stranded DNA, and single-stranded binding proteins keep it unwound. Different DNA polymerases are used to copy the lagging strand and the leading strand.

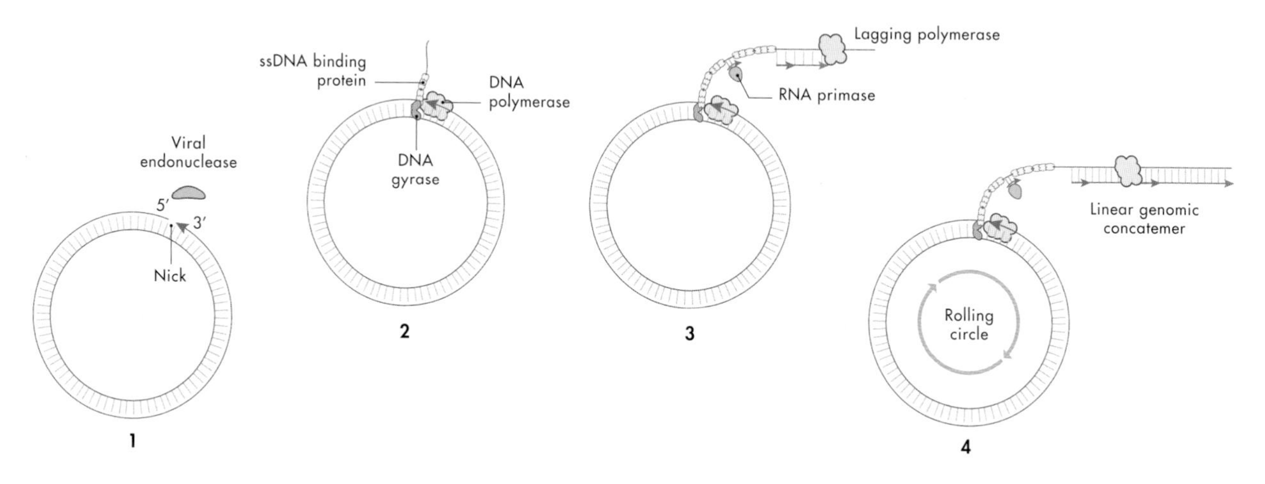

Rolling circle replication

Large DNA viruses and most bacterial DNA viruses replicate by a rolling circle mechanism, where the DNA is copied around a circular version of the genome. The enzymes are the same as shown in the diagram above.

1. DNA replication begins at specific locations in the genome, called "origins." A viral endonuclease creates a nick in the origin of replication.

2. The replication machinery assembles, with the DNA polymerase on the 3′ extremity.

3. The DNA polymerase and associated factors begin to proceed to a strand-displacement synthesis, producing a concatemer linear single-stranded DNA with one genome copy per turn of replication. On the concatemer strand, Okazaki fragments (see diagram above) are elongated after sequential RNA primer synthesis by the primase, thus turning it into dsDNA. The concatemer strand RNA primer is removed and Okazaki fragments ligated.

4. The replication forks go on and produce a long linear concatemer, which will be processed into linear genomes and encapsidated.

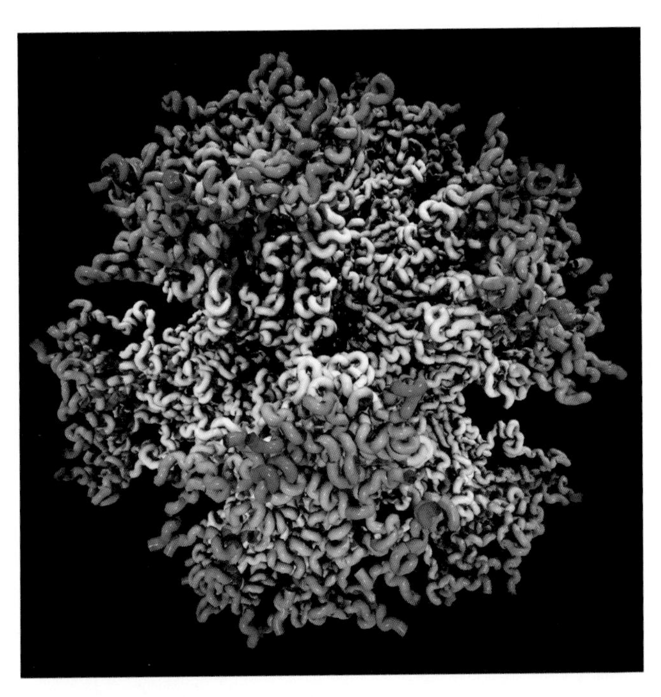

← Structure of the penton particle of human adenovirus 3.

↓ Artist's rendition of a human herpesvirus, drawn from a transmission electron micrograph.

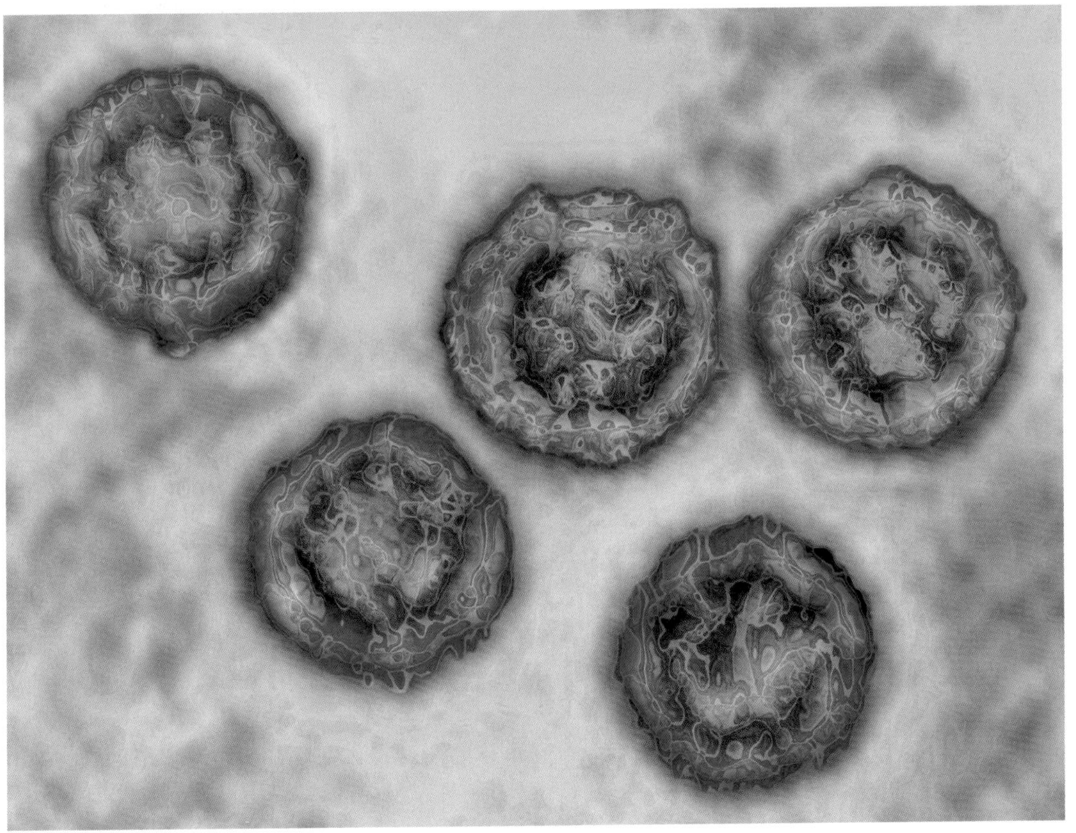

Adenovirus strand-displacement replication

Strand-displacement replication copies only one strand at a time. This then releases a single-stranded DNA that is in turn copied into the double-stranded DNA. Adenovirus, shown here, has a protein bound to the 5′ end of the genomic DNA (TP). It uses pre-terminal protein (pTP) to prime the DNA synthesis. The pTP binds to the polymerase (pTP-Pol) to guide it to start the DNA synthesis (step 1). The polymerase also acts as a topoisomerase to unwind the DNA and the DNA is kept single-stranded by single-stranded DNA binding protein (green, ssDB; step 2). After the first strand is completed the double-stranded intermediate DNA can be recycled to make more first strands (step 3). Once enough first strands are made they circularize using short complementary sequences at the ends of the genome (step 4). Then the second strand synthesis begins in a similar process to step 1 (steps 5, 6, and 7). In large DNA viruses such as the poxviruses a loop from the end folds back to prime the DNA synthesis, making long strings of multiple genomes that are processed by viral enzymes into genome lengths.

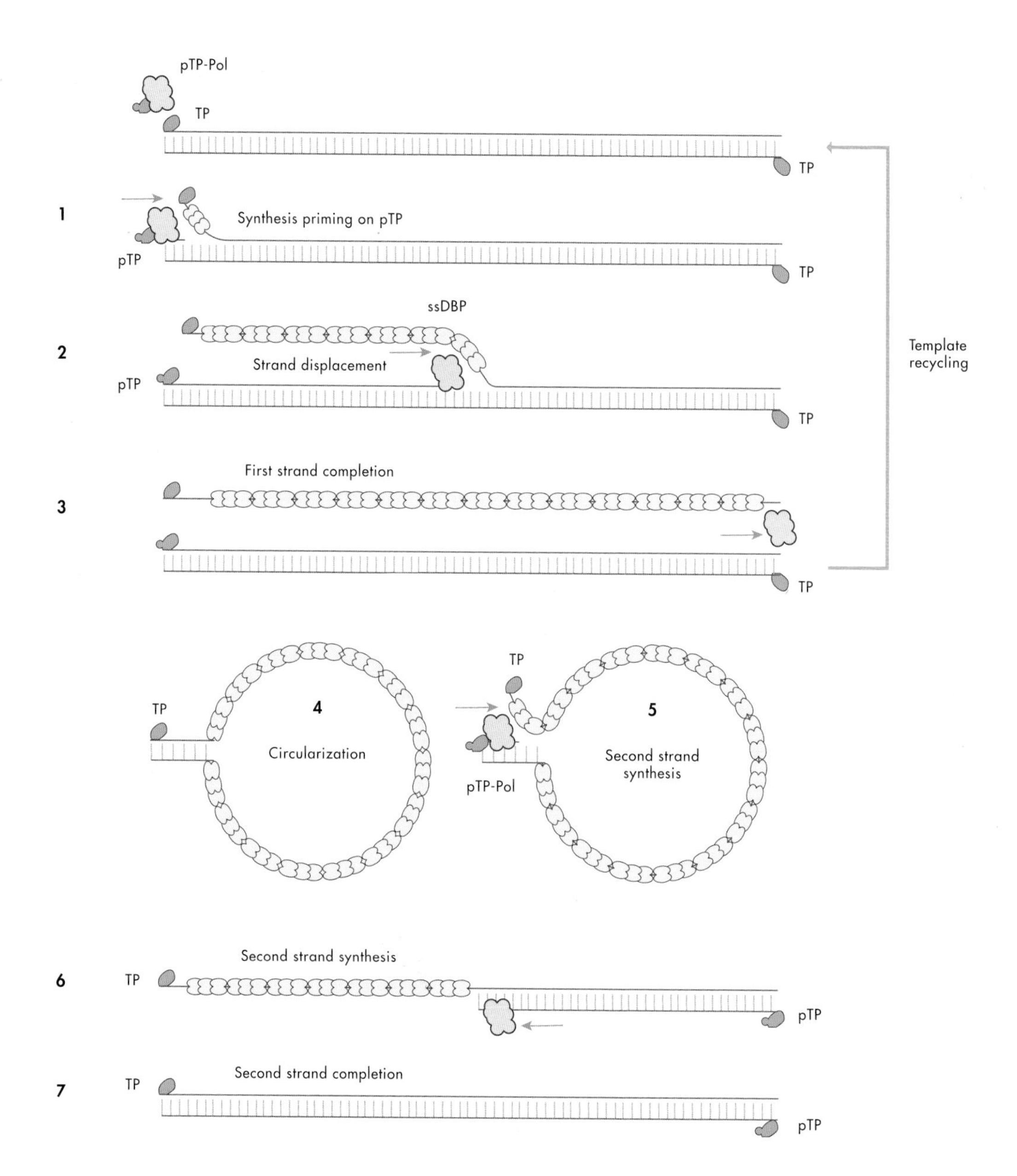

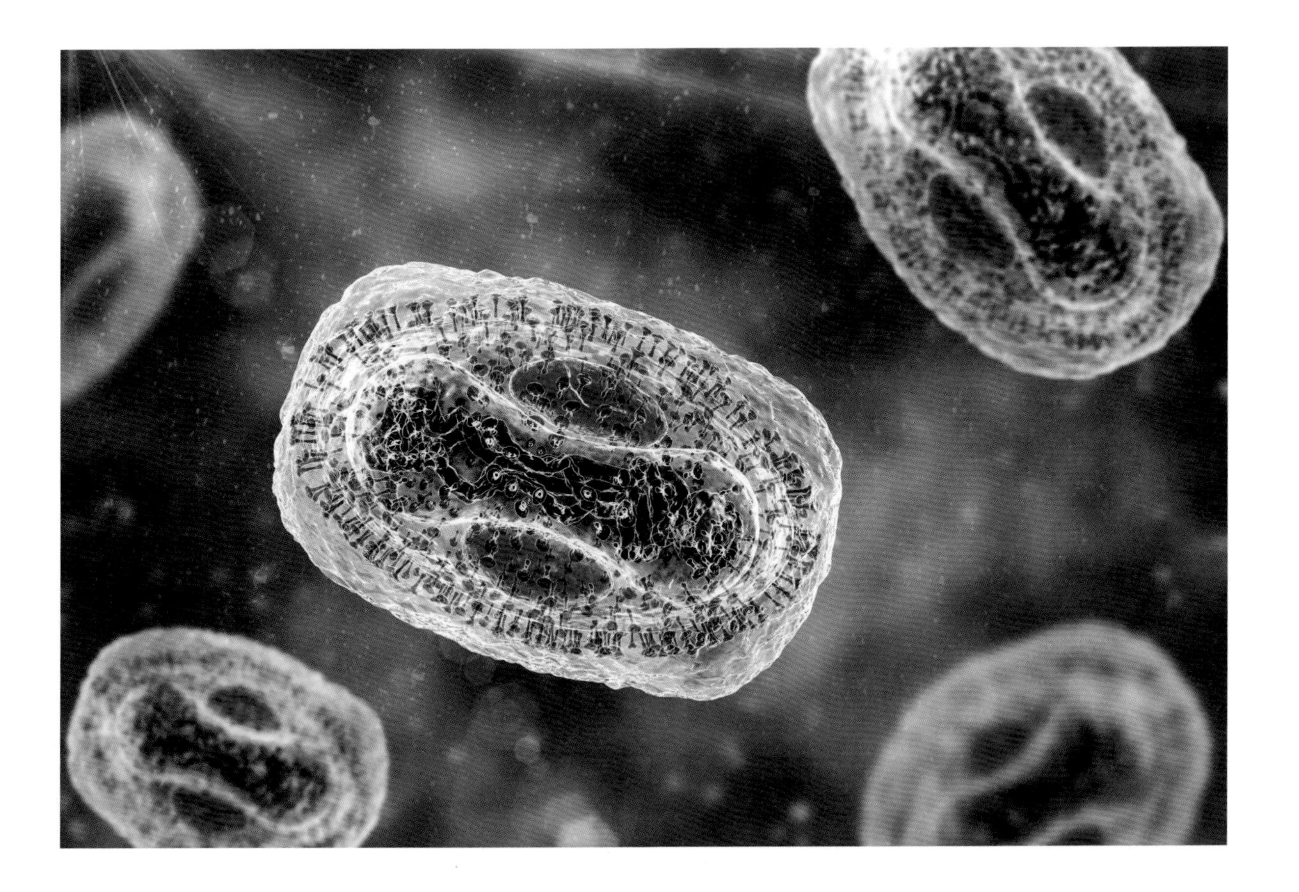

↑ Computer-assisted illustration of the monkeypox virus. All pox viruses are closely related and are very similar in structure.

Copying double-stranded DNA is the most complex replication process, so it makes sense that many viruses use the host enzymes for this. However, some double-stranded DNA viruses have their own sets of enzymes for this process. The poxviruses are examples of this. They replicate in the cytoplasm of the host cell, using about 14 of their proteins in the process. The central enzyme for copying DNA is called DNA polymerase. The poxvirus DNA polymerases are related to the polymerase enzymes of eukaryotic cells, but are substantially different. Some evolutionary biologists have speculated that the poxvirus replication enzymes are older than cellular ones, and that the nucleus of eukaryotic cells originated from a pox-like virus that was engulfed by an early protocell.

SINGLE-STRANDED DNA VIRUSES

Possibly the most abundant viruses on Earth, single-stranded DNA viruses are found in all kingdoms of life. They are also ancient—sequences of one geminivirus are found in the genome of tobacco and, based on its distribution in related plants, the virus is probably more than a million years old. Other sequences unique to single-stranded DNA viruses are found in many other hosts, including mammals, insects, fungi, and bacteria.

Most single-stranded DNA virus genomes are circular and are also very small—these viruses are among the smallest known (see, for example, porcine circovirus; page 50). The circular genome is copied into a double-stranded form by a host DNA polymerase. This is then used to make messenger RNA to produce the viral proteins, and to make more genomes.

Many circular genomes replicate by a rolling circular mechanism (see diagram opposite); these are collectively known as CRESS viruses (see page 50). The virus protein known as Rep makes a nick in one strand of the double-stranded circle. The Rep protein acts as a primer for the host polymerase to begin making single-stranded DNA, which continues around the circle until the copy is complete. In some viruses the enzyme keeps going, resulting in a string of genomes all linked together, called a concatemer. These are then cut into individual genomes, and the circle is closed by a host enzyme. The newly made genome is ready to repeat the process or to be packaged into a new particle. The virus uses many host enzymes for this process, but not for the same purposes as those required by the cell—for example, the host polymerase is usually used to make double-stranded DNA. The geminiviruses are CRESS viruses that infect plants (see page 212). Even though they use the host enzymes during their replication, which should have proofreading functions, they have large amounts of variation, similar to RNA viruses. For a virus, a lot of variation can be an advantage because it makes the virus more flexible for infecting new hosts.

Single-stranded DNA viruses with linear genomes, such as the parvoviruses (see page 268), replicate by a related method called rolling hairpin replication. In these viruses the single-stranded DNA ends fold back on themselves by base-pairing to form a hairpin. This creates a primer on which the DNA synthesis can begin.

← Pepper plants infected with a geminivirus, showing typical bright yellow symptoms.

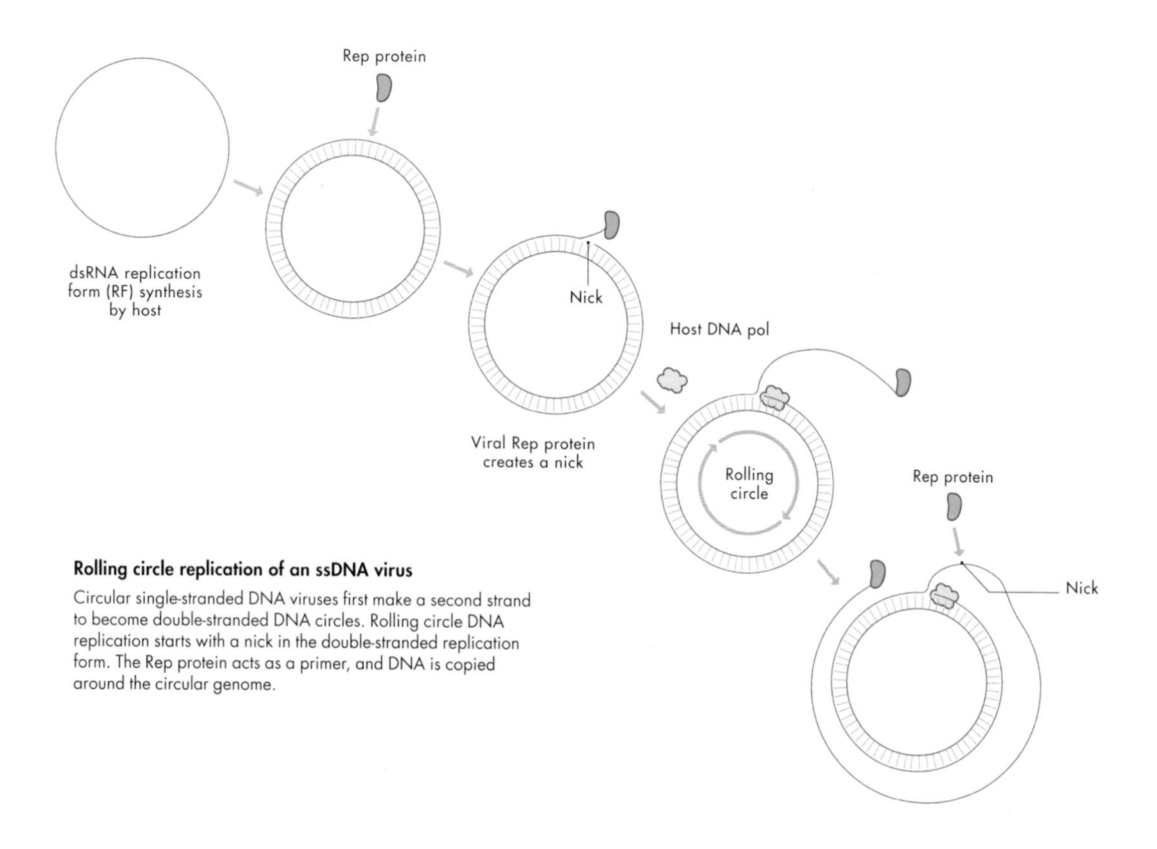

Rolling circle replication of an ssDNA virus

Circular single-stranded DNA viruses first make a second strand to become double-stranded DNA circles. Rolling circle DNA replication starts with a nick in the double-stranded replication form. The Rep protein acts as a primer, and DNA is copied around the circular genome.

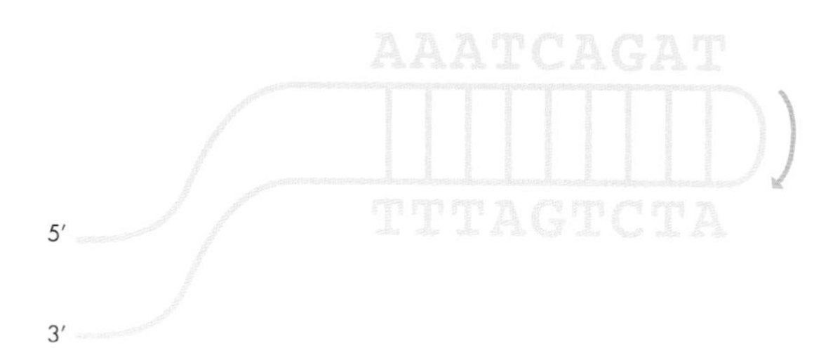

Hairpin loop for priming replication

Single-stranded DNA viruses with linear genomes make a hairpin that acts as the primer for the DNA polymerase. This is made from complementary nucleotides at the end of the genome, which fold back on each other. The hairpins can form because they contain inverted repeat nucleotide sequences. The actual hairpins are usually much longer—a simplified version is shown here to illustrate how they fold. After the strand is copied to the end, the other end is filled in.

RNA viruses

RNA viruses encode their own enzymes for replication, called RNA-dependent RNA polymerases. They copy RNA into RNA—unlike the host RNA polymerases that are DNA-dependent RNA polymerases and copy DNA into RNA. Different from DNA polymerases, RNA polymerases do not require a primer, which makes replication simpler. Most RNA-dependent RNA polymerases do not have the proofreading functions that DNA polymerases have, so they are more prone to making mistakes. An exception is the polymerase of coronaviruses, which can correct some mistakes.

Messenger RNA carries the code for DNA from a cell's nucleus to its cytoplasm, where it is read by ribosomes, which make proteins. Transfer RNA (tRNA) acts as a link between messenger RNA and the amino acids that are used to make the proteins. Messenger RNAs in cells have structures at their 5' ends called caps. These are added by enzymes in the cell, and they provide protection by making the RNAs recognizable as belonging to the cell. At their 3' ends, the messenger RNAs usually have a string of adenine residues, called the poly-A tail. Viruses use these structures, too, but some have a protein at the 5' end and some attach a transfer RNA to the 3' end.

RNA viruses have different strategies for making messenger RNA from their genomes for translation into proteins. A common strategy is called one RNA,

← Tobacco plant infected with tomato spotted wilt virus, a (–)RNA virus that infects plants and insects, and is related to animal viruses.

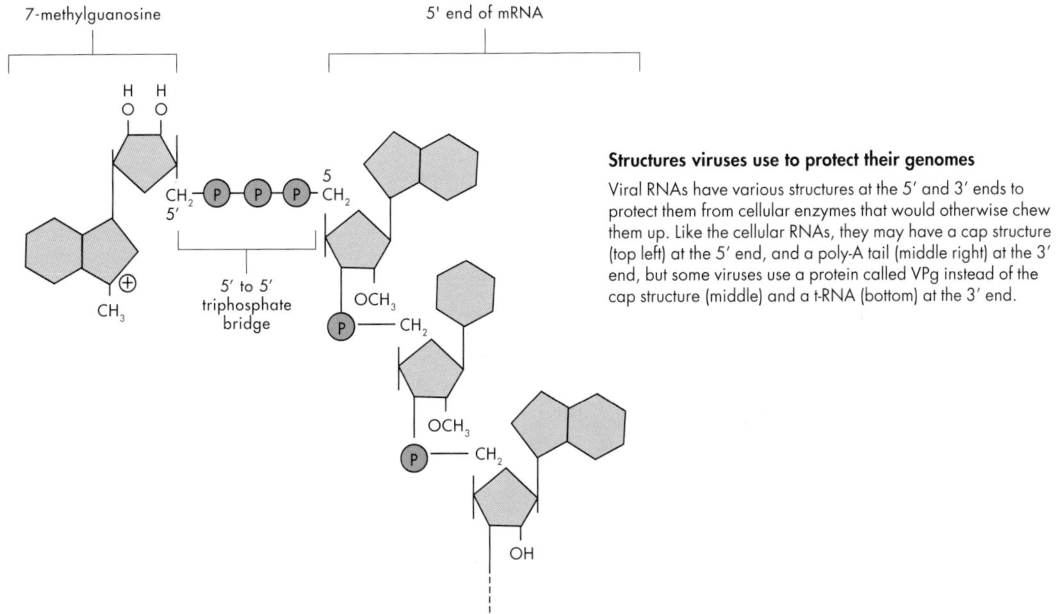

Structures viruses use to protect their genomes

Viral RNAs have various structures at the 5′ and 3′ ends to protect them from cellular enzymes that would otherwise chew them up. Like the cellular RNAs, they may have a cap structure (top left) at the 5′ end, and a poly-A tail (middle right) at the 3′ end, but some viruses use a protein called VPg instead of the cap structure (middle) and a t-RNA (bottom) at the 3′ end.

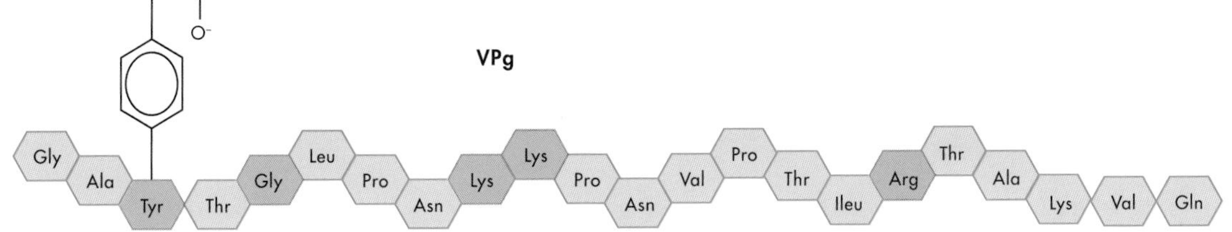

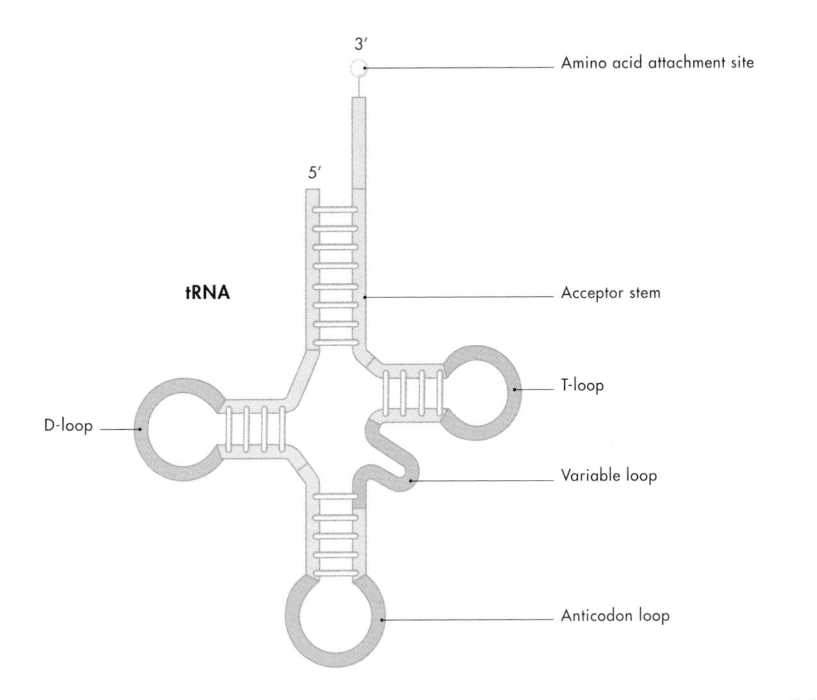

one protein, whereby each protein is encoded on a separate segment of the genome—influenza virus (page 252) is one example of a virus that uses this strategy. Another strategy is to make one large polyprotein from a single RNA molecule. The polyprotein is then cut up by the virus into the correct sizes for the individual proteins. Enteroviruses such as polio (page 210) and the rhinoviruses (page 128) use these strategies, as do many plant viruses. A third strategy is to make smaller RNA molecules from the genome, known as subgenomic RNAs, which act as messenger RNAs. Tobacco mosaic virus (page 96) uses this strategy. Some viruses, including cucumber mosaic virus (page 214), use a combination of these strategies. Finally, some negative stranded RNA viruses make their messenger RNAs directly from the genomic RNA by starting and ending the RNA copying process at different points across the genome, much like DNA templates for making messenger RNA. The genomic RNAs have specific sequences of nucleotides that act as signals to tell the enzymes where to start and stop, just as the messenger RNAs have signals indicating where protein synthesis starts and stops (see table on page 12).

Many RNA viruses replicate in complex structures that they induce the cell to make out of membranes, called viroplasms. These provide a safe place for the virus to replicate without any interference from the cell, and also keep all of their necessary enzymes together in one place rather than floating around freely in the cell's cytoplasm.

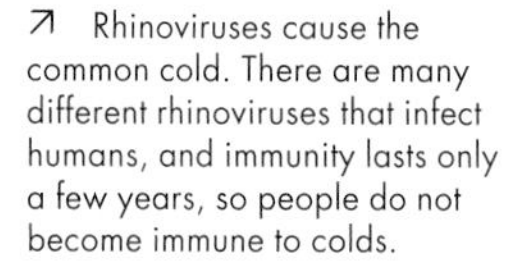

↗ Rhinoviruses cause the common cold. There are many different rhinoviruses that infect humans, and immunity lasts only a few years, so people do not become immune to colds.

→ Poliovirus can cause poliomyelitis, a severe disease that affects the nerves and can result in paralysis. Some people fully recover from this disease, but others are left with lifelong disability.

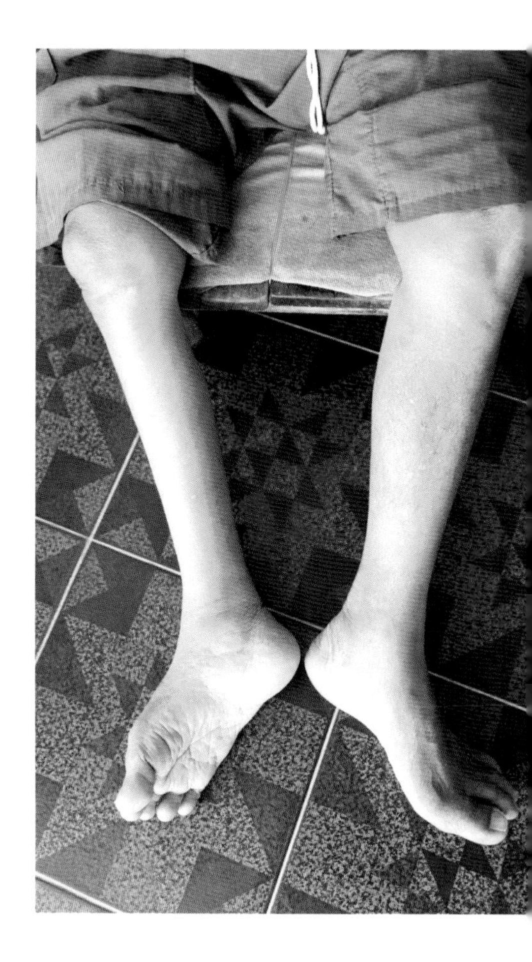

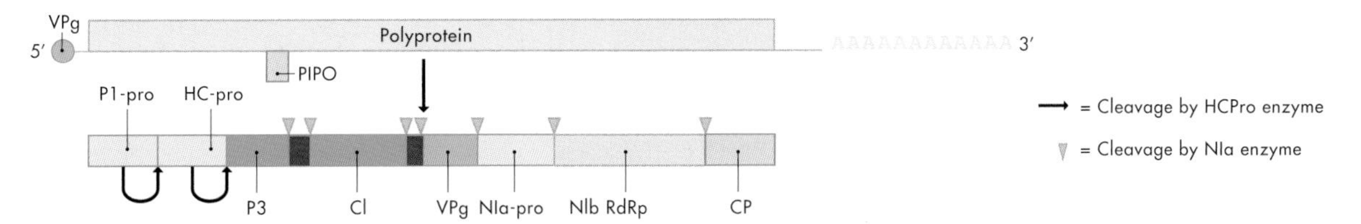

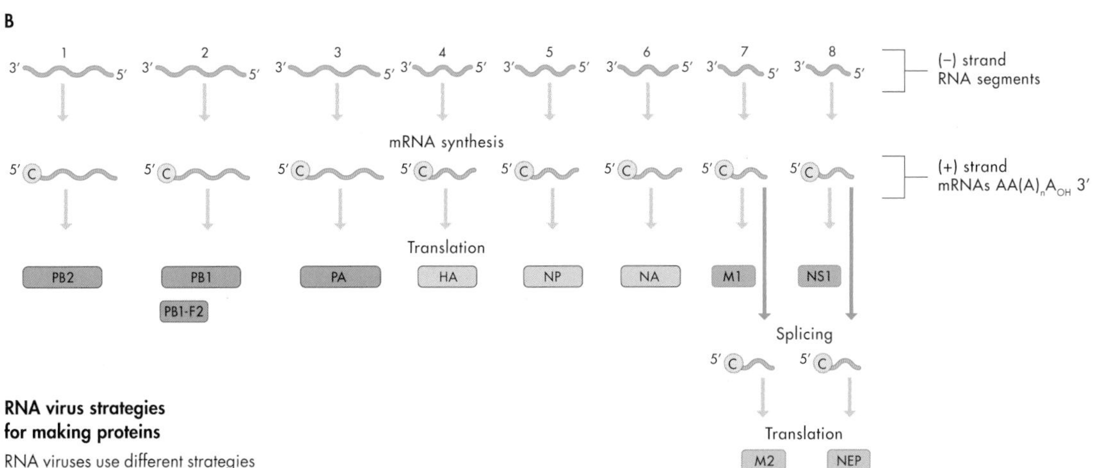

RNA virus strategies for making proteins

RNA viruses use different strategies to make their proteins.

(A) Plum pox virus, a (+) sense virus, makes one large protein, which is cut into functional proteins (indicated with abbreviations) by enzymes that the virus makes. There is an additional protein, PIPO, that is not part of the polyprotein. It is made when the polymerase making the mRNA for the polyprotein skips a nucleotide (polymerase slippage).

(B) Influenza virus, a (–) sense virus, mainly uses the one-RNA-one-protein strategy. Most of the eight segments encode only a single protein, although segments two, seven, and eight make two.

(C) Tobacco mosaic virus uses the subgenomic RNA strategy. Signals in the RNA, called promoters, tell the polymerase where to start making the RNA. It also uses "leaky" translation to maximize the proteins it can make. A codon that is usually a stop codon, signaling the end of translation, is occasionally read as a code for an amino acid, and the protein synthesis proceeds.

(D) Rhabdoviruses, another (–) sense RNA virus, make their messenger RNAs for each protein directly from the infecting genome.

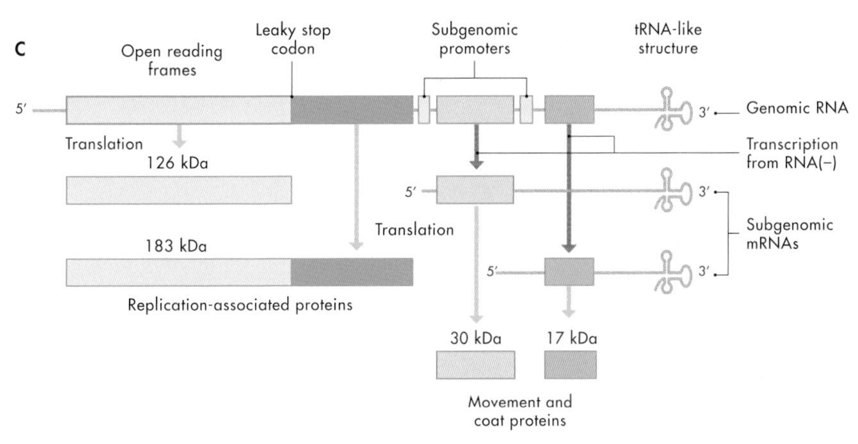

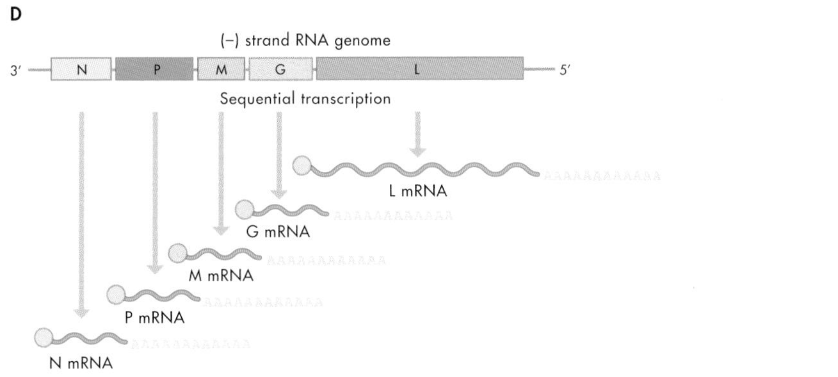

DOUBLE-STRANDED RNA VIRUSES

These viruses carry their polymerase in the virus particle. They do not fully uncoat and stay inside the core virus particle, making messenger RNAs and complete RNA copies of the genome, and then push these out into the cell. This process may have evolved because cells do not make large double-stranded RNAs, and as a result these trigger several host immune responses. The virus's strategy of keeping the genome in the particle hides it from the cell.

Double-stranded RNA virus replication

When double-stranded RNA viruses are copied, the 5′ end of the RNA may be naked, capped, or linked to a VPg. The genomic RNA is first copied to make the messenger RNAs and the pre-genomic RNA inside the viral capsid. The polymerase is usually found at the point in the capsid where five coat proteins come together. Once the viral proteins are made by the cell from the messenger RNAs, the single-stranded pregenome is packaged in new virus particles, and then the second strand is made to produce the double-stranded RNA genome. As in most RNA viruses, the virus life cycle takes place in a structure inside the cell called the viroplasm, which is surrounded by a membrane derived from the host cell.

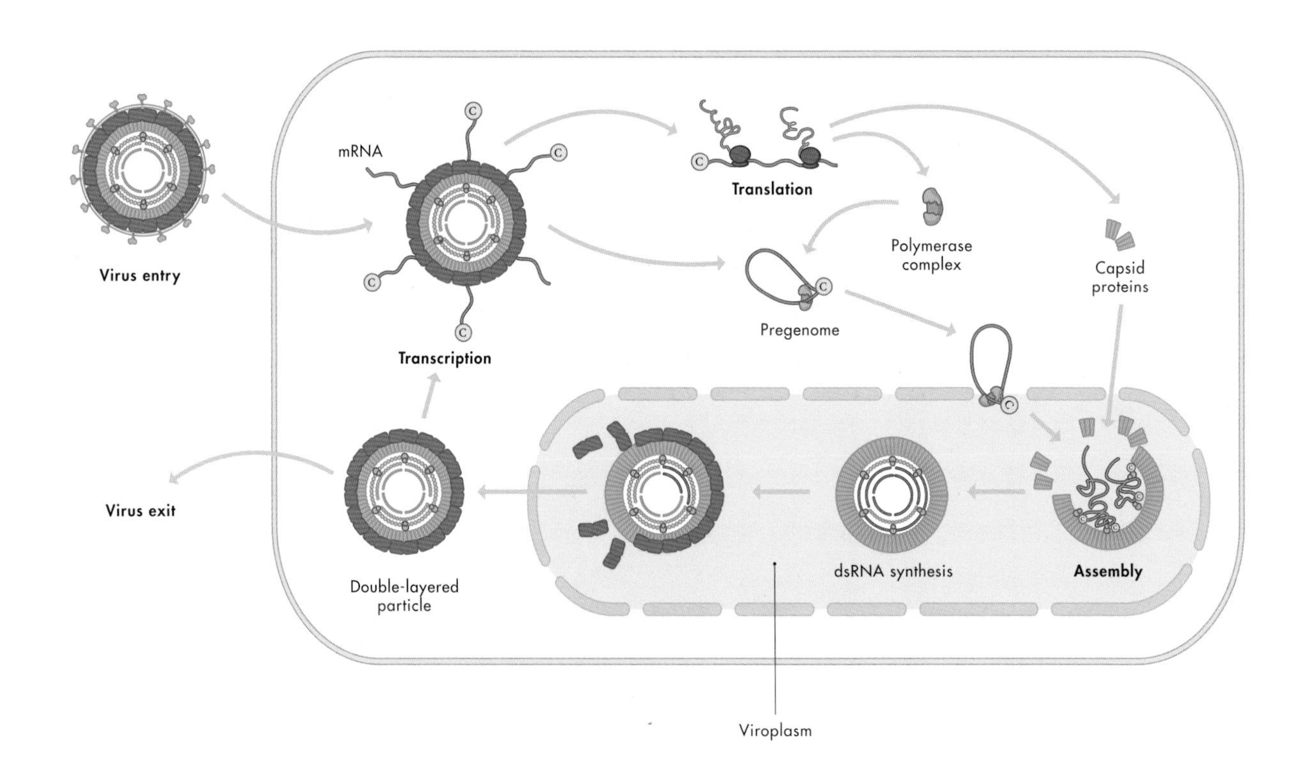

Double-stranded RNA viruses usually follow the one RNA, one protein rule. Some of the more common human double-stranded RNA viruses have 10 or 11 RNAs, which are all packaged into a single particle. For many plant and fungal double-stranded RNA viruses that have simpler genomes (see pages 238 and 246), each segment is packaged into its own particle. Single-stranded plant viruses use this strategy too. For the virus, the smaller particle size is easier to make and can move more easily between cells. However, to establish an infection, the virus has to move multiple virus particles into one cell.

The messenger RNAs made by the virus are translated into proteins by the host ribosomes. These proteins allow the virus to complete its life cycle, replicating and packaging new genomes into new particles. However, the viruses package a pregenome, a single-stranded version of the genome. Once the pregenome is safely packaged into a new particle, it copies its second strand to make the double-stranded genome.

↑ Most raspberry plants grown in North America are infected with raspberry latent virus, a double-stranded RNA virus that does not cause any symptoms.

SINGLE-STRANDED RNA VIRUSES

These viruses come in two varieties: plus sense (+) and minus sense (−). The (+)RNA viruses have genomes that can act as messenger RNAs directly. They do not need to carry the polymerase they need in their particle, because they can make it directly from the genomic RNA. This has been a bonus for virologists, because the genomic RNA can be used by itself to establish an infection. Many studies looking to understand the genetics of (+)RNA viruses have used clones of viruses that can be mutated in the lab. For example, if you make a deletion in a viral gene and then see what happens during an infection, then you can tell what the gene normally does. This is called reverse genetics. Another approach may be to take a gene from one virus and put it into another virus that behaves differently to see if you can add a new property (called gain-of-function genetics). These tools are very powerful, and have helped in our understanding of thousands of viruses and viral genes. They are not risk-free, however, especially when used with viruses that cause serious diseases. As a result, scientists take strict precautions to contain engineered viruses.

Plus-sense RNA virus replication

Replication of a (+)RNA virus usually occurs in the cytoplasm of the host. The infecting RNA is able to act as a messenger RNA, so the first proteins are made directly from the genome (translation). The antisense strand is then made using viral enzymes and host factors. When the antisense strand is made, there is likely a temporary double-stranded form before the new (+)RNA is made, either for creating further messenger RNAs or for new genomes. These viruses can reach very high levels of genome replication, because each infecting genome can be used to make many antisense strands, and then each antisense strand can make many plus strands.

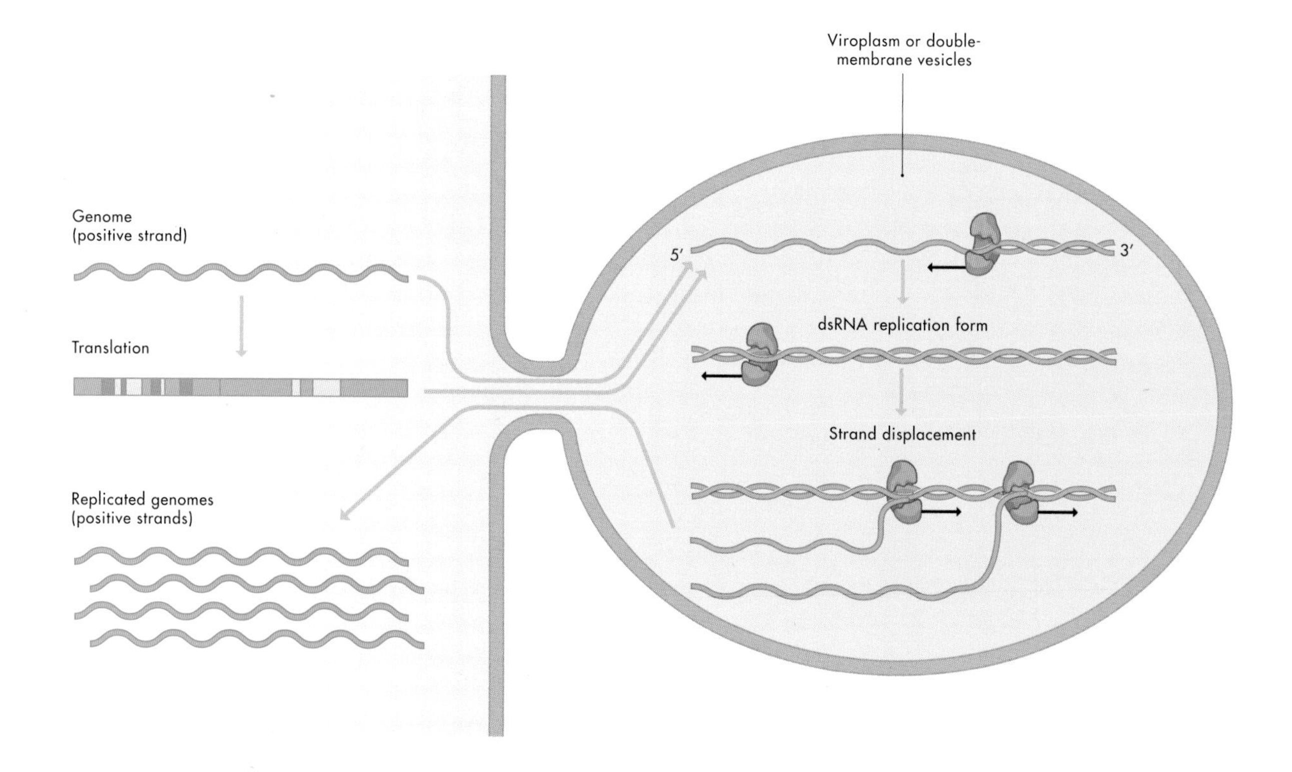

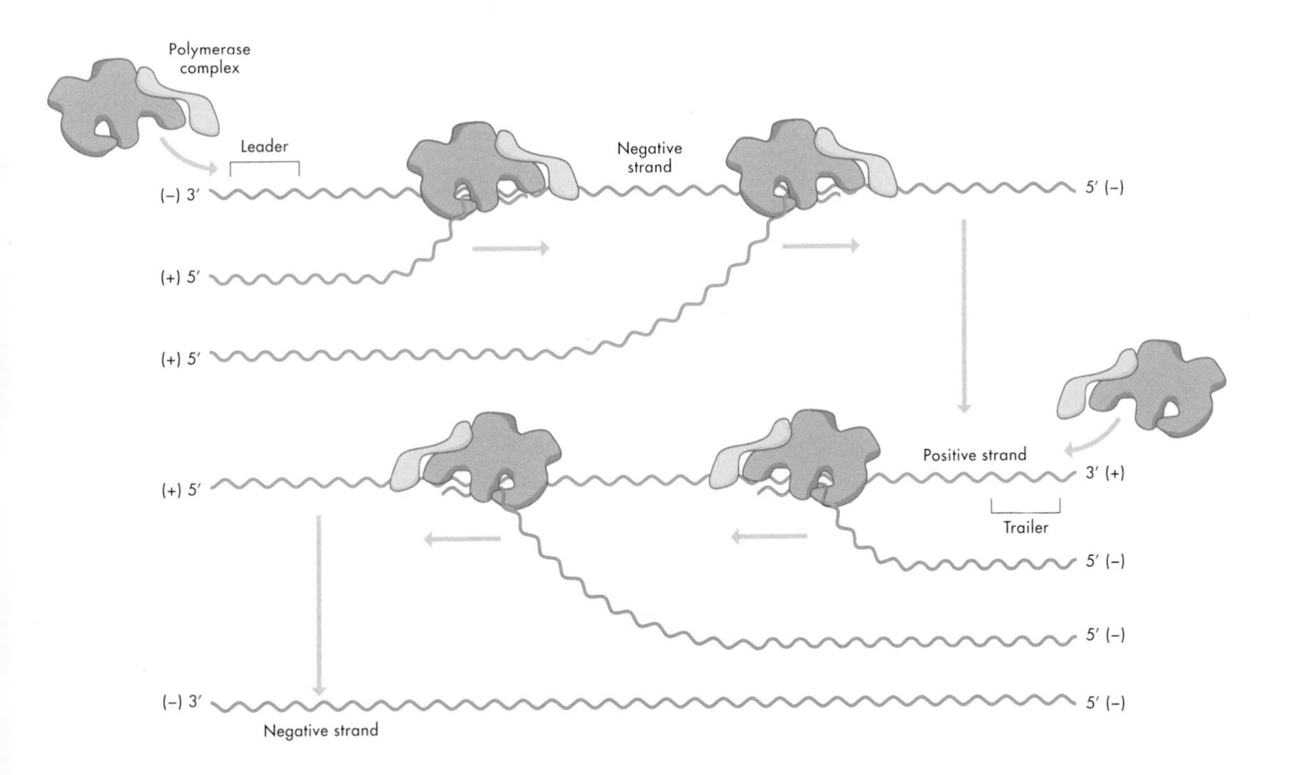

Single-stranded RNA viruses copy their genome by making first one strand and then the other, resulting in an intermediate that is double-stranded. This process happens inside the viroplasm, which keeps the double-stranded RNA from being detected by the cell. The double-stranded form is temporary, the goal of the virus being to make more copies of its single-stranded genome.

The (−)RNA viruses have a genome that cannot act as a messenger RNA, but has to be copied first into the complementary strand. These viruses have to carry their polymerase with them, just like the double-stranded RNA viruses. Cloning for these viruses can be done, but it is much more complicated than for (+)RNA viruses because the virus has to be provided with its polymerase separately.

Minus sense RNA virus replication

With (−)RNA virus replication, the genome cannot be directly translated into protein. The virus has to make the plus-sense strand, so like double-stranded RNA viruses it must also take its polymerase with it. The plus-strand synthesis begins at the 5′ leader to make either messenger RNAs or new genomes. The negative-sense genome is then made from the (+)RNA and the new genomes are packaged.

Retroviruses and pararetroviruses

Some viruses convert their genome between RNA and DNA. An example is the retroviruses, which have a single-stranded (+)RNA genome and carry two copies of this in the virus particle. They also carry their unique polymerase, called reverse transcriptase. This enzyme can copy RNA into DNA, a process early molecular biologists thought was impossible until it was discovered in the 1970s.

RETROVIRUSES

Retroviruses have envelopes and enter the cell by fusion with the cellular plasma membrane. Once inside the cell, the inner particle moves to the nucleus and copies its RNA genome into double-stranded DNA. This DNA is then inserted into the genome of the host cell, where it remains for the rest of the cell's life and in the offspring of that cell. If this event happens in a germ-line cell (e.g., a sperm cell or egg cell), the virus becomes a permanent part of the host genome. This is why so many retrovirus sequences are found in genomes (8 percent of the human genome is retroviral). Most of the time these viruses do not infect germ-line cells, so the integrated virus is not passed on to the offspring of the host.

Once the retrovirus DNA is integrated into the host cell, it acts much like any other gene. The host enzymes make messenger RNA from the integrated virus, which they then splice. Splicing is a very common process in messenger RNA. Cellular messenger RNA is made with parts that are removed, called introns, and parts that are kept, called exons. The introns are cut out by cellular enzymes to make the mature messenger RNA. This splicing allows the single RNA to be converted to several different messenger RNAs, so that it can make all of the necessary proteins for the virus. Some of these proteins are polyproteins, which are cut into the required functional parts by the virus digestive enzymes, called proteases. These proteases are unique to retroviruses, and have been used as targets for antiretrovirus drugs. The virus genome is also made from the integrated DNA. This is then packaged into new virus particles, with two genomes per particle. The virus core particles bud through the host plasma membrane and acquire their new envelope on the way out.

Researchers have used retroviruses as vectors for studying genes of interest. A gene can be cloned into the virus in the lab, and then moved into cells to study them. The most common virus for this purpose is a mouse virus, Maloney murine leukemia virus. Retroviruses have also been proposed for use as gene therapy, to provide a good copy of a mutant gene for people with genetic diseases. There is more about the beneficial uses of viruses on page 234.

Retroviruses are common in many different types of animal life. Although there are currently no known active retroviruses in fungi, protists, and plants, parts of these viruses are found in the genomes of these organisms, indicating their presence in them in the past.

Replication of retroviruses

Retroviruses enter the host cell by fusing their membranes and releasing the inner core, which contains two copies of the genome. The single-stranded RNA is converted to double-stranded DNA by the viral enzyme reverse transcriptase. The virus core then moves to the nucleus and the double-stranded DNA (pink) is integrated into the host genome (blue). From there, the messenger RNAs and the new genomes are made just like cellular messenger RNAs. The viral proteins are made in the cytoplasm and used to assemble new virus particles.

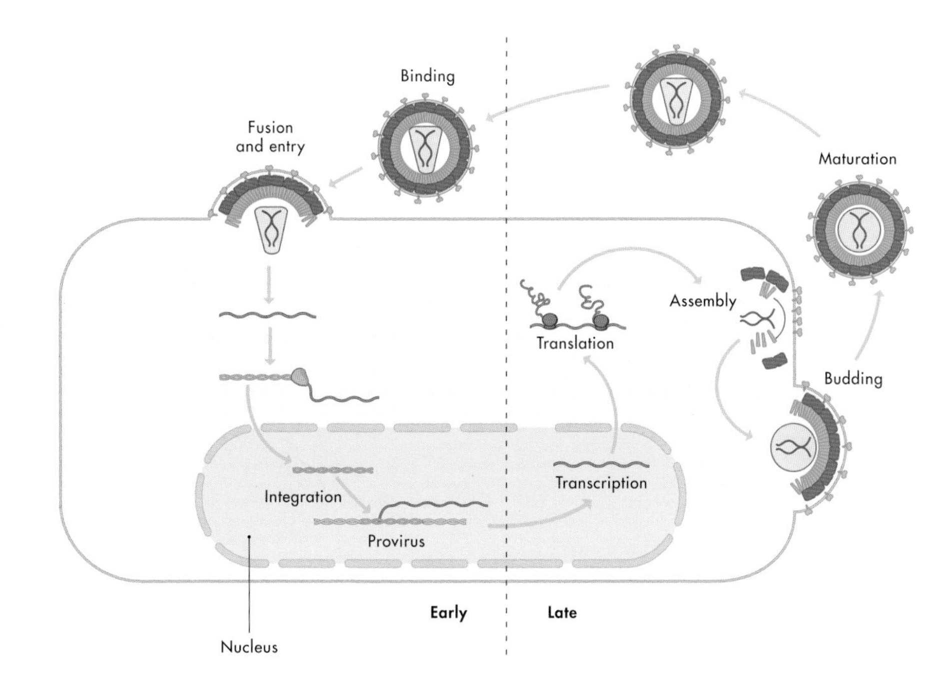

RNA splicing

When DNA is translated into RNA it contains parts that are used for translation into protein (exons) and parts that are not (introns). The introns carry other information related to how and when a gene is used. They have to be removed from the RNA before it can be used as a mature strand of mRNA, a process carried out by ribonucleoproteins that are part RNA and part protein. Most eukaryotic genes have introns and some viruses use this strategy too.

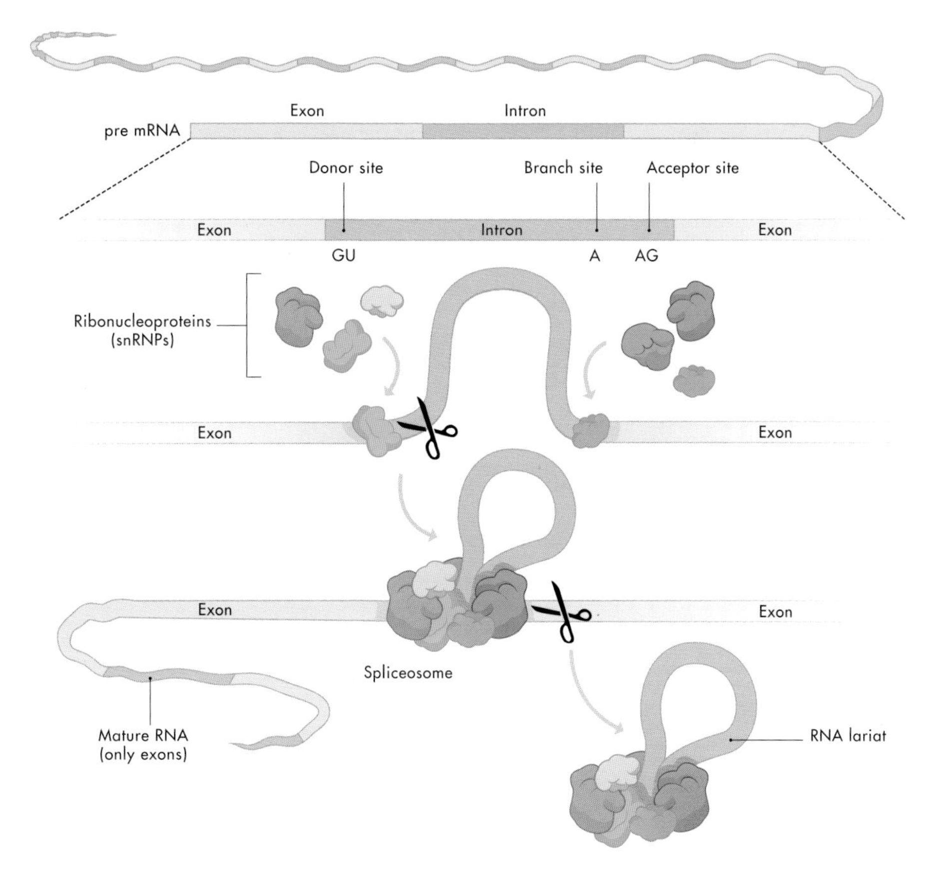

PARARETROVIRUSES

While they share some similarities with retroviruses, pararetroviruses differ in that they package their genomes as DNA or as a DNA–RNA hybrid. They are found commonly in plants, and a few, such as the hepadnaviruses, can be found in humans and other animals.

Although some pararetroviruses are found integrated in the host genome, this is not usually a requirement for their replication. Instead, they enter the host nucleus as double-stranded DNA, which is then converted to a "minichromosome." The viruses use cell proteins called histones (which are normally attached to chromosomes) as a template, and the

Replication of a pararetrovirus

Caulimoviruses are plant-infecting pararetroviruses (Type VII). Once the virus enters the cell, the DNA genome is released (1). The viral genome enters the cell nucleus and is converted to fully double-stranded DNA (2). The genome complexes with the host histone proteins (3). The host enzymes produce two mRNAs (4), one of which makes the P6 protein (5), while the other makes all the other viral proteins (6). The reverse transcriptase (RT) makes the RNA pregenome (7), and the second strand (DNA) is made (8) and packaged (9).

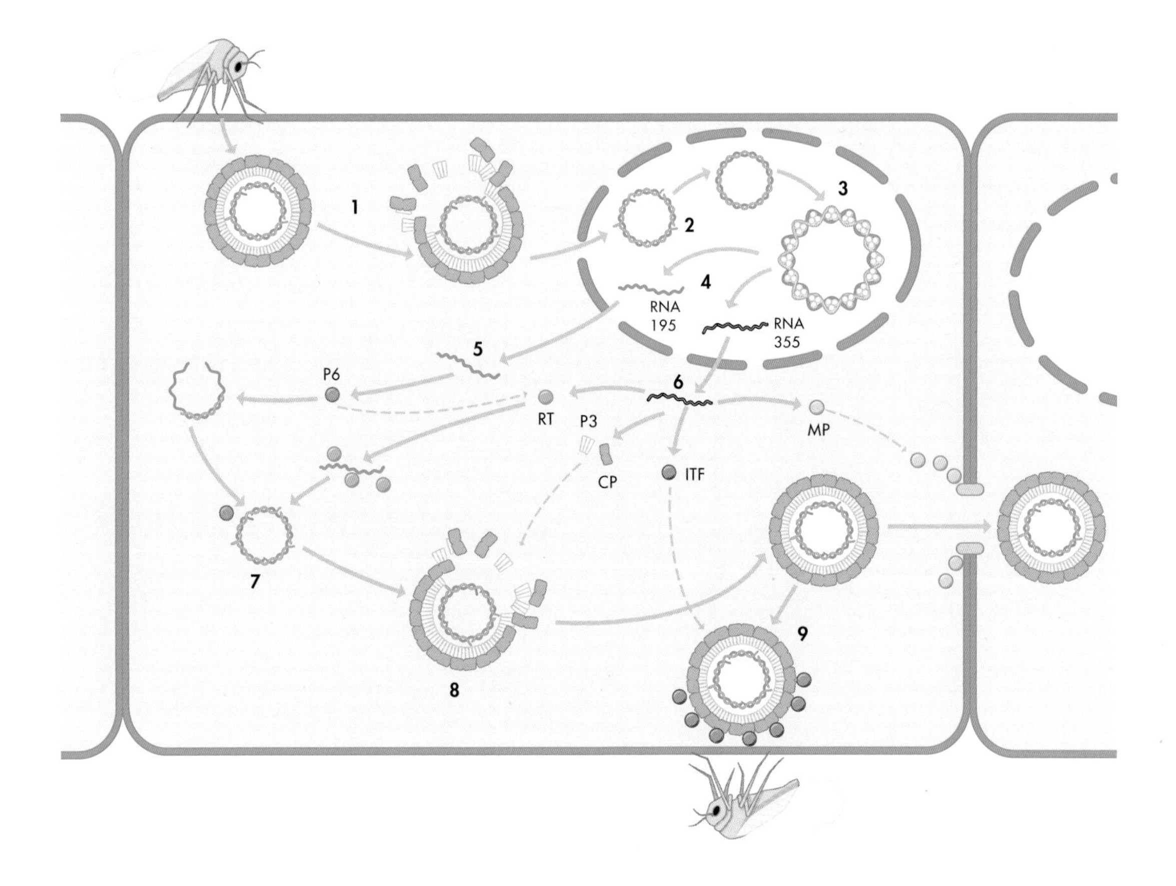

messenger RNAs are then made from this, similar to the process in retroviruses. A full-length RNA is made as a pregenome, and the virus then converts this to its DNA genome. In some pararetroviruses the pregenome is completely converted into double-stranded DNA, but in others it is only partially converted, so the packaged genome is a hybrid molecule that is part DNA and part RNA.

Integration of retroviruses and pararetroviruses can have serious consequences for the host. In some cases the virus can disrupt an important gene, while in others it can activate a gene when it integrates nearby, leading to cancer. Many viruses also carry oncogenes, or cancer-causing genes, so integration can lead to cancer by providing an oncogene in the integrated virus, or by activating a gene in the host genome.

← Cauliflower mosaic virus is a pararetrovirus that infects several different crop plants. The distortion symptoms on these cabbage leaves are caused by this virus.

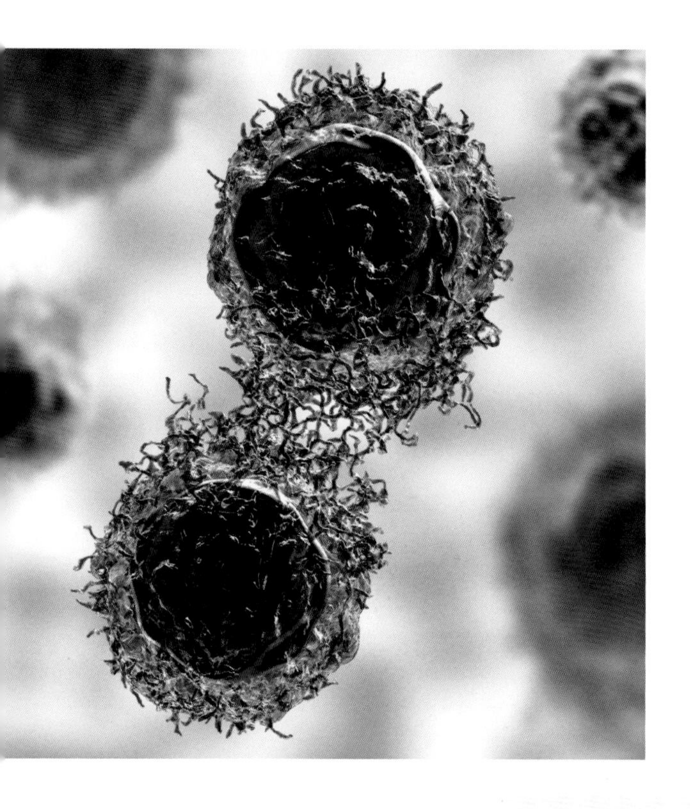

For all viruses, the critical process of replication can take place only inside of a living cell. Viruses need to get all the basic components for this process from their host, including nucleotides, amino acids, machinery to synthesize proteins, and often enzymes. Some viruses replicate to such high levels that the host cell is overwhelmed—for example, it is estimated that millions of acute plant viruses are reproduced by each cell that is infected. On the other hand, some viruses make many fewer copies of themselves and may go unnoticed by the host—for example, the plant persistent viruses (see page 238) make fewer than 500 copies per cell.

↑ Cancer cells from a Burkitt lymphoma, a cancer caused by Epstein–Barr virus.

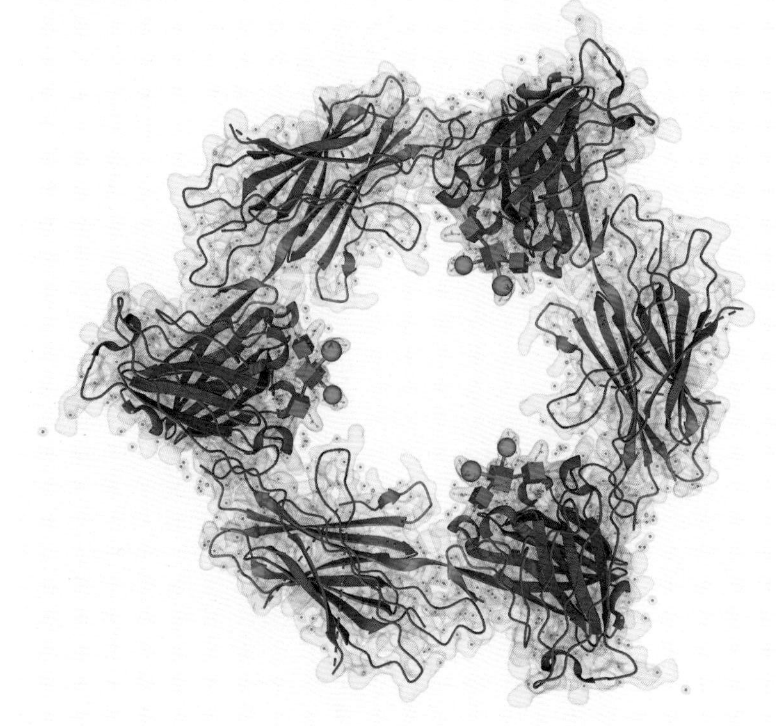

→ Ribbon diagram structure of BARF1, a cancer-causing gene from Epstein–Barr virus. This herpesvirus causes mononucleosis in humans, and is implicated in several human cancers, including Burkitt lymphoma and throat and stomach cancers, but the role of the oncogene has been shown clearly only in cells grown in culture.

REPLICATION OF VIROIDS AND SUBVIRAL ENTITIES

Viroids are small, infectious RNA molecules that are most often found in plants, such as the potato spindle tuber viroid (page 60). They are divided into two major families, the Pospiviroidae and the Avsunviroidae. Viroids are replicated by the host enzyme, called RNA Pol II, which is used by the host to make messenger RNA from DNA, or by a chloroplast RNA polymerase. The two families both use a rolling circle method of replication, but they use different strategies to resolve the long concatemers of their genomes. Members of the Pospiviroidae family use a host enzyme called RNase III, whereas the Avsunviroidae have a structured RNA within their genome called a ribozyme. Ribozymes are RNA molecules that behave like an enzyme (enzymes are always proteins). By convention, the genomic strand of a viroid is called the plus strand, but viroids do not encode any proteins.

A number of other small RNAs or DNAs, called subviral entities because they are found associated with viruses, replicate in a similar manner to viroids. An example is the hepatitis deltavirus (page 58).

Viroid replication

Replication of viroids occurs by different pathways depending on the viroid family. The Pospiviroidae replicate in the host cell nucleus, using the host RNA polymerase II (RNA pol II), whereas the Avsunviroidae replicate in the chloroplast (a cell structure in plants responsible for photosynthesis) using the host nuclear encoded chloroplastic RNA polymerase (NEP). In both cases the circular genome is copied by a rolling circle method. In the asymmetric pathway the concatemers of antisense RNA are copied into concatemers of (+)RNA, which are then cut to genome length by the host enzyme RNase III. The genomes are made into circles by the host RNA ligase. In the symmetrical pathway, the antisense strands are cut into genome length by the viroid encoded ribozyme, an RNA structure that can cut itself. This is circularized and copied into plus-strand concatemers, which are also cleaved by the viroid ribozyme.

Asymmetric pathway (e.g., family Pospiviroidae)

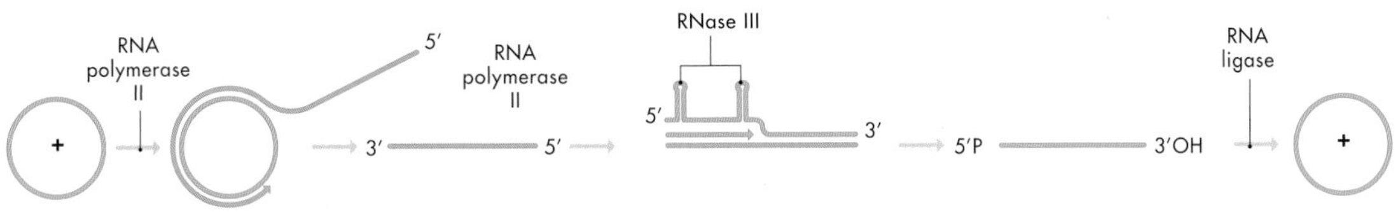

Symmetric pathway (e.g., family Avsunviroidae)

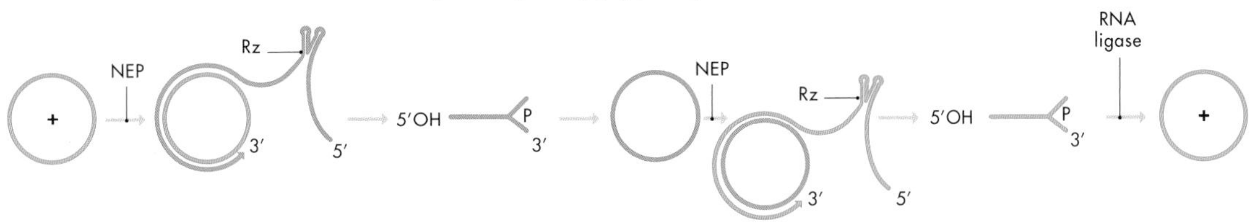

Fowlpox virus

Serious pathogen of domestic birds, easily controlled by vaccination

GROUP	I
FAMILY	Poxviridae
GENUS	Avipoxvirus
GENOME	Linear, single-component, double-stranded DNA of about 290 kilobases, encoding about 260 proteins
VIRUS PARTICLE	Enveloped, brick-shaped, about 360 nm by 250 nm
HOSTS	Chickens, turkeys
ASSOCIATED DISEASES	Fowlpox
TRANSMISSION	Mosquitoes, inhalation
VACCINE	Live attenuated related pigeonpox virus

Fowlpox is related to the viruses that cause smallpox and monkeypox in humans, and to a number of similar poxviruses of birds. The disease takes two forms, depending on how it is transmitted. Birds that get the virus from a mosquito have a mild form and usually recover fully, while birds that inhale the virus from other infected birds often die.

Like all poxviruses, fowlpox replicates in the cytoplasm of infected cells, using its own DNA-dependent DNA polymerase. Most other large DNA viruses use the host's enzymes for replication, and those that infect eukaryotes replicate in the nucleus of the infected cell. The genes of fowlpox virus are divided into "early", "middle", and "late" genes, depending on how long after initial infection they are active. The products of early genes start showing up about 30 minutes after the cell is infected, whereas late genes may not be active until up to two days after infection. The virus doesn't fully uncoat until the middle genes become active. The late genes code for proteins that make up the virus particle itself, called structural proteins.

Some strains of fowlpox virus contain an entire retrovirus within their genome. This is a unique example of a retrovirus endogenizing in another virus rather than the host genome. It is not known how this happens. The strains of fowlpox that contain this retrovirus genome are more likely to be linked to a lymphoma, a form of cancer, in the infected birds.

Most commercially grown birds are vaccinated against fowlpox, but the virus can show up in backyard chickens. Once a flock is infected, there is no treatment, although any spread may be contained by exterminating the mosquitoes that transmit the virus.

→ Ribbon diagram of two proteins involved in the replication of fowlpox virus, complexed with the viral DNA, which can be recognized in the center as a double helix.

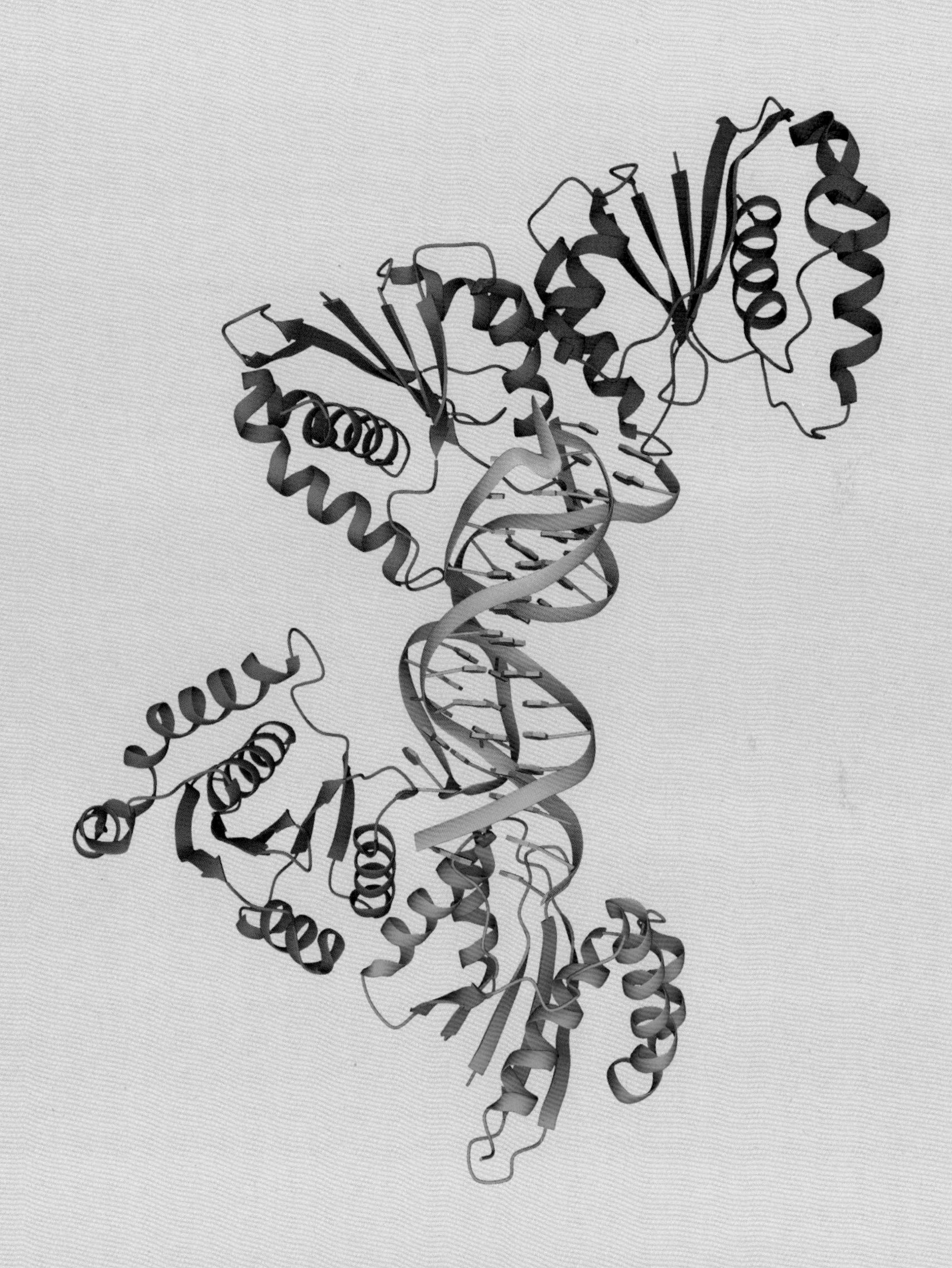

TTV1

Torque teno virus 1

Major part of the human virome

GROUP	II
FAMILY	Anelloviridae
GENUS	Alphatorquevirus
GENOME	Circular, single-component, single-stranded DNA of about 3,900 nucleotides, encoding four proteins
VIRUS PARTICLE	Icosahedral, about 30 nm
HOSTS	Humans, but related viruses are found in many mammals
ASSOCIATED DISEASES	None
TRANSMISSION	Unknown

Torque teno virus 1 (TTV1) was first described in the late 1990s, in a patient who had undergone a liver transplant. It is sometimes called transfusion transmitted virus, but in fact it is ubiquitous in humans and is not associated with any disease. However, the level of the virus can vary significantly, making up about 10 percent of the virome in a healthy individual but up to 65 percent in people who have had immunosuppression therapy in preparation for an organ transplant.

The genome of TTV1 is quite variable, and this has led to studies of the virus in different human populations. It is found in populations from both rural and urban areas, and is more similar among people in a single community, implying that it is spread within a local population. Some people have multiple variants of the virus, and many people have antibodies to the pig version of TTV. It is not known how the virus spreads, and it may use multiple routes of transmission.

TTV1 can also be found in the environment, in water supplies, wastewater treatment facilities, and hospitals. This indicates that it is very stable and can withstand various environmental stresses.

With the high degree of variation in levels of TTV1 between individual hosts, scientists have proposed its use as a sort of human fingerprint for studying human migration patterns and also in forensics. It has also been proposed as an indicator of human fecal contamination of groundwater.

→ Computer-generated image of Torque teno virus.

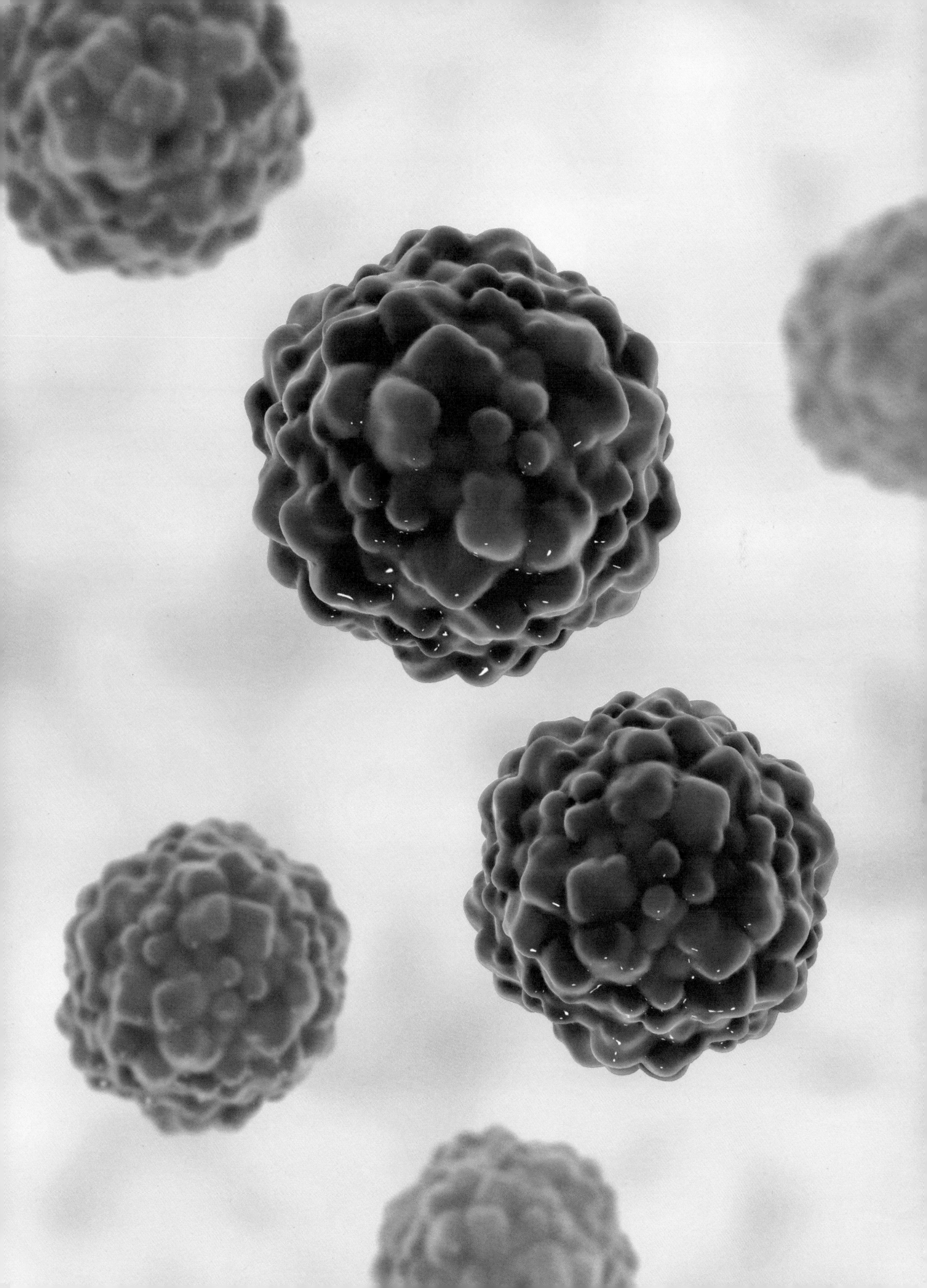

RRSV

Rice ragged stunt virus

Plant virus that infects its insect vector

GROUP	III
FAMILY	Spinareoviridae
GENUS	Oryzavirus
GENOME	Linear double-stranded RNA in 10 segments of about 26,000 nucleotides, encoding 13 proteins
VIRUS PARTICLE	Non-enveloped, double icosahedron with outer spikes, about 70 nm in diameter
HOSTS	Rice, other grasses, planthoppers
ASSOCIATED DISEASES	Ragged stunt disease
TRANSMISSION	Leafhoppers

Rice ragged stunt virus (RRSV) could be considered an insect virus that uses plants as its vector. It infects both its rice and leafhopper hosts, but causes serious disease only in the plants.

To infect hosts in two very different kingdoms—plants and insects—RRSV has to have the tools to get into them. In insects, the virus is also transmitted to the offspring, so it must overcome another barrier to get into the ovule.

Rice plants infected with RRSV are stunted and their leaves eventually become twisted and jagged. The virus doesn't kill the plant, but the yield of rice is greatly reduced. Control of the disease is difficult, because pesticides that kill the leafhoppers often do more harm than good. Not only are the chemicals toxic to humans and wildlife, but they also kill the predators of the leafhoppers.

Like all double-stranded RNA viruses, RRSV does not uncoat once it enters a host cell. Instead, it carries the enzymes it needs to make RNAs in its particle, and the new RNAs are synthesized and pushed out of the particle into the host cell. These RNAs are used to make proteins (messenger RNAs) and also pregenomes, which are then packaged in newly made virus particles. Once inside the new particle, the second strand of RNA is made to complete the double-stranded genome.

The reoviruses infect mammals, insects, plants, and fungi. The prefix "reo-" stands for "respiratory enteric orphan," after the mammalian reoviruses, which are often asymptomatic. No disease is associated with them, so they were called orphans—viruses without a disease. We now know that most viruses do not have an associated disease.

→ Rice ragged stunt virus causes a severe disease in rice plants, although it doesn't usually kill them. Like most plant viruses, it can't be cured, and the best strategy is usually to remove infected plants.

TMV

Tobacco mosaic virus

Virus that started the whole story

GROUP	IV
FAMILY	Virgaviridae
GENUS	Tobamovirus
GENOME	Linear, single-component, single-stranded RNA of about 6,400 nucleotides, encoding four proteins
VIRUS PARTICLE	Non-enveloped rigid rod about 300 nm long and 18 nm in diameter
HOSTS	Tobacco (*Nicotiana* spp.) and many related plants
ASSOCIATED DISEASES	Mosaic disease, necrosis
TRANSMISSION	Mechanical

Tobacco mosaic virus (TMV) was the first virus to be discovered. It was found in the sap of tobacco plants showing a mosaic of light and dark green areas in the leaves. Researchers found that it could be transmitted to other plants via the sap of infected plants, but knew it was not a bacterium because it was small enough to pass through a 0.2 μm filter.

TMV is a typical (+)RNA virus, in that the genome can act directly as a messenger RNA. The 5' end of the viral genome has a cap structure, and the 3' end has a tRNA structure (see diagram on page 77). The genomic RNA is translated into two distinct proteins, because it has a special sequence that is occasionally read as an amino acid instead of signaling for the ribosome to stop translating. TMV makes two smaller RNAs (subgenomic RNAs) that act as messenger RNAs for the coat protein and for the movement protein that helps the virus move between cells in the plant.

Many major advances in the study of virology and molecular biology resulted from studies of TMV. It was the first virus to be crystallized, which allowed a more detailed understanding of its structure. It was also the first virus to be visualized in an electron microscope. It was important in understanding genetic code, or how RNA encodes amino acids to make proteins, and it was part of the first demonstration that RNA is a genetic material. It was the first virus to have a gene transferred to a plant. The coat protein of TMV was transferred to tobacco plants by genetic engineering, with the result that these plants were resistant to infection by the virus.

Many tobacco are resistant to TMV because they have a gene that results in the virus-infected cells being killed. This causes small spots of necrosis on the leaves, called local lesions. The virus does not spread outside of these spots. However, the effect is sensitive to temperatures—if the infected plant is held at high temperatures (above 82 °F/28 °C), the gene is not effective and the virus spreads. If the plant is then brought down to a lower temperature, the entire plant becomes necrotic and collapses.

→ Computer-generated cutaway structure of tobacco mosaic virus, showing the coat protein in blue and the RNA in orange.

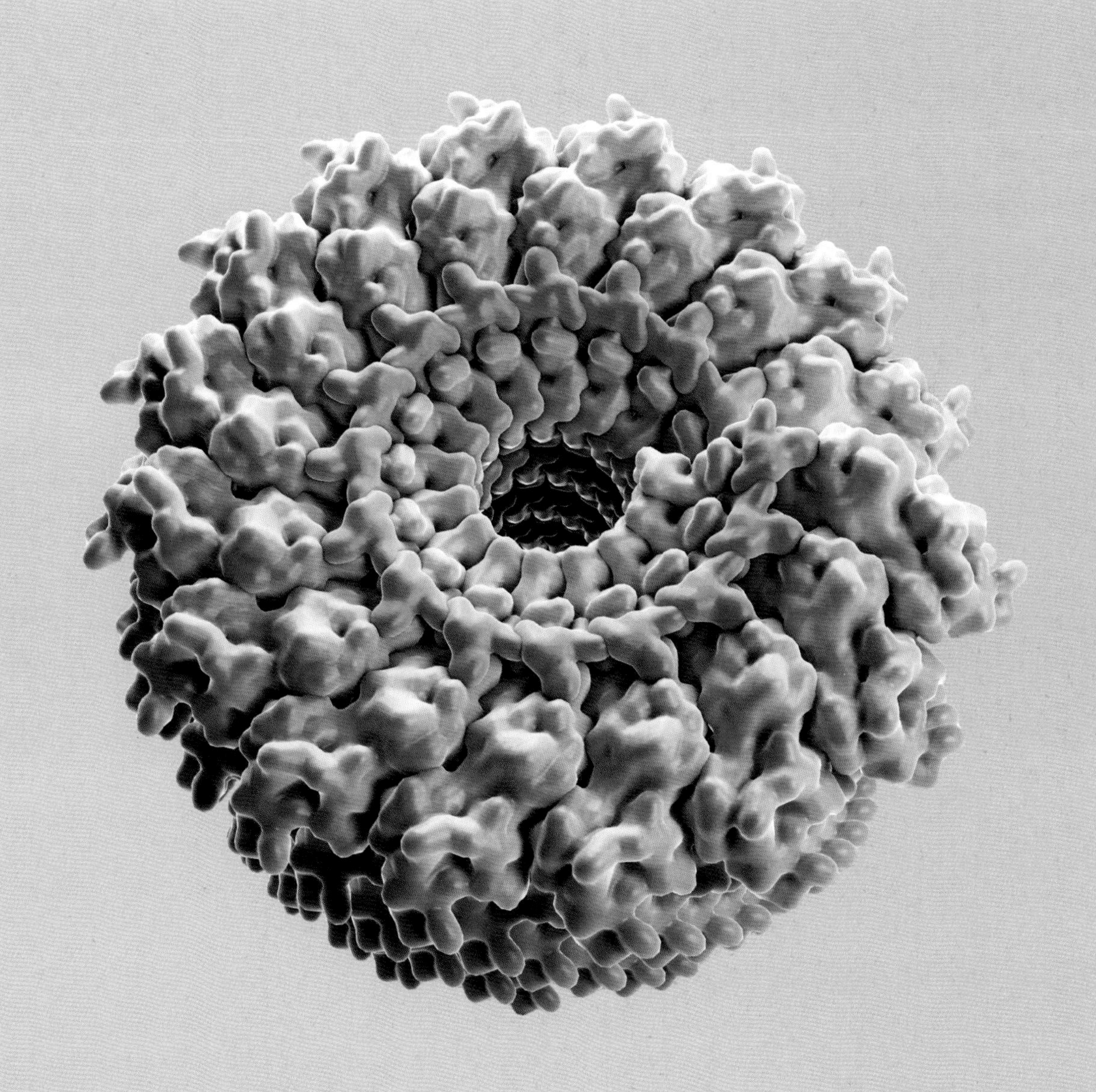

Lyssavirus rabies

One of the most feared viruses in the world

GROUP	V
FAMILY	Rabdoviridae
GENUS	Lyssavirus
GENOME	Linear, single-stranded RNA of about 11,000 nucleotides, encoding five proteins that are subsequently cleaved into functional units
VIRUS PARTICLE	Enveloped, bullet-shaped particle, about 180 nm long and 75 nm wide
HOSTS	Mammals, experimentally in birds and reptiles
ASSOCIATED DISEASES	Rabies
TRANSMISSION	Bite wounds
VACCINE	Inactivated virus

Lyssavirus rabies, also known as rabies virus, is rare in humans in North America and Europe, but is found more frequently in parts of the world where pets are not vaccinated. The virus is transmitted by bites and the disease it causes leads to madness and death.

In humans, the initial virus infection occurs months before any sign of disease, making it very difficult to determine the source. Fever and headache are usually the first symptoms, but they progress to an inflammation of the brain. The disease is essentially always fatal. There was one documented case of a young girl in Wisconsin, USA, who in 2003 survived rabies after an extensive treatment, including an induced coma, now known as the Milwaukee protocol. However, although there have been a few reports of other cases of human survival using this protocol, they are not well documented, and in general the protocol has been abandoned as ineffective.

The vaccines for rabies are very effective, and in many parts of the world most pets are vaccinated. People whose work could place them in contact with rabid animals, including veterinarians and wildlife researchers, can also be vaccinated. The initial progression of the disease is so slow that people can be effectively vaccinated after a known exposure. Use of immune sera—made in animals, including horses and sheep—was once the only post-exposure treatment, and involved a series of painful injections. This treatment is sometimes still used in conjunction with the vaccine.

The major source of rabies is from wild animals, including bats, Raccoons (*Procyon lotor*), skunks, and wild canines. Birds can also be infected but do not show any symptoms. Human infection is often from bats, whose bites are rare but often go unnoticed. Unlike many other human viruses that can be acquired from bats, rabies does cause disease in bats, although it is not usually fatal.

→ Colorized transmission electron microscope image of a section of tissue infected with rabies virus (in red) that is displaying cellular inclusions (in blue).

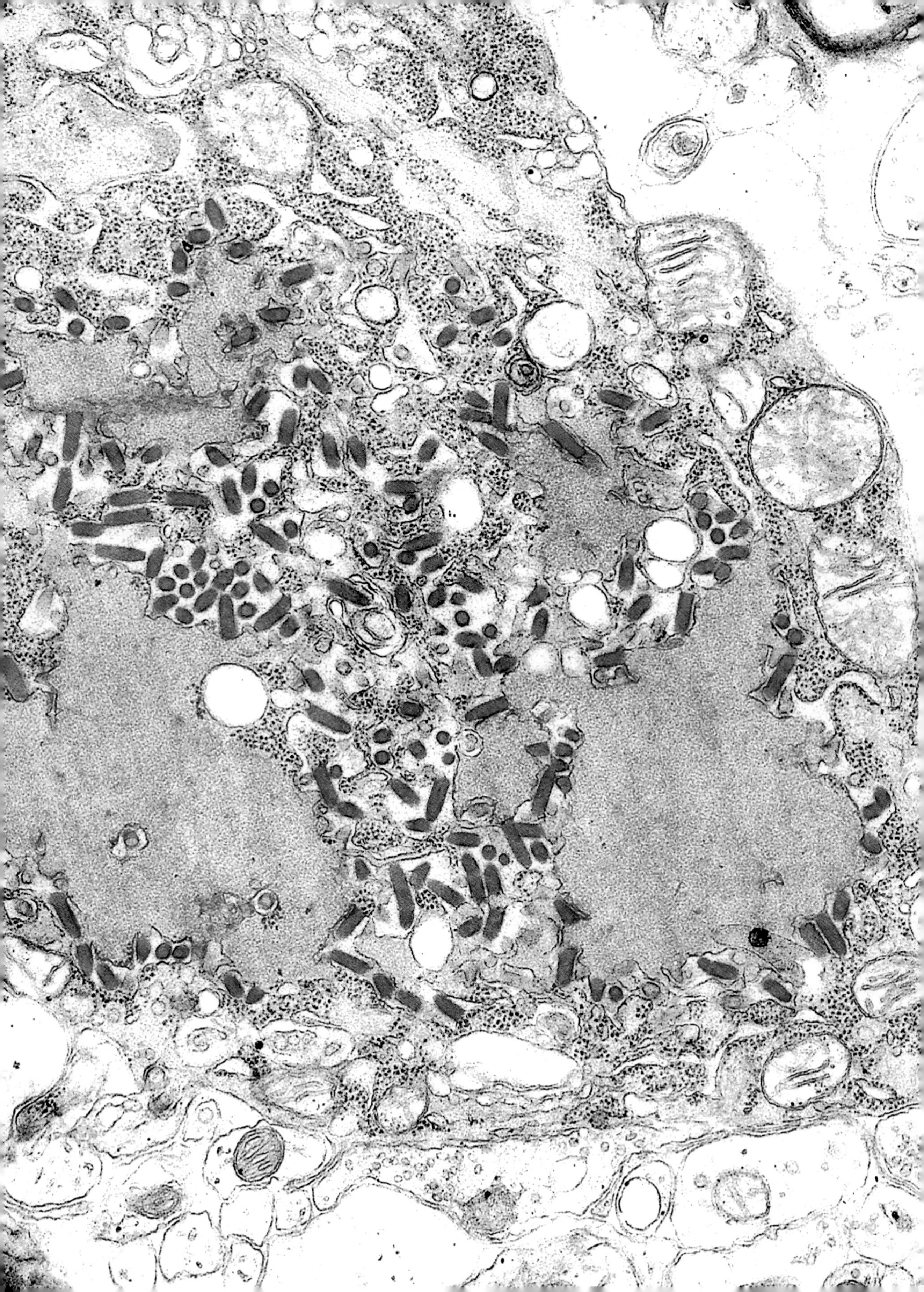

RSV

Rous sarcoma virus

First retrovirus known to cause cancer

GROUP	VI
FAMILY	Retroviridae
GENUS	Alpharetrovirus
GENOME	Linear, single-stranded RNA of about 7,200 nucleotides, encoding four proteins
VIRUS PARTICLE	Enveloped, with a spherical core of about 90 nm
HOSTS	Birds
ASSOCIATED DISEASES	Tumors
TRANSMISSION	Experimental
VACCINE	None

Rous sarcoma virus (RSV) was discovered more than 100 years ago by the American pathologist Francis Peyton Rous (1879–1970). He found that a cancer in chickens could be transmitted to other chickens by injecting them with an infected extract. Because Rous passed the extract through a filter that excluded larger microbes, he concluded that a virus was responsible for the tumors. In 1966, many years after his discovery, Rous was awarded a Nobel Prize for his work.

RSV is a typical retrovirus, with a small RNA genome that is packaged in duplicate. The RNA has a 5' VPg and a 3' poly-A tail. After conversion to DNA and integration into the host cell genome, the messenger RNA is made for the first polyprotein, and a readthrough to make the Pol protein that is the reverse transcriptase. The messenger RNA for the second polyprotein is made by splicing the same messenger RNA to remove the first protein-coding regions.

The study of RSV also led to the discovery of oncogenes. These genes are found in some retroviruses and in cells, and are involved in causing cancer. They signal the cell to make other proteins, including growth factors (cancer is essentially unregulated growth of a cell). After the discovery of RSV, scientists found other retroviruses—including tumor-causing viruses—in many animals. However, it was not until 1977 that the first retrovirus was found in humans. No cancer-causing retroviruses are known in humans, but there are a few other types of viruses that do cause human cancers, including herpesviruses (page 154) and papilloma viruses (page 124).

In spite of the fact that RSV has been essential in our understanding of retroviruses and the link between viruses and cancer, its natural history is poorly studied. Antibodies to RSV are found in most chicken flocks, but the birds do not develop tumors unless they are exposed to chickens that have been experimentally injected with extract from a tumor.

→ The structure of icosahedral viruses is made up of arrangements of the capsid proteins in arrays of five (pentamers) or six (heptamers). Shown here is the structure of an RSV pentamer derived from cryogenic electron microscopy analysis.

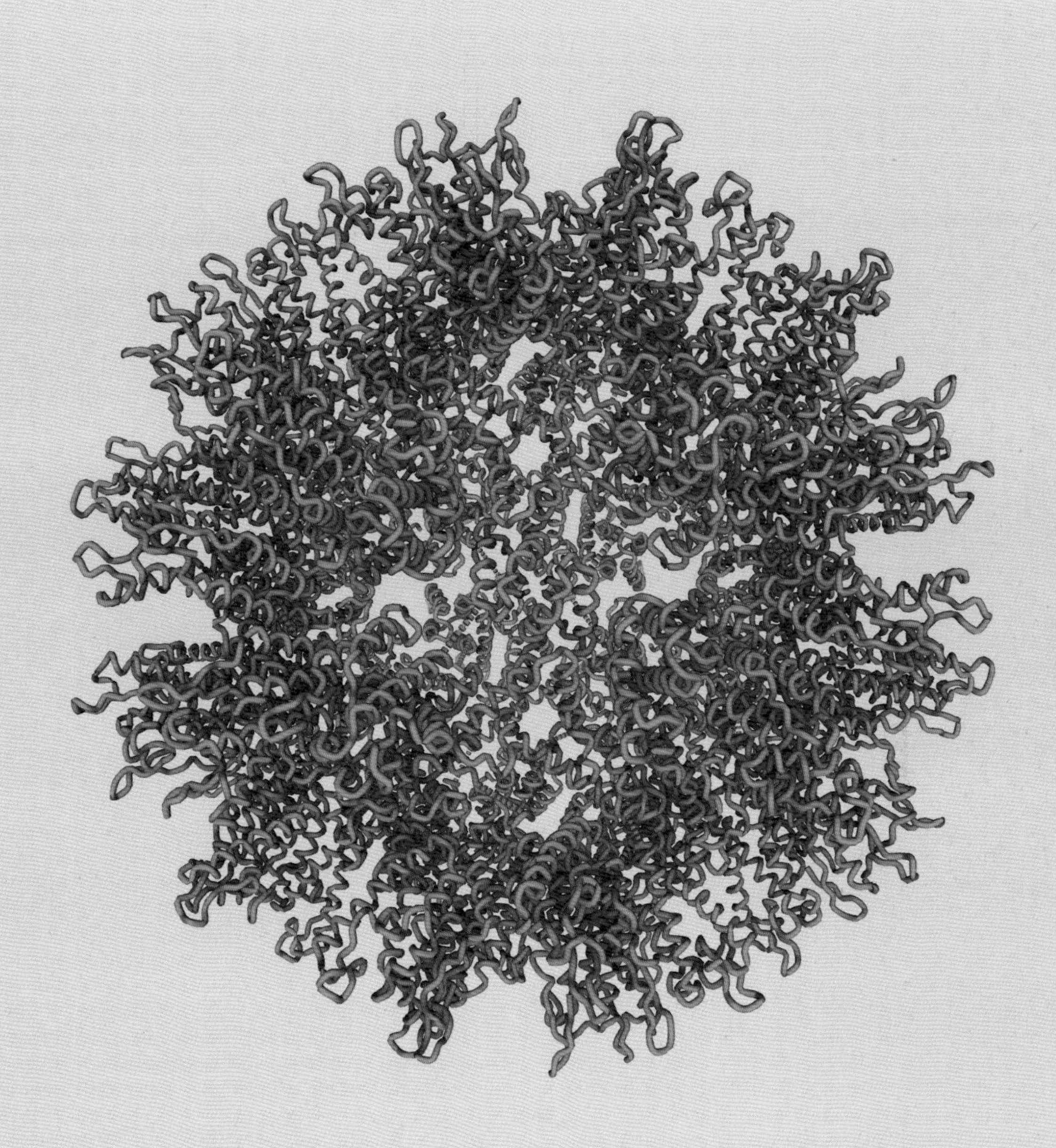

RTBV

Rice tungro bacilliform virus

A virus that needs help for its transmission

GROUP	VII
FAMILY	Caulimoviridae
GENUS	Tungrovirus
GENOME	Circular, single-component, double-stranded DNA of about 8,000 nucleotides, encoding four proteins
VIRUS PARTICLE	Non-enveloped, bacilliform-shaped, about 130 nm long and 30 nm in diameter
HOSTS	Rice and related grasses
ASSOCIATED DISEASES	Stunting, discoloration of the leaves, loss of tillers
TRANSMISSION	Leafhoppers

Rice tungro bacilliform virus (RTBV) is a typical pararetrovirus, turning its DNA genome into RNA in the nucleus of the infected cell, and then reverse transcribing the RNA into DNA that is packaged as new virus particles. The virus is transmitted by leafhoppers, but this requires the presence of another virus, called rice tungro spherical virus. RTBV can infect rice by itself, but causes only a mild disease and is poorly transmitted.

Tungro means "degenerated growth" in a Filipino dialect. Rice tungro disease was first described in the Philippines in the 1950s, and was originally thought to be a problem of plant nutrition. The viral nature of the disease was discovered about a decade later. It is considered one of the most serious rice diseases in Southeast Asia, where rice is a very important staple food. In the rice field, it is hard to distinguish infected plants from those that are under forms of stress such as insect damage, other diseases, drought, or heat stress. Early test methods for the virus used an antibody generated in rabbits that could bind to the virus particles. A polymerase chain reaction (PCR) test is often used to test leaf samples now, similar to tests used to detect many other viruses, including human viruses such as SARS-CoV-2.

Control of tungro disease is mainly through the use of insecticides that kill the leafhoppers. However, this is both costly and environmentally unsound, and the insects often evolve resistance to the chemicals. The International Rice Research Institute, located in the Philippines, maintains about 80,000 rice cultivars, many of which have been found to be resistant to the leafhoppers. However, as with the chemical treatments for insects, these cultivars often do not survive under the pressure of heavy insect loads and evolution. There has been little success in finding cultivars that are resistant to the virus, but genetic engineering has produced better results for resistant rice.

→ Rice plants infected with rice tungro bacilliform virus, showing typical stunting and yellowing.

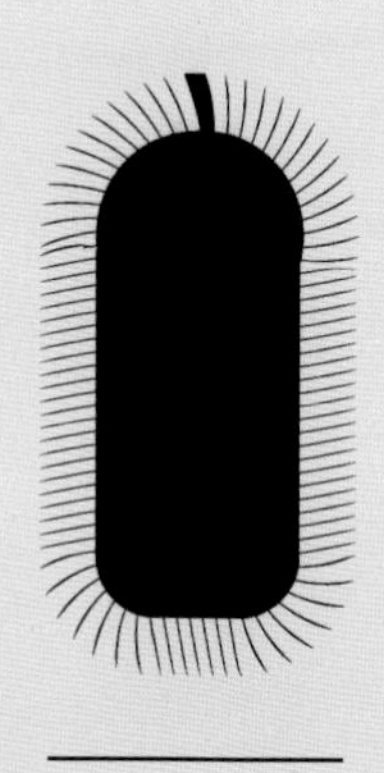

HOW VIRUSES GET AROUND

Introduction

Viruses have many different ways of entering and exiting their hosts, and moving between cells within the host. These methods depend on the virus structure and size, the type of host, where the host lives, and whether the host can move around.

During the early stages of the COVID-19 pandemic, much was said about how the virus might move between hosts. There were some reports that the virus could stay on surfaces for as long as 24 hours, and that people needed to disinfect everything that could have come into contact with another person. A lot has been learned since those early days, including that just because viral RNA can be recovered by very sensitive methods it does not mean there is any infectious virus present. One of the hallmarks of viruses with an envelope, like SARS Co-V2, is that they are very unstable in the environment. Knowing the ways that viruses get around is important for the understanding of necessary measures that can protect from infection.

CHALLENGES VIRUSES ENCOUNTER WHILE ENTERING AND EXITING THEIR HOSTS

Various types of hosts have different barriers that viruses must overcome. In this table the 'yes' means the host has this characteristic, and the 'no' means it does not. In some host categories there is variation. For example, many animals are motile, but not all animals; some protists have cell walls and some do not.

	Animal	Plant	Fungus	Protist	Archaea	Bacteria
Cell wall	no	yes	yes	yes/no	yes	yes
Motile	yes/no	no	no	yes/no	yes/no	yes/no
Airborne	yes	no	no	no	no	no
Waterborne/foodborne	yes	no	no	yes	yes	yes
Vectored	yes	yes	no	no	no	no
Vertical	yes	yes	yes	yes	yes	yes
Integrated	yes	yes	yes	yes	yes	yes

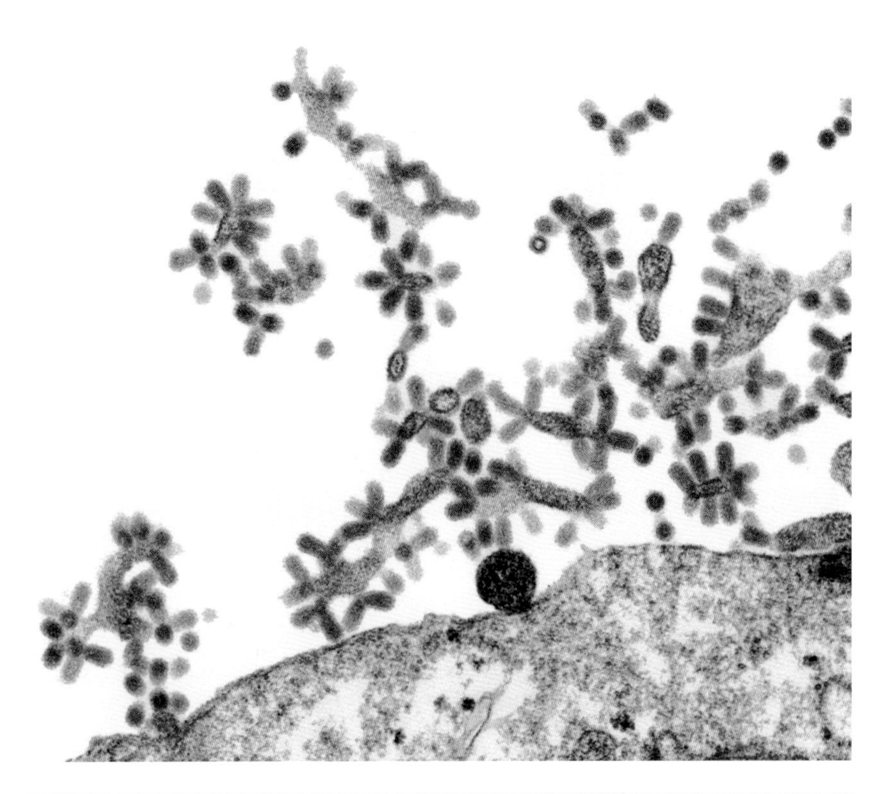

↖ Colorized electron micrograph of influenza virus (orange) exiting an infected cell (green).

↙ A colony of pea aphids (*Acyrthosiphon pisum*) on a plant. Pea aphids transmit many plant viruses.

Cell walls and membranes

All cells are surrounded by a membrane, called a plasma membrane, which is mainly made up of fats and proteins and is easily penetrated. In the animal kingdom cells have only a plasma membrane and no walls. In contrast, the cells of organisms in other kingdoms (including plants, fungi, bacteria, and some protists) have walls on the outside of their membranes that are rigid and made up of different stable compounds that are hard to penetrate (see images on pages 10 and 11). Archaeal cells have a different outer surface, comprising mostly proteins.

Viruses may also be surrounded by a membrane, which in their case is usually called an envelope. This envelope is helpful when a virus infects an animal cell because it contains proteins that recognize the proteins on the outside of the host cell, acting like a key to the host's lock. The envelope and the host cell membrane fuse, allowing the virus to get inside the cell. When enveloped viruses exit the cell, they bud through the host cell membrane, taking it with them as their new envelope but with their own proteins inserted into the membrane. Some viruses bud through different cell membranes, such as the one around the nucleus. Viruses that don't have envelopes still bind to the animal cell surface by protein recognition, but they are taken into the cell by other methods such as endocytosis (cell drinking). Many of these viruses cause the cell they are infecting to burst open, killing it and releasing the virus.

Most viruses that infect hosts other than animals do not have envelopes, because they have to penetrate a cell wall. There are a few exceptions, however, including some plant viruses that also infect insects—having an envelope helps them enter the insect cells. There are also some bacterial and archaeal viruses that have envelopes.

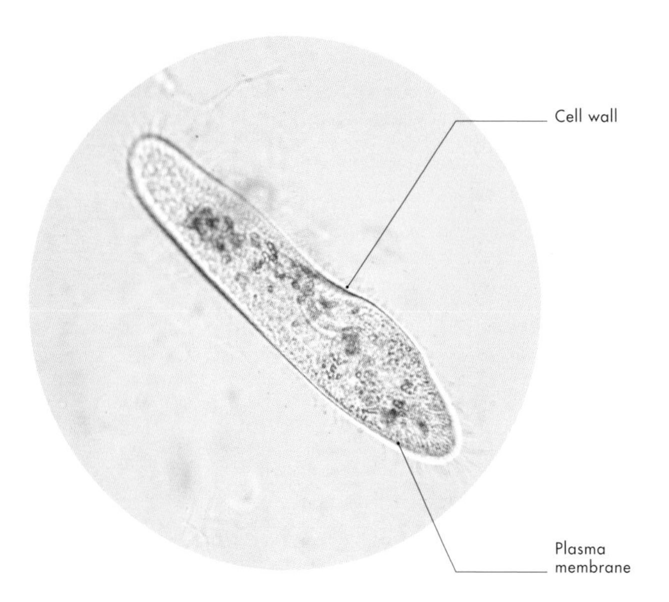

← *Paramecium* spp. are single-celled eukaryotes, with a cell wall and a plasma membrane surrounding the cell.

Entry by membrane fusion

An enveloped virus has exterior proteins that bind to specific receptors on the surface of a cell. The virus and cell membranes fuse, and the virus capside is released into the cell to initiate the infection.

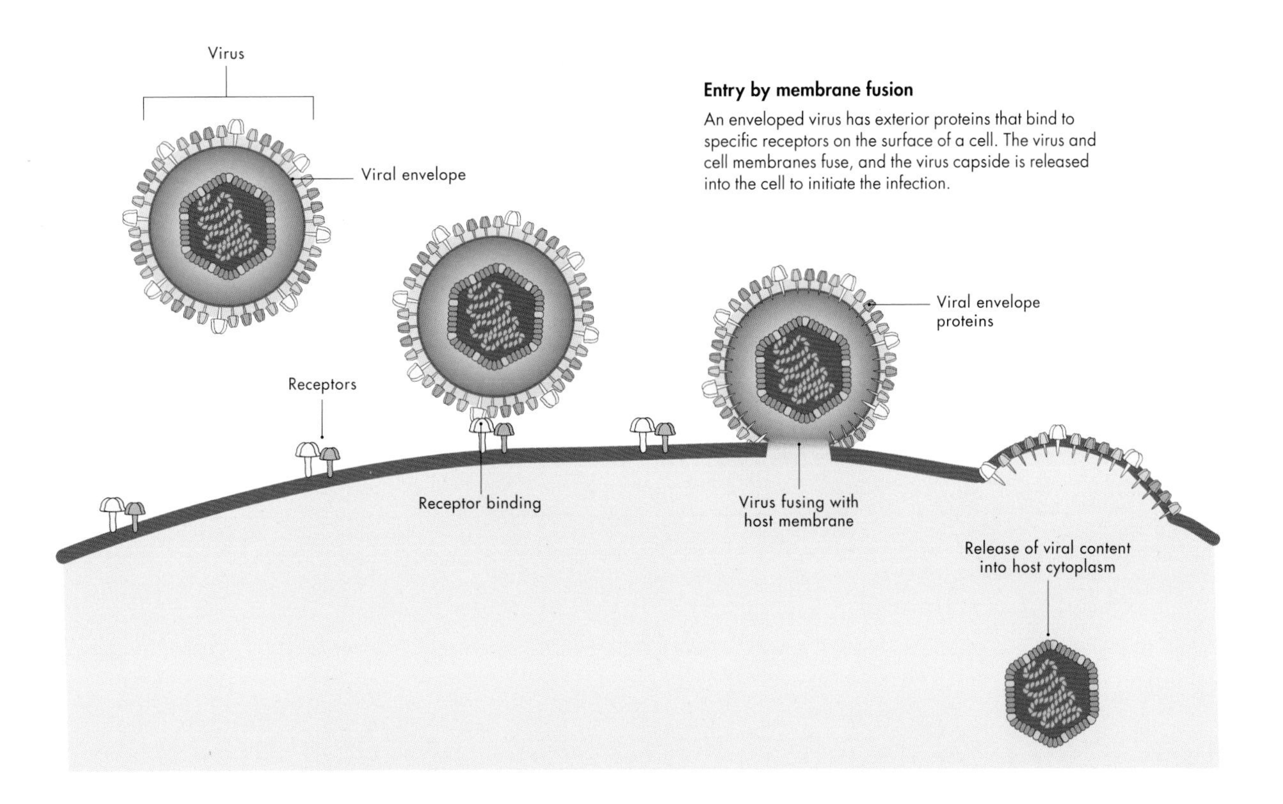

Entry of a non-enveloped virus into a cell

The virus exterior proteins attach to the cell membrane receptors and the cell engulfs the virus. The cell membrane forms a structure around the virus called a vesicle. Once fully inside the cell, the vesicle and virus capsid break down and the viral genome is released.

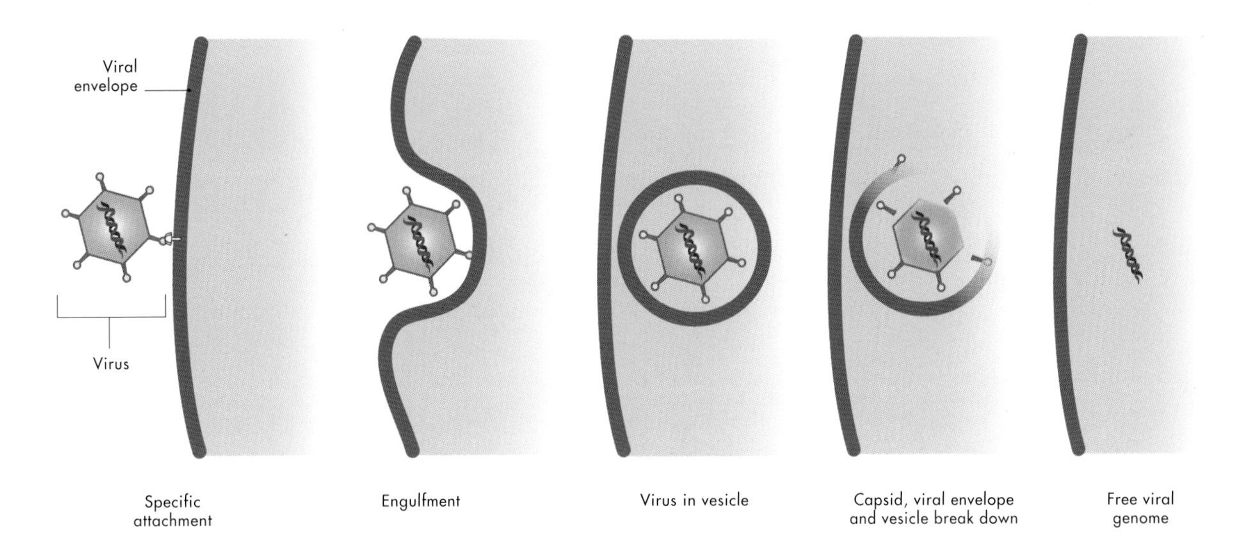

Breaking through the wall

Viruses use different methods to break through cell walls. In the case of plant viruses, many are transmitted by insects that feed on plants. While they are feeding, the insects make a hole in the cell wall and deposit the virus. The virus exits the plant in a similar way, being taken up by an insect feeding on the plant.

Cell wall structures in plants

Plant viruses move between plant cells by structures in the cell wall called plasmodesmata. Viruses make proteins that either bind to the outside of the virus to help it move through these structures, or bind to the plasmodesmata to make it larger.

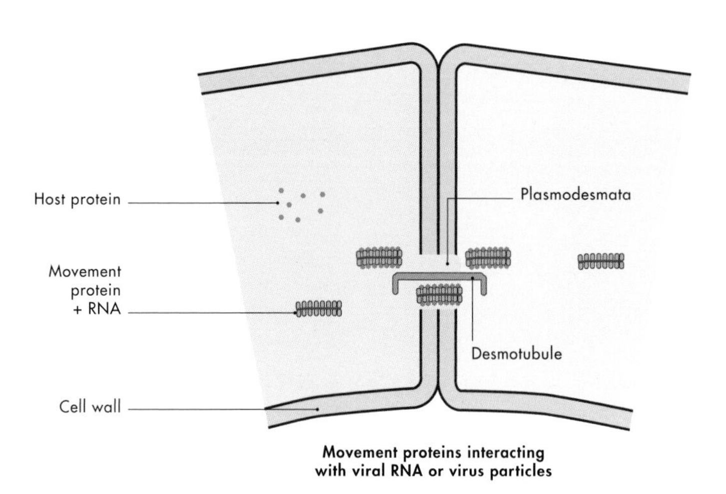

Movement proteins interacting with viral RNA or virus particles

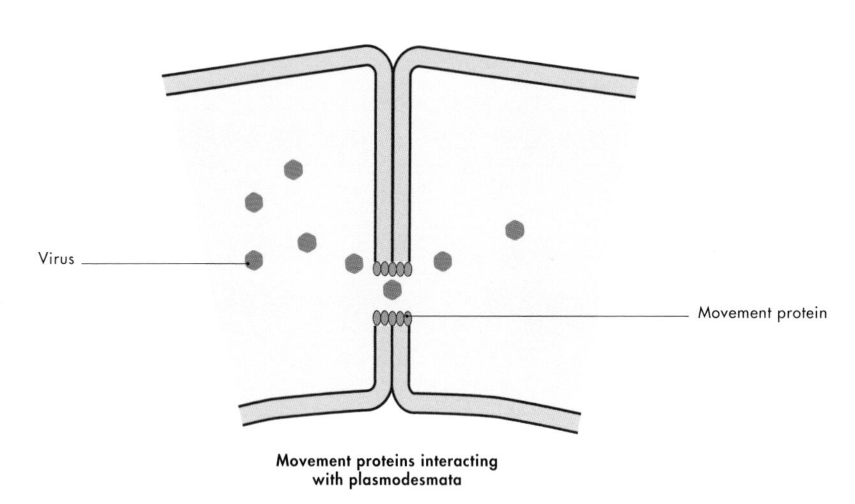

Movement proteins interacting with plasmodesmata

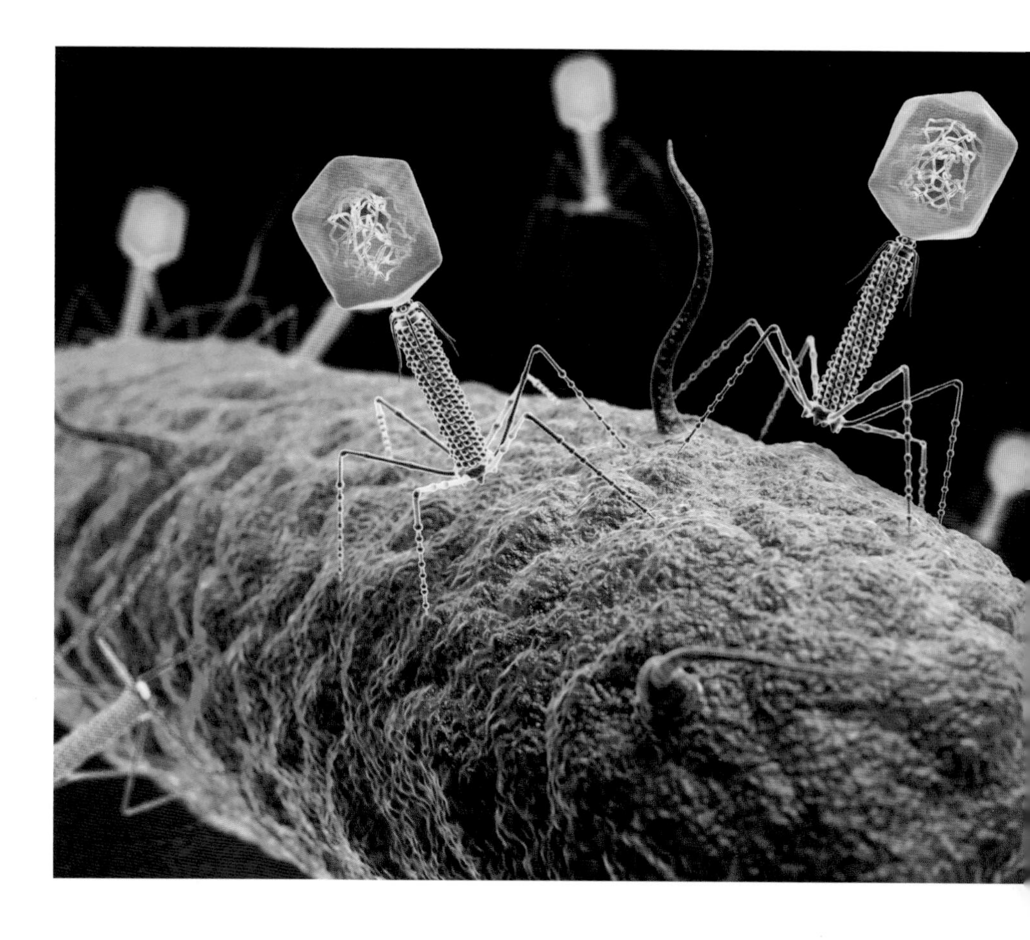

→ Bacterial viruses such as T4 bacteriophage have landing gear to allow them to attach to the cell. They then inject their DNA genome into the bacterial cell.

Plant viruses move between cells within a plant using the existing connections between the cells, called plasmodesmata. These highly regulated structures form channels between the cells, and most viruses make proteins that help them pass through. Plant viruses also move through the phloem and xylem, the tissues in plants used to transport water and nutrients.

Fungal viruses are transmitted by connections that occur between different individual fungi, a process called anastomosis. This can occur only between closely related fungi, but since very similar viruses are found in different fungi, it seems likely that the viruses have other ways of getting around that scientists don't yet know about. Fungi have pores in their inner walls similar to plant plasmodesmata, but not as tightly regulated, which viruses use to move between cells.

Bacterial viruses often land on their host cells and then inject their DNA into the host cell with a syringe-like apparatus. Some archaeal viruses use this method too. Others use complex structures that allow them to attach to host cells and pass their DNA genomes into the host. Most bacterial viruses and some archaeal viruses cause the host cell to burst open once it is full of virus, releasing the virus to infect other cells. However, sometimes the viruses integrate into the genome of the host cell. When this happens, the virus protects the host from infection by similar viruses, and all the cells generated by subsequent cell division carry the original virus.

VERTICAL VERSUS HORIZONTAL TRANSMISSION

The process by which a virus moves directly from a parent to an offspring is called vertical transmission. In the case of humans and other animals, this usually means that the virus is transmitted to the offspring either before birth or during the birthing process. Although viruses are found in human semen, it isn't known if they can be transmitted to an egg during fertilization.

In plants, vertical transmission can take place through either the ova (female) or pollen (male). Many plant viruses are transmitted only vertically. These viruses are passed on for many generations, and perhaps even thousands of years. Since the plant embryo is already infected, either by the ova or the pollen, every cell in the plant has the virus. Many fungal viruses seem to have a similar, mainly vertical, transmission, and bacterial or archaeal viruses can also be transmitted in this way. In these cases the virus moves via cell division.

Horizontal transmission is when a virus moves from one individual to another. There are many ways this can happen, but they all require some kind of contact, either direct or indirect.

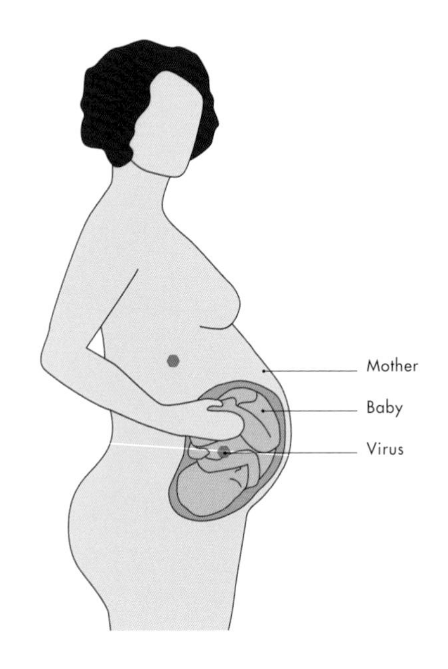

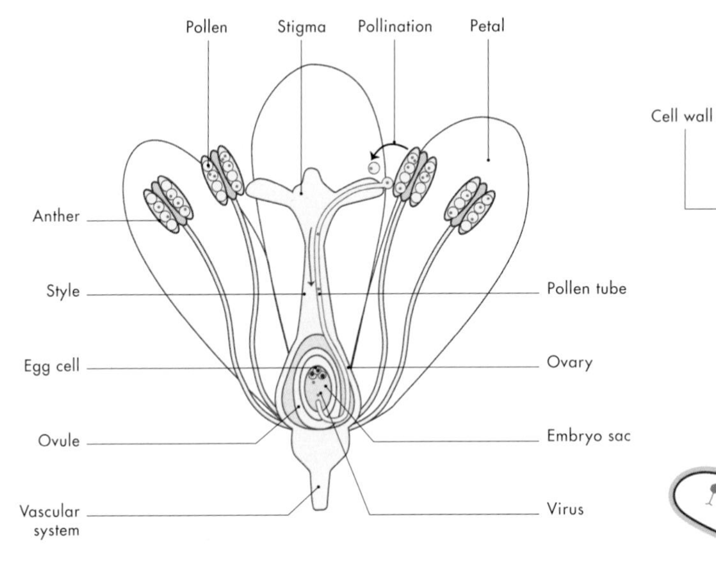

Vertical transmission

Vertical transmission occurs when a mother passes a virus to an unborn child, a plant ova or pollen grain passes a virus to the fertilized egg, or an asexual cell like a bacterium divides by cell division.

HOST-TO-HOST CONTACT

Most animals move around and come into contact with other animals, giving viruses the opportunity to infect a new host. Some marine animals are sessile, meaning they do not move around but stay attached to something for their entire adult lives, but land animals generally move. Plants and fungi don't move around except in their seed or spore stages, but even then they are not really active. Viruses of these hosts therefore have to find ways to get from one organism to another. In the case of bacteria, archaea, some fungi, protists, and sessile animals, these organisms live in environments that are aquatic or contain water or another liquid that acts as a medium for viruses to move around in.

The stability of a virus has a big impact on how it gets around. For example, viruses with envelopes are very susceptible to drying out, and cannot survive very long outside their hosts. These viruses therefore require more direct contact to move from one host to another, or the use of a vector (see page 117). Viruses that are very stable may survive for long periods of time—sometimes many years—outside a host. For example, tobacco mosaic virus is stable in many environments, including water and the guts of humans and other animals. It does not infect the animals but just passes through them. Most enteric (fecal-oral) viruses that are transmitted on food or in water are very stable. An example is canine parvovirus, which has been

→ A sneeze releases thousands of droplets that spread through the air, how far depending on temperature and humidity. If the sneezer is carrying a respiratory virus, these droplets are likely to be infectious to anyone breathing them in.

reported to survive in soil for many years. Unvaccinated dogs are at high risk from being infected by this virus because it is hard to eliminate it from the environment.

If a virus infects only one kind of host, the way it moves between hosts will differ from viruses that infect more than one kind of host, where the different hosts have to come into contact with one another. For example, bats carry many viruses, some of which may infect humans, but bats and humans don't come into contact with each other very often, so transmission is rare. In North America, the rabies virus in humans comes from bats, so human rabies here is extremely rare. However, in parts of the world where dogs are not vaccinated against rabies, human rabies is more common because humans do have frequent contact with dogs. Some viruses have evolved to infect very different groups of organisms—for example, there are viruses that infect both plants and insects, and insects and vertebrate animals.

Animals that move around can transmit their viruses by direct contact with other animals, through touching or more intimate contact. The latter includes the exchange of body fluids that can occur during medical procedures, drug injection, sexual contact, and animal bites; breathing in viruses through the air after they are released from another host; or eating or drinking food or water that is contaminated by viruses. To prevent spread of a virus, different protocols are required depending on the transmission method.

↓ Transmission of fecal–oral viruses such as hepatitis A virus can occur via food. Sometimes the food is contaminated in the field or during harvest, but it can also be contaminated during preparation. It is important for food workers to take precautions to prevent this by washing fresh produce to remove existing contamination, and by washing their hands and wearing gloves to prevent further contamination.

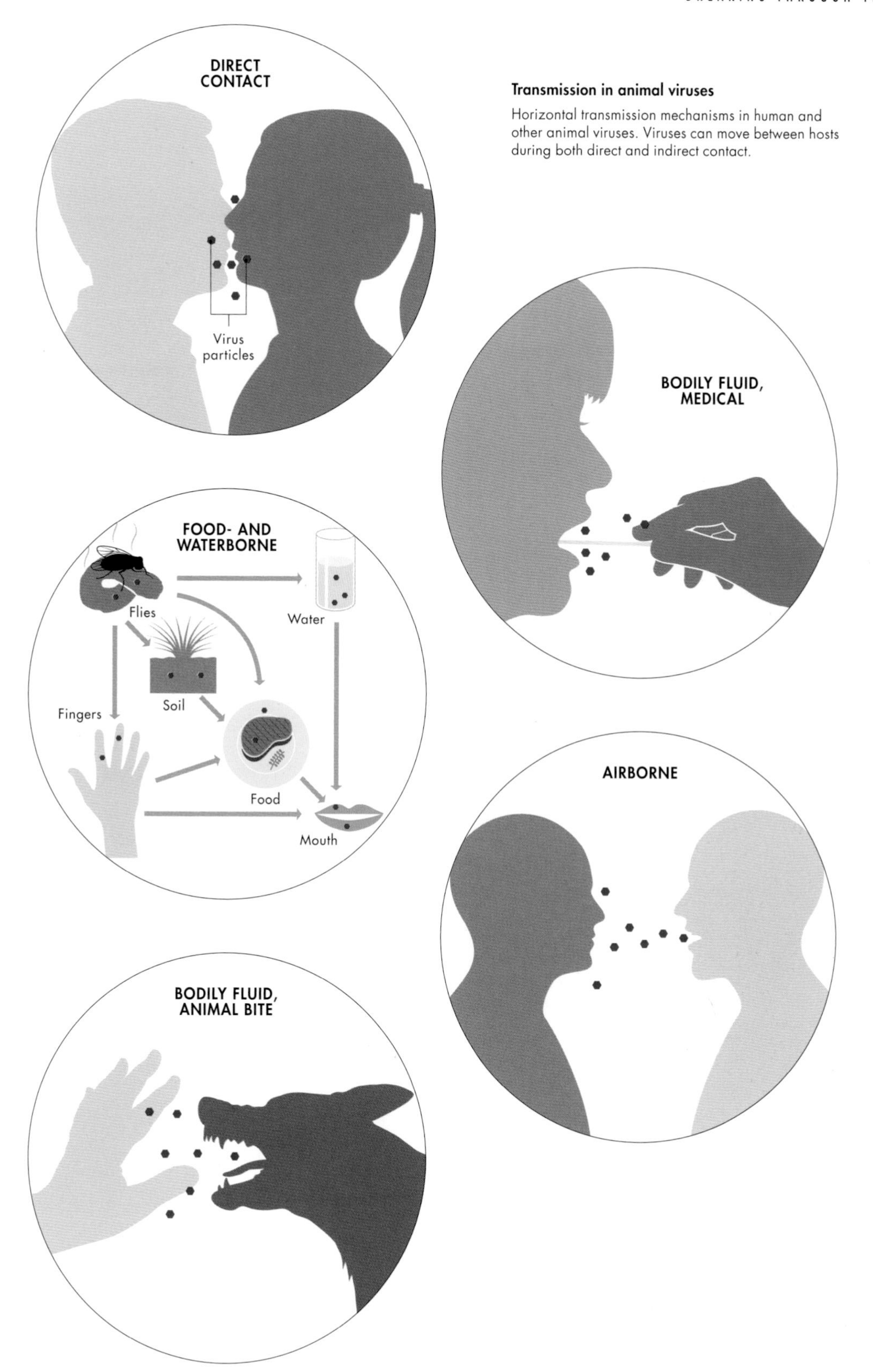

Transmission in animal viruses

Horizontal transmission mechanisms in human and other animal viruses. Viruses can move between hosts during both direct and indirect contact.

In many cultures it is traditional to shake hands as a gesture of trust, a tradition that dates back to at least the ninth century BCE in Babylon. However, handshaking can be a great way to transmit respiratory (airborne) or enteric viruses from one person to another. During the COVID-19 pandemic, people began to avoid shaking hands, replacing this tradition with other forms of greeting such as the elbow bump. Good handwashing practices can also reduce this type of transmission.

Airborne viruses such as influenza and SARS-CoV-2 are found in very small droplets released by the infected host, often through sneezing or coughing, although even talking or singing can release them. The droplets may then be inhaled directly from the air if two hosts are inhabiting the same airspace. Virus-containing droplets travel different distances through the air, depending on the air temperature, how dry the air is, if the air is moving or stagnant, and how large the droplets are. Wearing a mask can help reduce this type of transmission, both by reducing release of the droplets by an infected person and intake of the droplets by an uninfected person. Masking is especially important in indoor spaces, where the air does not move much. Many people who have worn masks to prevent the spread of SARS-CoV-2 also found that they avoided getting influenza or common colds.

The virus-containing droplets also land on many surfaces, so touching these and then touching the face can allow the virus to be inhaled. The best way to avoid this type of transmission is by washing or sanitizing hands and avoiding touching the face. Wearing a mask can indirectly help people avoid touching their face.

Food- and waterborne viruses generally infect the gut. They are passed from the gut of a human or animal to contaminate food or water, and are then ingested by the next host through eating or drinking. Viruses that are transmitted in this way are common in some crowded situations, as seen in epidemics of Norwalk virus on cruise ships.

← Many people wore face masks during the 1918 influenza pandemic to prevent the spread of the airborne virus, which is transmitted by inhaling tiny droplets released by infected individuals.

Vectors

Many viruses are moved around among hosts by another agent, called a vector. They are most often insects or other arthropods. The most common vectors of viruses of humans and other animals are mosquitoes, although ticks, mites, and midges can also carry animal viruses.

Some very important human pathogens are moved around by mosquitoes including dengue virus, yellow fever virus, West Nile virus, chikungunya virus (see page 130), and Zika virus. Most animal viruses that are transmitted by mosquitoes also infect the insects themselves. Some manipulate the feeding patterns of the mosquitoes, making them probe more frequently and thereby enhancing virus transmission.

Plant versus animal virus transmission

Aphids take up and transmit plant viruses via their mouthparts. On the other hand, mosquitoes take up viruses in their mouthparts, but they can release them from their gut where they enter the bloodstream of the host through the wound made by the mosquito.

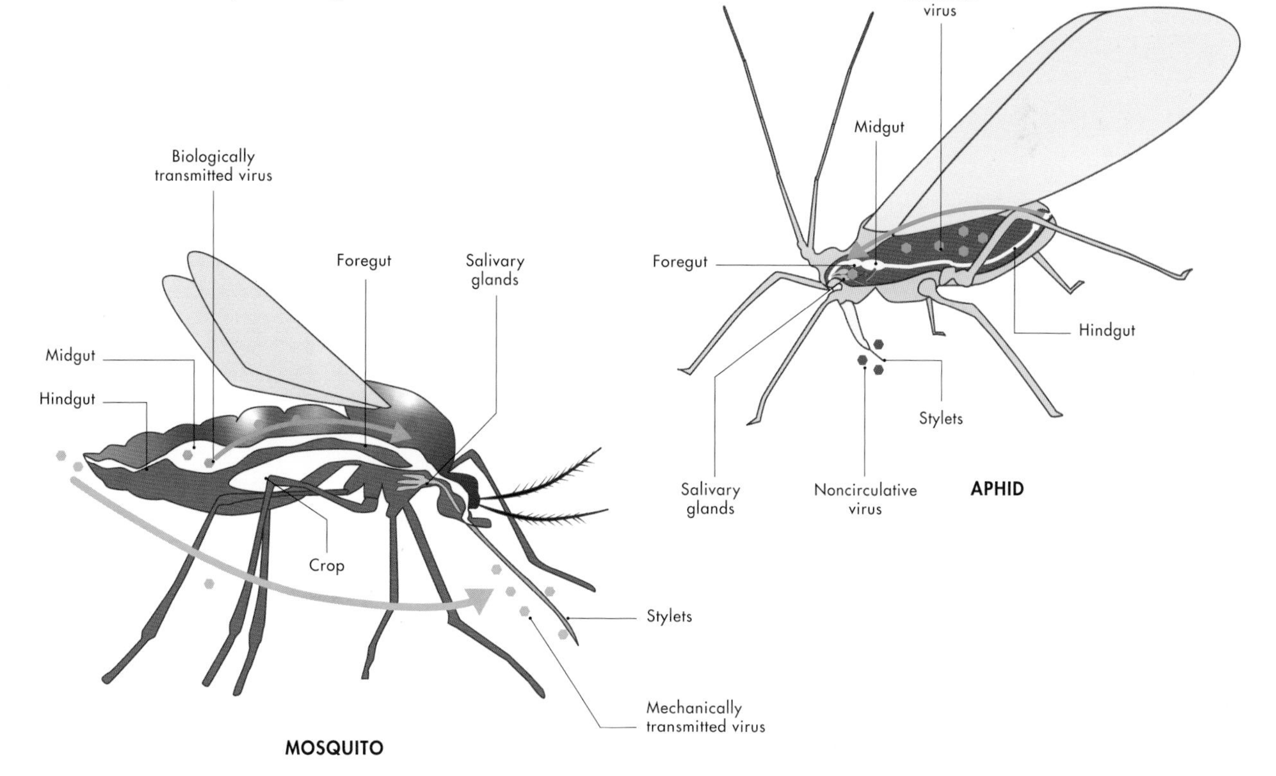

Plant viruses use sophisticated methods to ensure their transmission by insects. Many manipulate the host plants to make volatile compounds that insects can smell and are attracted to. Once the insect is drawn to the plant, settles on it, and starts feeding, the virus may further manipulate the plant to make chemicals that the insect doesn't like. This will make the insect move off to another plant, carrying the virus with it. Another way viruses make infected plants attract insects is through color changes in the host. Aphids in particular are attracted to the color yellow, and virus-infected plants often have a yellow look. Some viruses can also increase the number of offspring produced by insects that feed on infected plants.

Most insect-transmitted plant viruses have specific relationships with their vectors, and this can vary a lot among different viruses. For example, cucumber mosaic virus (page 214) can be transmitted by nearly 400 different species of aphid, whereas barley yellow dwarf virus is usually limited to a single species of aphid. Small soil worms called nematodes can also act as vectors for plant viruses, as can fungi that live in the soil. Viruses transmitted in this way are sometimes called soilborne viruses, even though they don't usually enter the host directly from the soil.

Some plant viruses are transmitted during grazing by herbivores, which break open the cells of plants as they chew them. In some insects that chew on plant parts, including beetles, viruses are transmitted in a similar way. Farm and gardening equipment can also spread plant viruses, although in this case the virus needs to be relatively stable. An interesting twist on vectors is that a plant can be a vector for some insect viruses. The insect deposits the virus while it feeds on the plant, and then an insect that subsequently feeds on the plant becomes infected.

All of the variation in how viruses get around indicates that the relationships between viruses and their hosts have existed for a very long time and have evolved in very diverse ways to overcome similar challenges: how to get into a host, how to move around it once inside, and how to get out.

← St. Augustine decline is a grass disease spread by lawnmowers. It is caused by panicum mosaic virus and is often found in the southern United States, where the heat-tolerant St. Augustine Grass (*Stenotaphrum secundatum*) is used for lawns.

→ Many vectors involved in the transmission of animal and plant viruses are insects (e.g., mosquitoes and aphids) or other arthropods (e.g., ticks and mites), but microbes can also be vectors, as can grazing animals and farm equipment in the case of plant viruses. A, tick; B, mosquito; C, flea; D, midge; E, aphid; F, thrips; G, cattle; H, dodder; I, nematode; J, treehopper; K, planthopper; L, long-tailed mealybug.

ANIMAL VIRUS VECTORS
PLANT VIRUS VECTORS
A
B
C
D
E
F
G
H
I
J
K
L

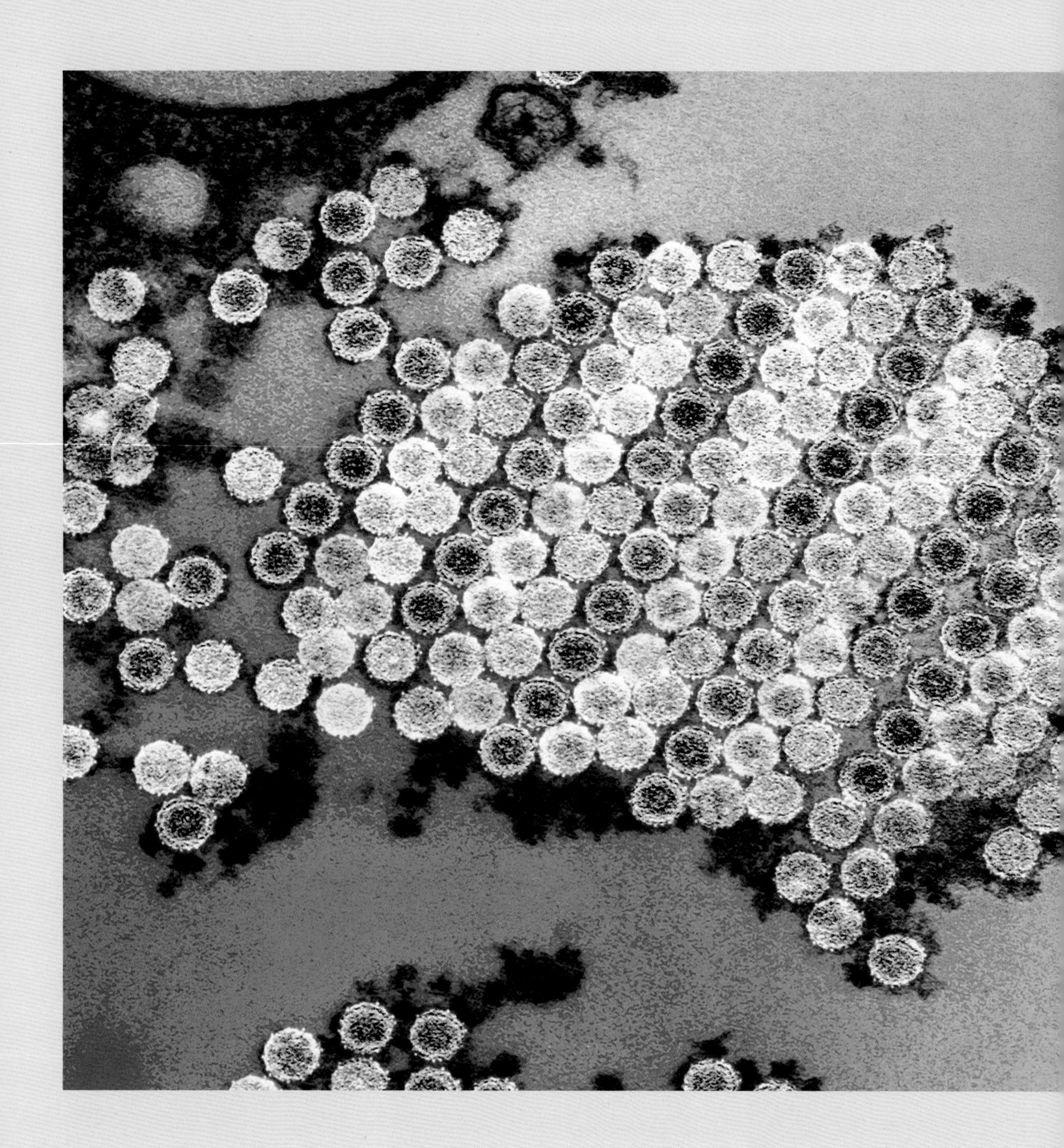

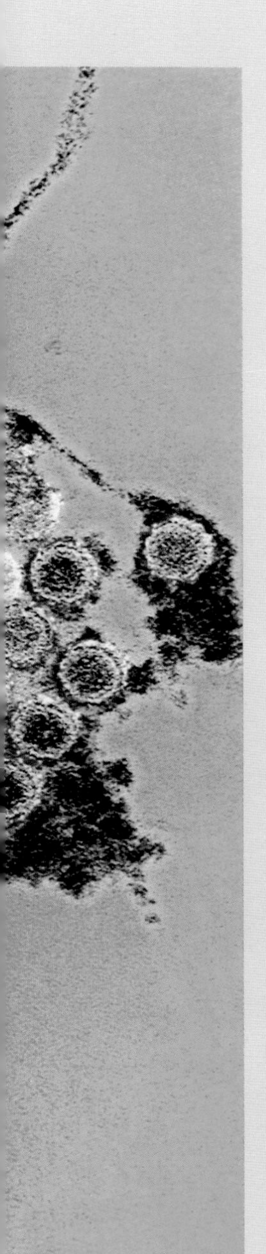

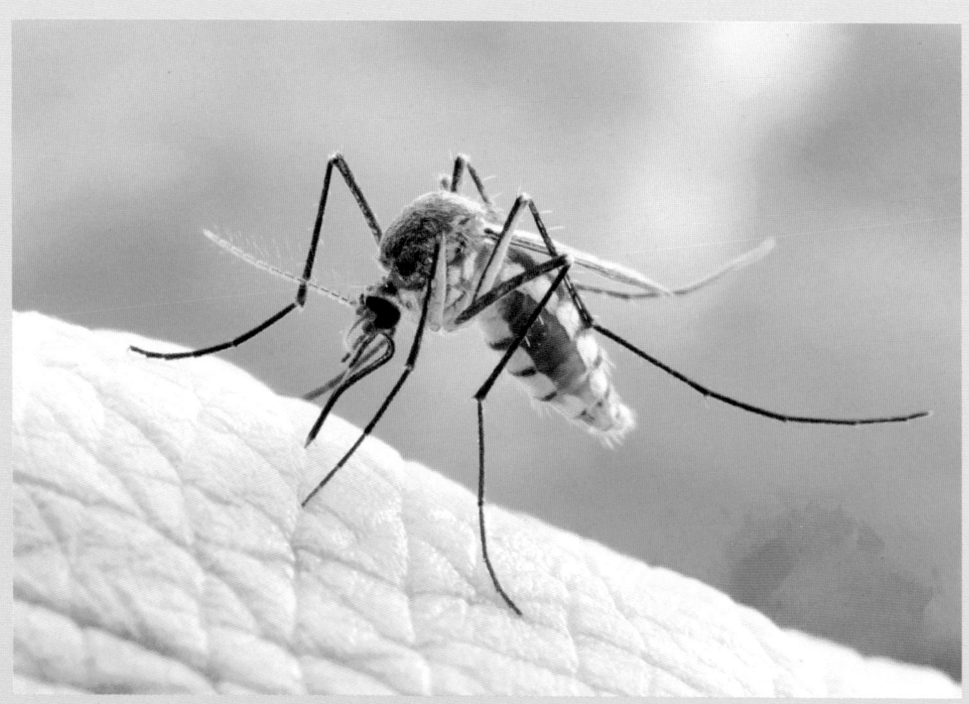

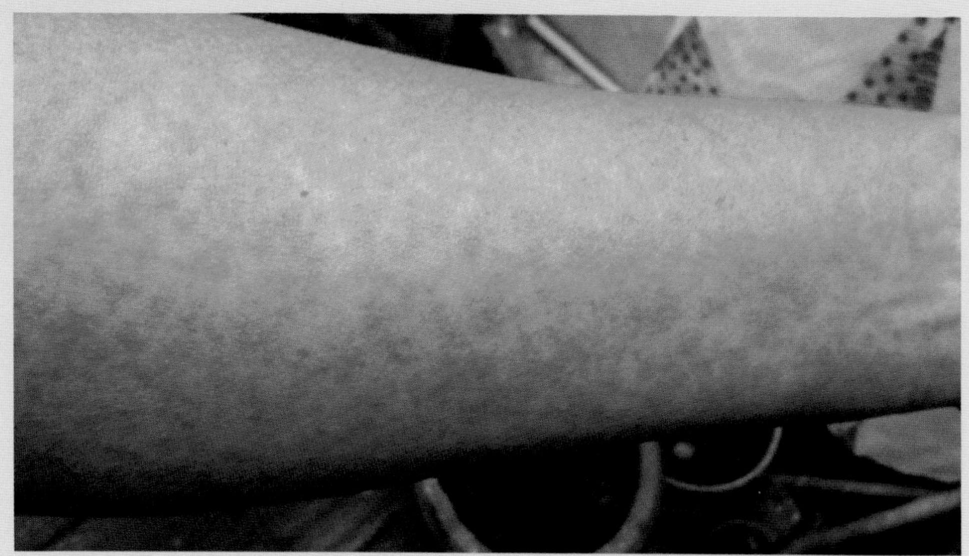

↑ Colorized transmission electron microscope image of Chikungunya virus.

↑↑ A mosquito taking a blood meal from a human.

↑ The Chikungunya virus can cause a rash, but more severe symptoms include painful arthritis, which can last for years.

HAV

Hepatovirus A

A food- and waterborne virus that causes severe outbreaks of hepatitis

GROUP	IV
FAMILY	Picornaviridae
GENUS	Hepatovirus
GENOME	Linear, single-stranded, single-component RNA comprising about 7,500 nucleotides, encoding 11 proteins via a polyprotein
VIRUS PARTICLE	Icosahedral, non-enveloped
HOSTS	Humans, wild primates, rodents in experimental conditions
ASSOCIATED DISEASES	Hepatitis A
TRANSMISSION	Water- and foodborne
VACCINE	Single-antigen or double-antigen injection

Hepatitis is an infection of the liver and is caused by several different hepatoviruses. Hepatovirus A (HAV) is common in some parts of the world, and is transmitted by contaminated food and water.

Shellfish farmed in or gathered from contaminated water is a common source of HAV infection. Outbreaks of hepatitis A in the United States, Europe, and Australia have often been linked to contaminated spinach or other greens, and smaller outbreaks have been linked to infected food service workers. Good handwashing practices among workers in the food service industry are critical to prevent the spread of the disease.

Young children who get hepatitis A rarely have any symptoms, but may be a source of HAV infection for other family members. Adolescents and adults can have serious symptoms, including fever, headache, nausea, jaundice, diarrhea, and fatigue. Unlike the hepatovirus C or hepatovirus B viruses, which establish chronic infections in humans, most people fully recover from HAV infection without any long-term effects. There is also some evidence that HAV infection prevents infection with hepatovirus C.

Preventing HAV infection in travelers used to involve injections of gammaglobulin derived from immune human blood, a painful process that also came with serious risks from receiving human blood products. The immunoglobulin injections now used are much safer and are thoroughly screened. However, they are rarely necessary today because a very effective vaccine for HAV has been available since the mid-1990s. Immunity, either from vaccination or infection, is very long lasting.

→ Cryo-EM structure of hepatitis A virus complexed with an antibody.

Alphapapillomavirus

A sexually transmitted virus that can cause cancer

GROUP	I
FAMILY	Papillomaviridae
GENUS	Alphapapillomavirus
GENOME	Circular, single-component, double-stranded DNA comprising about 8,000 nucleotides, encoding eight proteins
VIRUS PARTICLE	Non-enveloped icosahedron, 55 nm
HOSTS	Humans, monkeys
ASSOCIATED DISEASES	Genital warts; cervical, penile, anal, and tonsilar cancer
TRANSMISSION	Sexual
VACCINE	Antigens from several strains

Alphapapillomavirus is a complex of many different closely related viruses that cause warts. Some of these cause skin warts, plantar warts, or flat warts, all of which are common in humans and are mostly just an annoyance. However, a number of strains are transmitted by sexual contact.

In fact, human papilloma virus is the most frequent sexually transmitted infection in the United States, with about 40 million cases per year. Sexual transmission can result in genital warts, and cancer long after the initial infection. There are numerous types of the virus. The types associated with the highest risk for genital warts are 6, 11, 42, and 44, while those associated with the highest risk for cancer are 16, 18, 31, and 45.

A vaccine for alphapapilloma virus was first introduced in 2006. The current vaccine, which is highly recommended for young people between the ages of 11 and 26, protects against strains 6, 11, 16, and 18, covering the most important strains for genital warts and about 70 percent of the incidence of cervical cancer. This vaccine was the first one ever released against cancer. There are two other vaccines used in some parts of the world that protect against other strains.

→ The structure of human papilloma virus, drawn from cryogenic electron microscopy images to create a high-resolution structure.

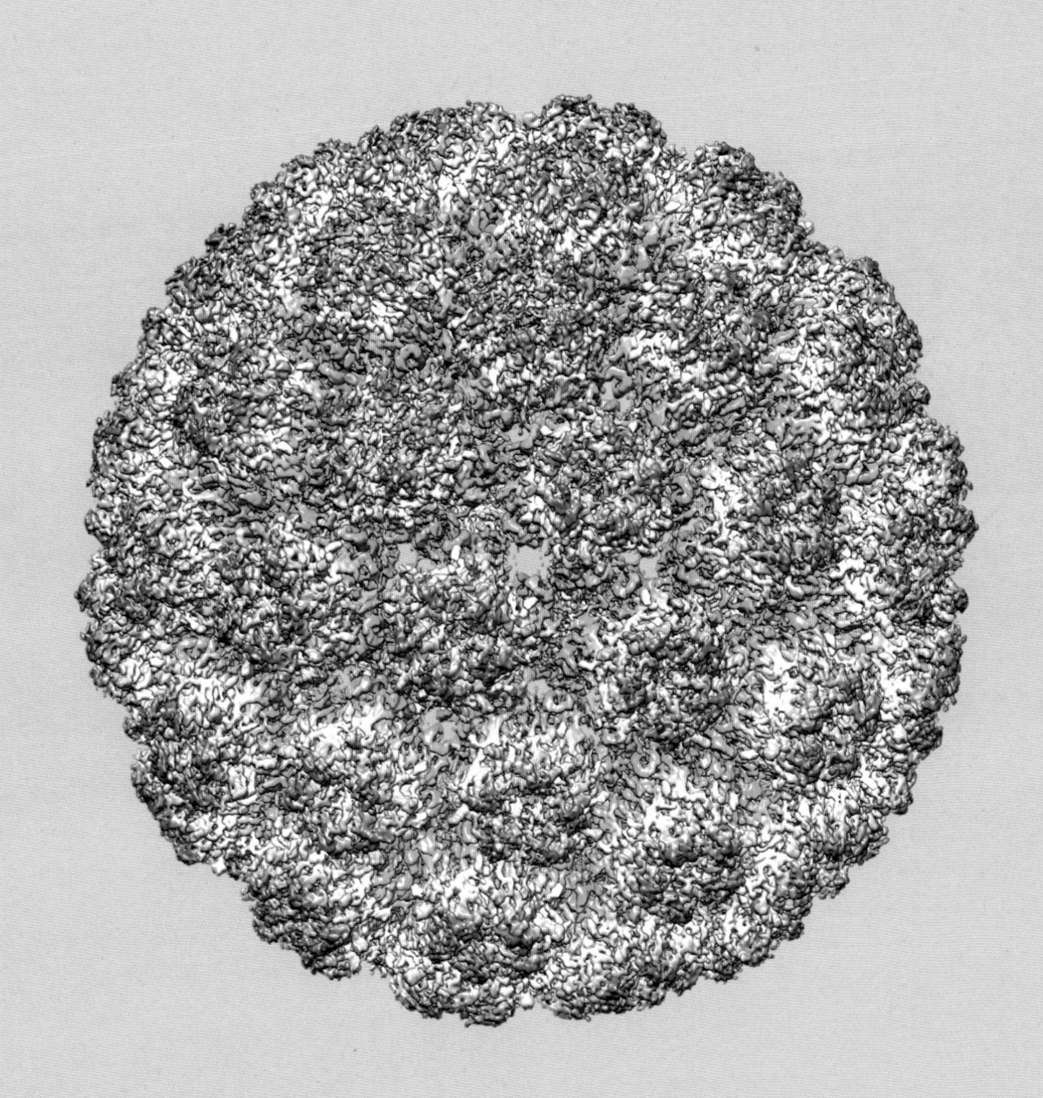

PMV

Panicum mosaic virus

A virus transmitted by lawnmowers

GROUP	IV
FAMILY	Tombusviridae
GENUS	Panicovirus
GENOME	Linear, single-stranded, single-component RNA comprising about 4,300 nucleotides, encoding six proteins
VIRUS PARTICLE	Icosahedral, non-enveloped
HOSTS	Turf grass, Switchgrass (*Panicum virgatum*), millet
ASSOCIATED DISEASES	St. Augustine decline
TRANSMISSION	Farm and garden equipment

St. Augustine decline, caused by panicum mosaic virus (PMV), is a common problem in lawns in the southern United States, especially those that are maintained by lawn care services. The virus is picked up from the infected plant sap during the mowing process, and then spread to other lawns that are mowed with the same equipment.

PMV can stay viable in plant debris for years, making it hard to get rid of once it is established within a system. The virus can spread from the debris to new plants by rain or wind, and it can remain on uninfected plants until they are wounded, when they have the chance to enter a plant cell.

PMV can support a satellite virus. This virus makes its own coat protein but can't replicate without the helper virus. In millet and Switchgrass (*Panicum virgatum*), the satellite virus can make the usually mild yellowing and stunting symptoms of St. Augustine decline much worse.

Scientists have proposed that some strains of PMV that cause very mild symptoms could be used as a sort of vaccine in Switchgrass against related viruses that cause more serious disease. In plant viruses this is called cross-protection and is a method that has been studied and used for decades in agriculture around the world.

→ Structure of panicum mosaic virus derived from chrystallography data.

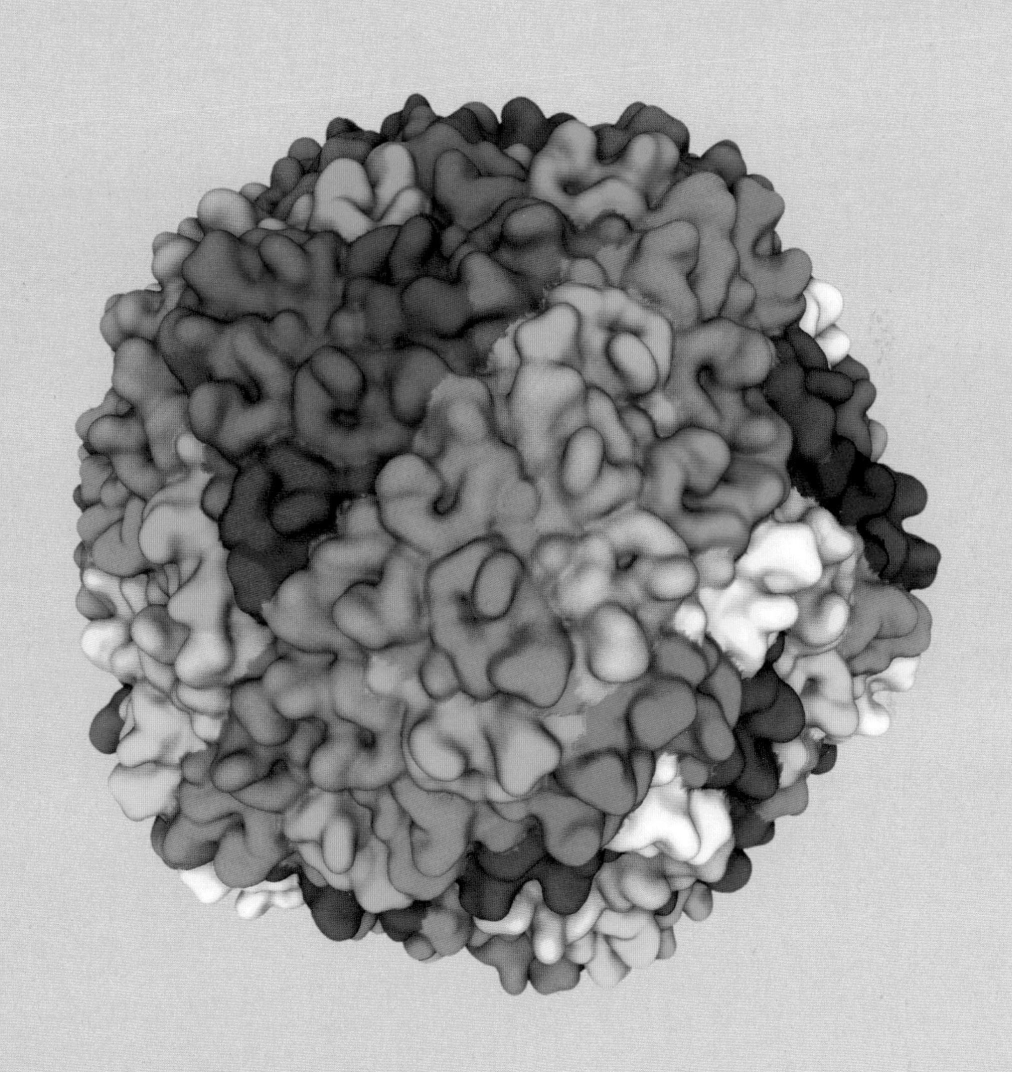

Rhinovirus C

A recent addition to the common cold viruses

GROUP	IV
FAMILY	Picornaviridae
GENUS	Enterovirus
GENOME	Single-stranded, single-component RNA comprising about 7,400 nucleotides, encoding 11 proteins via a polyprotein
VIRUS PARTICLE	Icosahedral, non-enveloped
HOSTS	Humans, other primates
ASSOCIATED DISEASES	Common cold, asthma
TRANSMISSION	Airborne
VACCINE	None available

A cure or vaccine for the common cold is still elusive, although prevention is possible. During the COVID-19 pandemic, many people noticed that they were not getting any colds. This is because they were wearing masks in public, one of the most effective measures for preventing infection by an airborne virus.

There are several species of rhinovirus, and many strains of each of them. Rhinovirus C alone has about 60 different strains, and there isn't much cross-reactivity by the immune system. This means that if you get one virus you will mount an immune response, but it probably won't protect you from other strains. In addition, immunity to rhinoviruses is not very long lived. All this makes it very difficult to develop a vaccine to prevent infection.

Rhinovirus C was discovered as recently as in the mid-2000s during routine screening for respiratory viruses after the first SARS-CoV epidemic. The virus was nearly impossible to grow in tissue culture, making it hard to study, but it is associated with the most severe rhinovirus infections, especially virus-induced asthma. It turns out that the human gene that codes for the receptor for rhinovirus C has two versions. The A version is found in all non-human primates and even in other animals with lungs, but it is rare in humans, who mostly have the G variant. The G variant is protective against severe infection with rhinovirus C, but rare individuals who have both alleles as the A variant are very prone to severe infection and virus-associated asthma. This is also linked to the development of chronic obstructive pulmonary disease in older adults. Chimpanzees (*Pan troglodytes*), which have the A/A genotype, often die from rhinovirus C infection.

→ A space-filling model of rhinovirus C drawn from x-ray crystallography data.

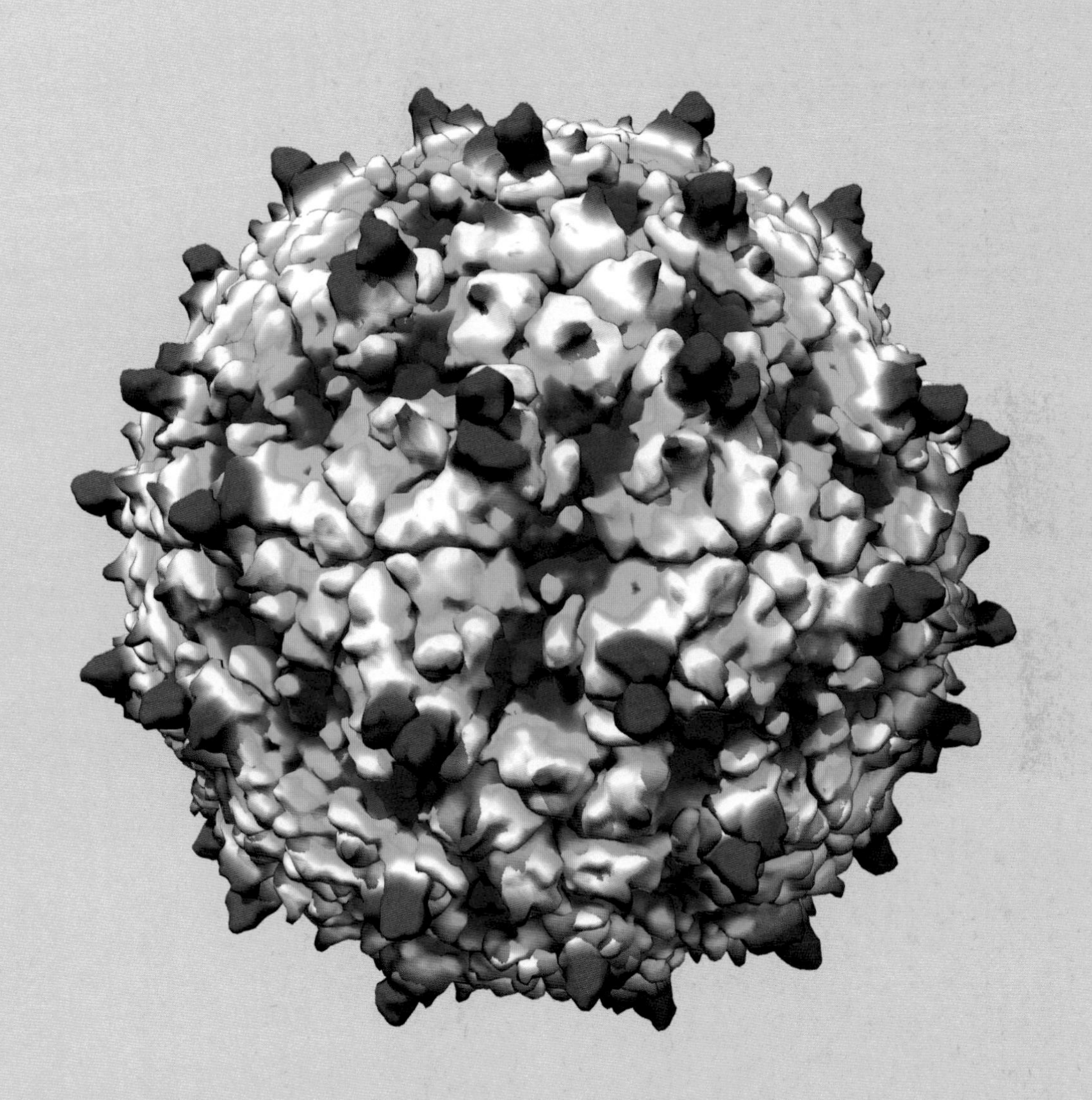

CHIKV

Chikungunya virus

A virus with an expanding transmission

GROUP	IV
FAMILY	Togaviridae
GENUS	Alphavirus
GENOME	Linear, single-stranded, single-component RNA comprising about 12,000 nucleotides, encoding nine proteins from a polyprotein
VIRUS PARTICLE	Icosahedral core with envelope
HOSTS	Humans, other primates, rodents, birds, mosquitoes
ASSOCIATED DISEASES	Chikungunya
TRANSMISSION	Mosquitoes in the *Aedes* genus
VACCINE	In development

Chikungunya is a serious illness that can cause long-term arthritis-like symptoms. It was first noted in parts of Africa in the 1950s, where it was transmitted to humans from wild primates by the Yellow Fever Mosquito (*Aedes aegypti*).

The Yellow Fever Mosquito is limited in range to tropical and subtropical climates. During the first decade of the twenty-first century chikungunya began showing up in Asia, and a decade later it arrived in the Americas. Then, rather suddenly, it was found in parts of Europe and North America with temperate climates.

Chikungunya virus (CHIKV) infects both mosquitoes and its primate hosts. In has to replicate in the gut of the mosquito in order to be transmitted to primates. Research has shown that the virus has mutated and is now being transmitted efficiently by another mosquito, the Asian Tiger Mosquito (*Aedes albopictus*). This mosquito has a much wider range, and has recently been found in many parts of Europe and North America. This means that the virus could readily emerge as a major problem in these parts of the world.

Both of the mosquito vectors of CHIKV are very well adapted to urban environments, laying their eggs in the humid microenvironments above standing water. Such environments are common in towns and cities, where rain collects in flowerpots and old tires, and the spread of the Asian Tiger Mosquito has been linked to the global trade in used tires.

→ Chikungunya virus particle drawn from cryo-EM data.

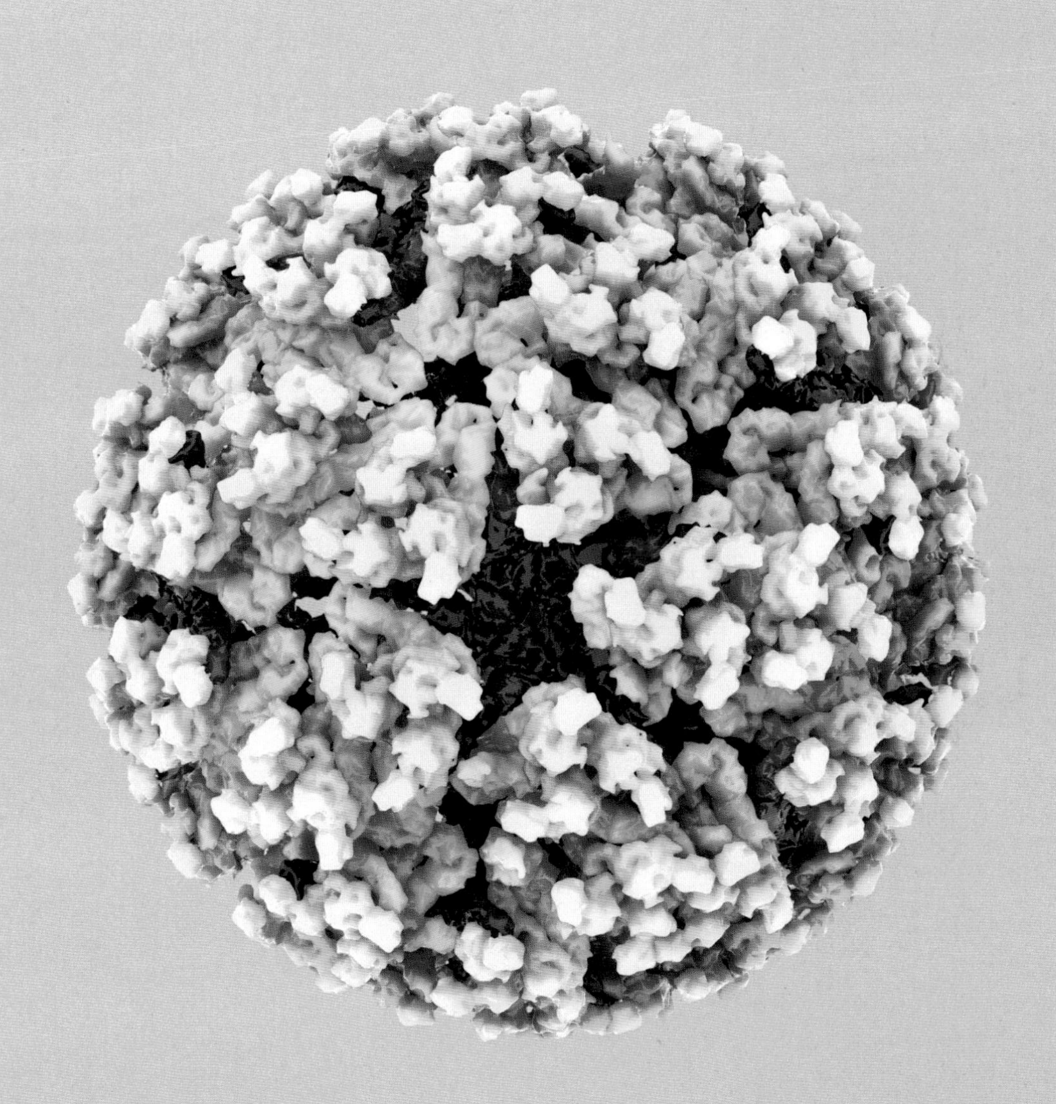

BPEV

Bell pepper alphaendornavirus

A naked virus with strict vertical transmission

GROUP	IV
FAMILY	Endornaviridae
GENUS	Alphaendornavirus
GENOME	Linear, single-stranded, single-component RNA comprising about 15,000 nucleotides, encoding one large polyprotein
VIRUS PARTICLE	None, naked RNA
HOSTS	Bell Pepper (*Capsicum anuum*) and some relatives
ASSOCIATED DISEASES	None
TRANSMISSION	Strictly vertical

Bell pepper alphaendornavirus (BPEV) is an unusual virus. Like all endornaviruses, it doesn't make a coat and is composed simply of its naked RNA genome. Single-stranded RNA is not very stable, so most endornaviruses have been discovered as the intermediate of replication, double-stranded RNA, which is very stable.

Plant endornaviruses are not transmitted between different hosts, except through the seeds. This vertical transmission means that they are passed on for many generations. Comparing these viruses in related plants can tell us a lot about the history of both the host and the virus. For example, a particular endornavirus affects all the "Japonica" Asian Rice (*Oryza sativa*) cultivars but not the "Indica" cultivars. These two cultivar types were domesticated about 10,000 years ago, from Wild Rice (*Oryza rufipogon*). Wild Rice itself has a related endornavirus that is about 24 percent different from the domesticated virus.

Scientists have compared BPEV affecting many different cultivars of Bell Pepper and some related peppers in North and South America. This has led to a deeper understanding of how the various types of peppers were domesticated in Mexico and South America. As with rice, peppers were domesticated about 10,000 years ago, but the wild ancestor of pepper does not have BPEV, indicating that the virus was introduced after domestication.

Genome of BPEV

The genome of BPEV is a single-stranded RNA, but it is always found as the double-stranded replicative intermediate. Here it is shown with an RNA-dependent RNA polymerase (RdRp) attached. There is a nick in the genomic strand of the virus.

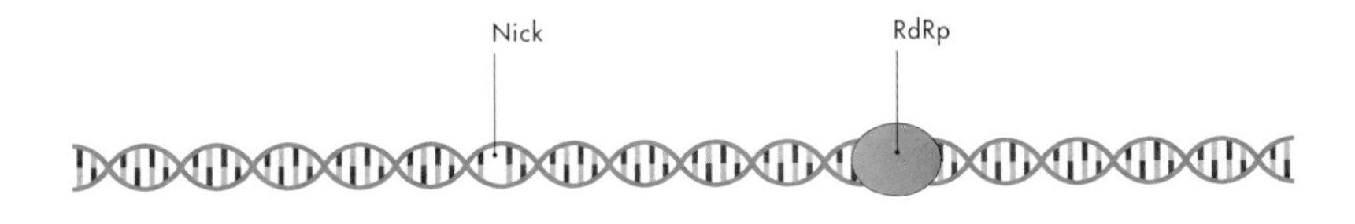

→ Bell Pepper (*Capsicum anuum*) plants are all infected with BPEV. The virus never causes any symptoms in the plants, but is just passed on to the next generation.

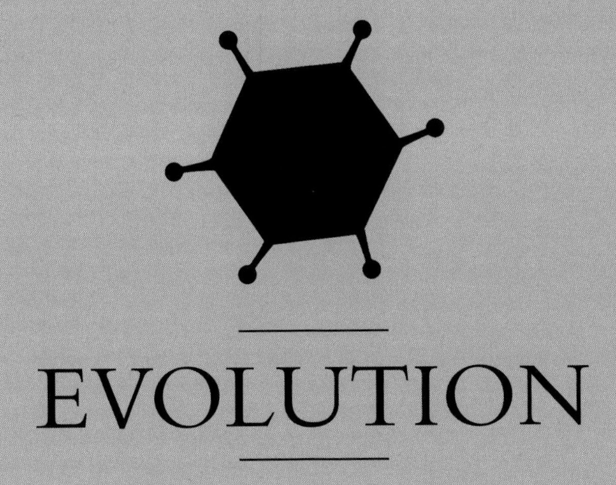

EVOLUTION

Evolution and natural selection

The word "evolution" usually makes people think of the English naturalist Charles Darwin and his seminal book *On the Origin of Species by Means of Natural Selection* (1859). Darwin's careful, detailed work, based on years of observations, provided critical insights into how the planet became populated with such a variety of species. However, he carried out all of his work before anyone had even heard of viruses, so made no mention of them. Darwin's work was also completed almost a century before the discovery of DNA and RNA as the genetic coding material for all living organisms.

In the era of molecular biology, evolution is discussed as the slow change of the genome over time as a result of mutations. A mutation is a change in the DNA or RNA genome. Mutations happen all the time, when a polymerase makes a mistake in copying DNA or RNA, or when an environmental factor such as a chemical or radiation makes a lesion in the DNA or RNA. These random changes can happen almost anywhere in the genome, although in cells they are more likely in active areas where transcription to RNA or copying of the DNA is taking place. In cells, most of these mistakes are corrected by enzymes that recognize them as such and cut out the bad DNA, replacing it with the correct DNA. Only rarely do mutations remain in the genome. To be passed on to the next generation, mutations have to occur in germ-line cells, such as eggs or sperm.

If mutations are passed on to the next generation, they may have a variety of effects. In most cases they are "neutral," meaning they don't have any effect that can be detected. Occasionally, however, they do have an effect, and whether or not they are carried on depends on whether that effect is positive or negative. If the effect is negative, the organism with the mutation won't be as competitive as its siblings that lack the mutation, and the lineage may die out. But if it is positive, the organism will be more competitive and the lineage may take over the population in time.

→ Charles Darwin (1809–1882) traveled around the world on board HMS *Beagle* as the ship's naturalist. On his journey, especially in South America, his observations led him to develop his hypothesis of evolution.

Mutations become important when food sources or the environment change. For example, let's say you have a mutation that allows you to tolerate heat better than most people. As the climate changes and temperatures increase, your offspring will be more likely to survive and prosper than those who cannot tolerate higher temperatures. However, if the climate were to get colder, your offspring would not have any advantage. This is the essence of natural selection, the theory of evolution described by Darwin.

Viruses evolve through natural selection too, just like all other gene-based entities. However, they face some different constraints. Viruses often have overlapping genes, so a single nucleotide mutation can impact more than one protein. In RNA viruses there is a lot of biological activity in the way the RNA folds, and this is dependent on the nucleotide sequence of the RNA. This means that natural selection can be important outside of the constraints of protein coding.

Accumulation of mutations during replication

On average, an RNA virus might make a mistake, or mutation, about once every time it copies its genome. A ssRNA virus may make several copies of the complementary strand, each with a potential mutation, and each of these can go on to make many copies of the genome. This leads to a very rapid accumulation of mutations.

(+) Strands

(–) Strands

Deleterious mutation

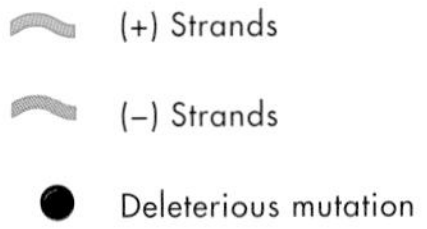

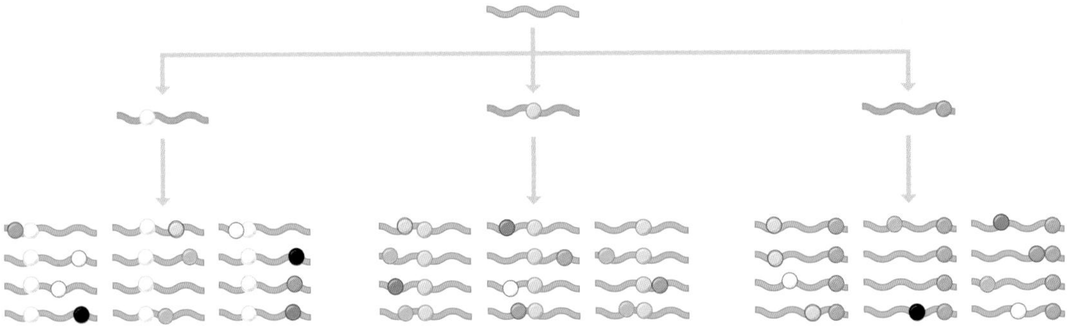

Another major difference between evolution of cellular life and that of viruses is the speed at which it occurs. There are several reasons for this. First, viruses have a very short generation time—in some cases less than a minute. Bacteria have generation times in the range of 30 minutes, while in humans a generation is about 20 years.

Secondly, most virus genomes are streamlined and most of the genome encodes proteins, and therefore many mutations can have an obvious effect. In humans, only 1–2 percent of the 3 billion-nucleotide genome is responsible for coding proteins, which is where mutations are usually most obvious. Other entities have even larger genomes: the largest known is the Japanese canopy plant, with 149 billion nucleotides.

Thirdly, viruses often make more mistakes when they copy their genomes, and they don't have the same means of correcting mistakes. Most RNA viruses make a mistake about once in every 10,000 nucleotides, whereas the mutation rate in eukarya is about once in every 1 million to 10 million nucleotides.

Fourthly, and finally, viruses that jump into novel hosts may show a very rapid rate of evolution as they adapt to a different environment. This type of dramatic change is uncommon in cellular life.

↓ An artist's rendition of a mutated coronavirus (in blue) exiting an infected cell.

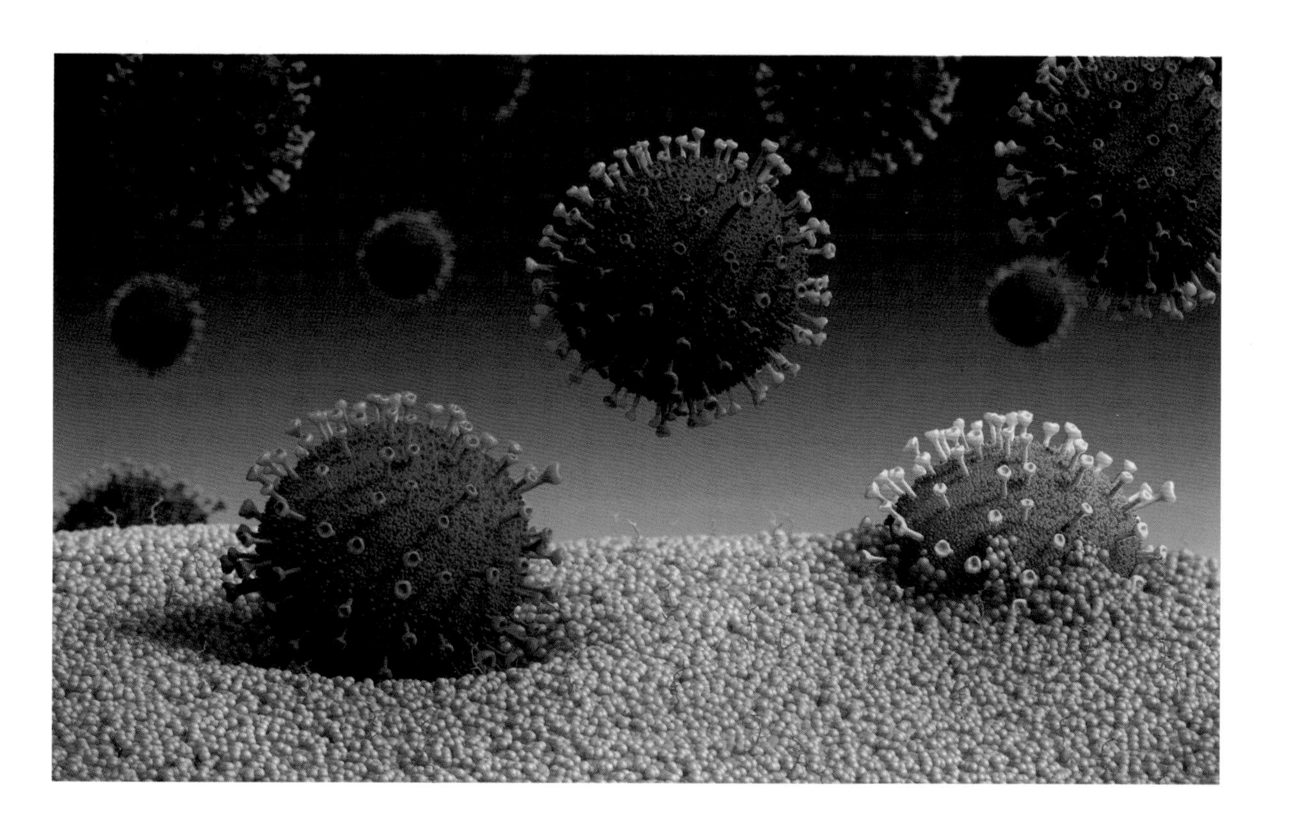

Fitness

The concept of biological fitness is simple: an individual that is more fit, that is well adapted and healthy within its environment, will produce more offspring and these will be more likely to survive, compared to an individual that is less fit. Measuring fitness is not so simple, however, and nor is understanding how fitness evolves.

↙ Fitness is often depicted as a landscape, like these mountains. The more fit the virus the higher the mountain peak it has climbed. A virus on the top of a steep peak like the one in the center may have a problem if its environment changes, because it will be hard for it to move, it will have to become very unfit first, whereas a virus on one of the peaks in the background can more easily move from peak to peak without losing too much fitness.

Clearly, if a virus mutates and that mutation makes it able to replicate faster or transmit better, then the virus would be more fit. However, there are limits to this. If the virus also makes its host very sick, then that can be detrimental to its fitness in the long run. For example, if you have a flu virus that replicates very rapidly such that you have a lot of virus coursing through your body, you are probably going to feel pretty bad and are most likely going to stay home in bed. That isn't very good for the virus, because you aren't going out and letting it spread to other hosts. Instead, it might be better for the virus to make you less sick. If this means the virus can't replicate to such high levels, it may evolve to some intermediate level, where it replicates well enough to make a lot of copies of itself, but not so well that it makes you too sick to get out and about.

Many viruses are most transmissible early in infection. If this stage happens ahead of symptoms then it is an advantage for the virus, as the host will be mingling with many susceptible hosts. It is possible that viruses become less transmissible over the course of an infection because of the accumulation of deleterious mutations. In this case, the virus population will not be robust enough to withstand the bottleneck that occurs during transmission (see page 145).

A virus that is lethal has an even bigger problem: if it kills its host before it has a chance to spread to a lot of new hosts, it will also die. It is likely that very lethal viruses such as Ebola have been infecting humans sporadically for a very long time. Ebola spends most of its time in an animal reservoir, and only rarely jumps into humans. Before people traveled around as much as they do now, the virus would have been self-limiting. It might have infected an entire small village and killed a lot of people there, but then it would run out of hosts and die out itself.

Experimental evolution

Rapid evolution in viruses means that the process can easily be observed, making them ideal subjects for studying how evolution works. Many studies have used bacterial viruses (also called bacteriophages) to study evolution, and these have demonstrated a couple of important points. First, when you experimentally evolve two virus lineages in the same environment, they end up having many of the same changes. This demonstrates the power of natural selection. And second, viruses can evolve to use different cell proteins as receptors to get into a cell. This has very important implications for how they adapt to new species.

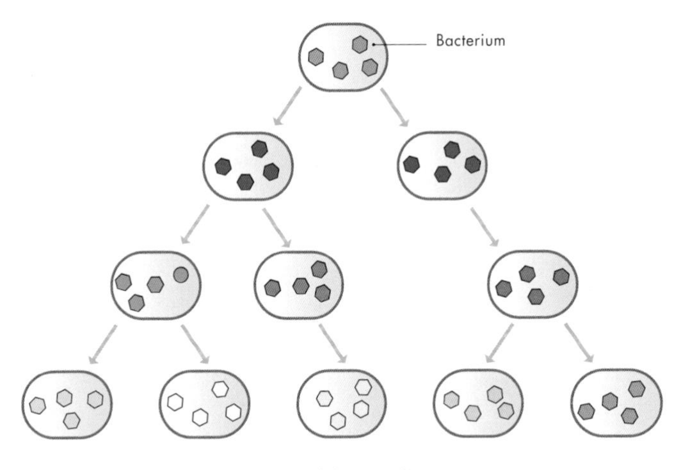

Viral evolution through bacteriophage generations. Each viral mutant is shown with a different color

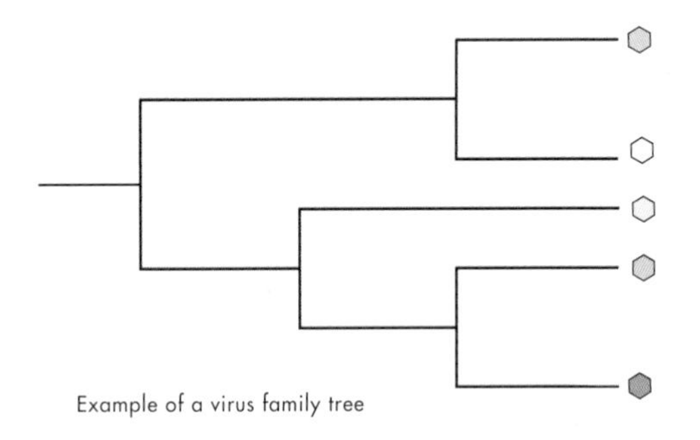

Example of a virus family tree

Studies of bacteriophage evolution have also been very important in developing better tools for the evolutionary analysis known as phylogenetics. This compares the genetic sequences of related entities and, using sophisticated computer algorithms, estimates how they are related to one another. This can be used to create a family tree. When we make a family tree of our ancestors, we know how everybody is related and so we have a known phylogeny. In a similar way, but starting with a clone, researchers allowed bacteriophages to evolve with a known relationship and then determined their genomic sequences. They then used these to generate a tree by various computer algorithms, allowing them to find a program that would re-create the known relationships.

Experimental evolution of a bacteriophage

When bacteriophages are experimentally evolved by infecting bacteria with a clone and the progeny then passaged in the scheme shown upper left, we know what the family tree looks like. When we are comparing viruses to see how they are related, we don't know that history and have only the end sequences. A correct tree, shown lower left, recapitulates the family tree.

Plant viruses have provided an excellent system for studying eukaryotic virus evolution. The first experiments in virus evolution were carried out in plants in the 1930s, before the RNA genomes and mutations were understood. Researchers passed tobacco mosaic virus and cucumber mosaic virus from plant to plant and followed changes in the virus symptoms, from a light and dark green pattern to a yellow mosaic pattern. Plant hosts make ideal experimental systems because they are easy and inexpensive to grow. In some cases, changes in symptoms can be observed in as little as 10 days. Plant virus studies have also shown that the type of host can make a big difference in how much variation there is in a virus population. For example, an infection of cucumber mosaic virus in peppers had high levels of variation, whereas the same virus in squash had much lower levels of variation. This implies that virus adaptation to new hosts or environments may happen more easily in some hosts.

SYMPTOM CHANGES IN EXPERIMENTAL EVOLUTION

The P6 strain of cucumber mosaic virus was one of the first experimentally evolved strains of a virus. The original strain caused a light and dark green mottle on tobacco leaves. The experiment involved inoculating plants, waiting a couple of weeks, and then taking a little tissue from the infected plants to inoculate a fresh plant. After the original strain was passaged in this way, a bright yellow strain appeared. The strain causing the symptoms on this tobacco plant was derived from P6 and shows the bright yellow coloration. However, it was continually passaged by researchers in greenhouses for many decades, and in this process it lost its ability to be transmitted by aphids—the way it gets around in nature. This occurred because there was no selection to maintain aphid transmission.

RECOMBINATION

Another feature of evolution that has been studied with viruses is recombination. All cellular genomes can recombine. In sexual life-forms, individuals receive two copies of each chromosome, one from each parent. These can recombine during replication of the genome, providing an enormous amount of variation in the offspring. In viruses this happens when multiple virus genomes are found in a single cell. Experimental studies to look at recombination show that it is very common in viruses, and can result in the emergence of new strains. Recombination is not usually random in viruses; in RNA viruses a lot of the recombination is driven by the way the single-stranded RNA folds on itself, making structures that help the polymerase that is copying the genome to jump to a different strand.

Recombination and reassortment

Virus recombination occurs when two strains of virus, represented here as Virus A and Virus B, are infecting the same cell. During the process of replication, the polymerase can jump from one genome to the other, resulting in recombinant viruses, represented as Virus BA and Virus AB. When the viruses have multiple components to their genome, these can also be mixed, a process called reassortment. Reassortment plays a major role in the evolution of influenza virus (see "The pathogens," page 248).

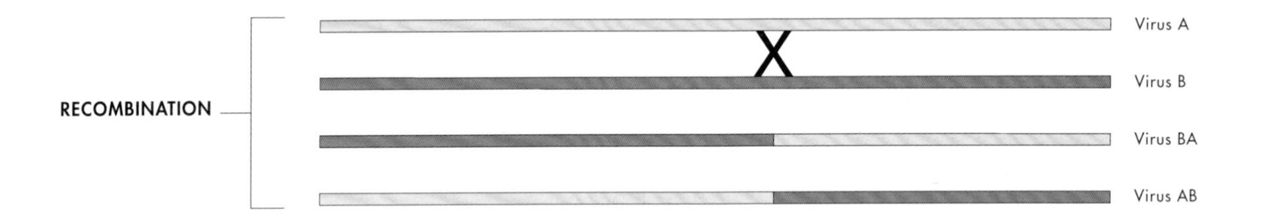

BOTTLENECKS

Bottlenecks occur when a population loses a lot of its genetic variation, usually due to a population crash. The most famous studies of population bottlenecks were carried out in Cheetahs (*Acinonyx jubatus*). These wild African cats went through a very severe population bottleneck thousands of years ago, resulting in very little genetic diversity. This has been made a lot worse for Cheetahs today because of human interference with their habitats, and their own low ability to produce offspring, and they are now considered to be at the brink of extinction.

Virus populations also go through bottlenecks, which probably occur during most transmission events and when the virus moves from its originally infected cell to the rest of the host. Because viruses evolve so rapidly, the effects of bottlenecks may be masked—they will generate a high level of variation after a single infection cycle, even when starting from a clone. Bottlenecks reduce the genetic diversity of a virus population, and probably occur during most transmission events.

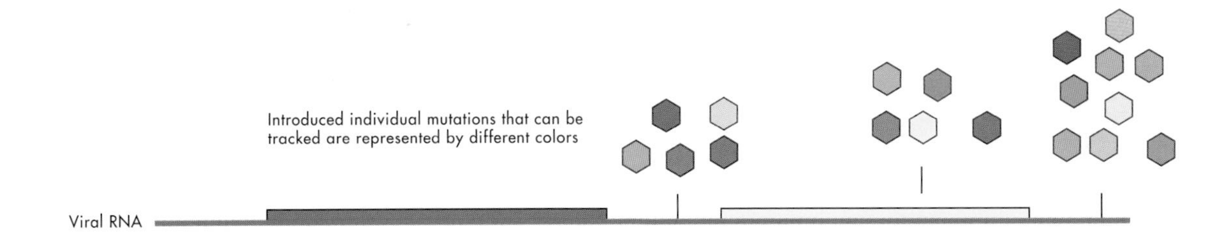

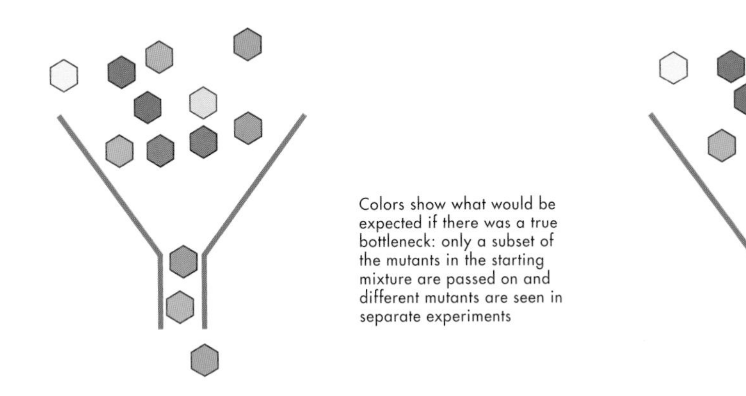

Bottlenecks during transmission or systemic infection

Representation of an experimental study of bottlenecks in virus infections. All the mutants were mixed and used to infect a host, then the resulting virus that had replicated and moved through the host was monitored. If the experiment is repeated, you would expect to see a similar number of mutants each time, but they would not be the same mutants because a bottleneck is completely random. In fact, this was observed when this experiment was carried out in plants using cucumber mosaic virus. Bottlenecks occurred during the systemic infection of the virus and during movement of the virus from plant to plant by the aphid vector of the virus.

Host–virus interactions in evolution

Viruses have played a central role in the evolution of their hosts. About 30 percent of the protein adaptations that make us human have been shaped by the viruses that have infected our ancestors. Some proteins involved in virus interactions were derived from Neanderthal portions of the human genome.

Viruses can impact evolution of hosts by facilitating movement of genes between species. This can happen when a virus integrates into the host genome, and then picks up a host gene when it leaves the genome and introduces it to a new host. This is called horizontal gene transfer. When genomes are compared across many species, there is a lot of evidence that this process has been very important in host evolution. There has also been frequent exchange of genes from viruses to bacteria and from bacteria to viruses.

In turn, hosts affect virus evolution in many ways. In a natural environment with a highly diverse set of hosts, viruses need to be more adaptable. In plants, for example, some viruses have a very broad host range—well over 1,000 species in some cases. For these viruses, having a lot of variation is an advantage. If the virus is transmitted by aphids, the next plant the aphid lands on may not be the same species as the one it acquired the virus from. In crops that are grown as monocultures, this is not important because the next plant is much more likely to be very similar to the one the virus came from. Monocultures of both plants and animals that are common in agriculture or in urban settings can allow a virus to evolve rapidly to be highly adapted to a single host. This may be why virus epidemics are quite rare in wild lands, but are much more common in agriculture and in human populations.

← Agricultural plants are generally grown as a monoculture, and as such are more susceptible to virus epidemics. Here, an expanse of soybeans in the midwestern United States is seen alongside a remnant of wild prairie where the plant diversity is very high.

Variants and escapees

The evolution of variants in virus populations is what gives rise to new strains that can have a different interaction with the host, or may be able to infect new hosts. A virus that generates many variants is likely to be able to infect a novel host—often called species jumping.

After a virus jumps species, a lot of variation can often be detected as it adapts to its new environment. This adaptability is very different for different viruses, and varies from host to host. One critical effect of this generation of variants is that the virus may rapidly evolve to escape the immune response of the host, whether this is due to infection or vaccination. This is discussed in more detail in the chapter on pathogens (page 248).

The population of a virus can be considered at many different levels. All of the various strains of the virus that are found across the world make up its global population. A subset of these strains is found in different communities around the world. Infection of an individual host will be through transmission from a variant or a few variants within the community, and then this will establish a new population within that host. Within a host there can be separate populations in different parts of the body or in different organs, and then within these organs there can be different areas where infection started and variants are again generated. Even individual infected cells can have separate populations, or populations in different compartments of the cell. In plant viruses, different populations of the same virus can be found in different branches of a plant.

← The cycle of virus transmission with a virus like Middle Eastern respiratory syndrome virus. The virus is carried by fruit bats, whose droppings are ingested by camels at watering holes. The virus is then transmitted to the human caretakers of the camels.

Virus population

Virus populations occur at many levels. The global population includes all the strains from all over the world. The individual host population includes all the virus variants in a single host. Different populations may also occur in different parts of the body. Within a site or organ there can be different foci of infection. There are also separate populations within individual cells, and even separate populations within different compartments inside a cell.

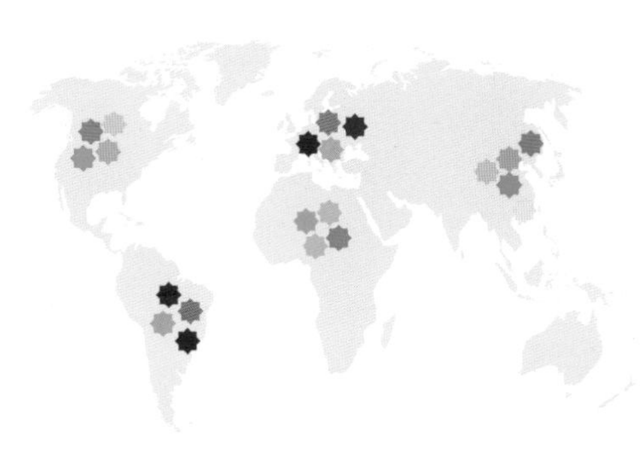

HOST POPULATIONS

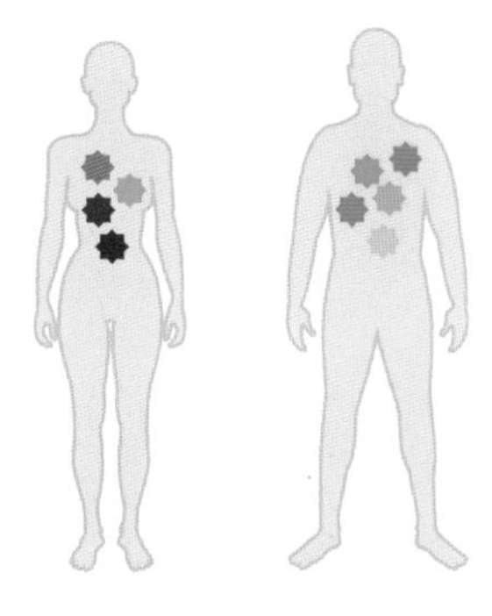

INDIVIDUAL HOSTS

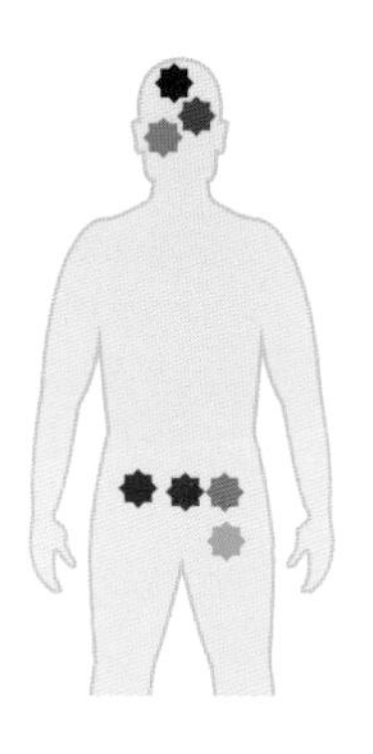

BODY SITES

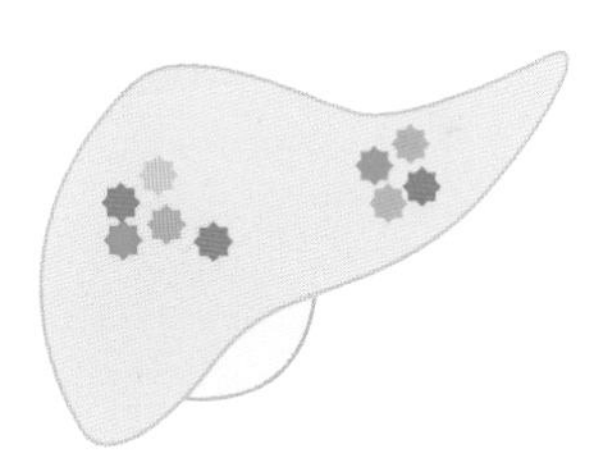

INFECTION FOCI

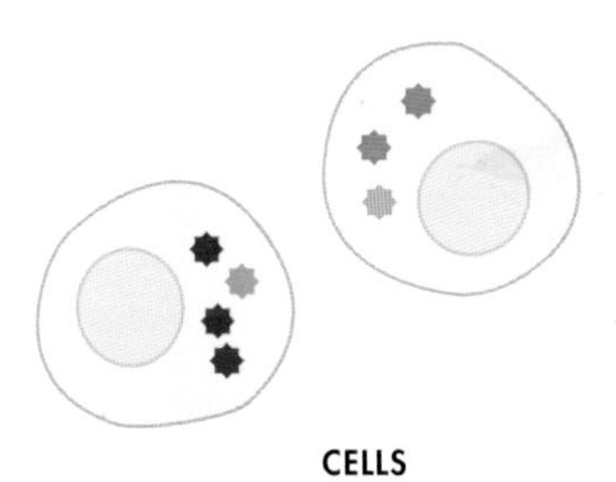

CELLS

REPLICATION CENTERS

Virus populations in different branches of a tree

In an experiment with plum pox virus, a plant virus that infects fruit trees, researchers inoculated very young trees with the virus and followed the populations for several years. After 13 years, different viral populations had evolved in different branches of the trees.

Highly variant viral populations, especially in RNA viruses, have been studied for decades. The earliest experimental studies in this area were carried out in RNA bacteriophages. One interesting effect of high levels of variation is that selection can act on the viral population as a whole rather than the individual. For example, there may be variants in the viral population that have mutations producing defective proteins, whereas other variants produce proteins that work better than the original protein. All the members of the population can make use of the better protein, even if their genome doesn't encode it. This is called complementation; different genomes within the same population can complement one another and provide extended functions. These variants are an advantage for the virus when it is infecting a host, but may be a problem when the virus is transmitted. This is because a lot of the variation will be lost through the transmission bottleneck and the best (most fit) genome may be not among the few that are transmitted.

DEEP EVOLUTION

Where do viruses come from? Comparing genomes of related entities provides a lot of information about the origins of life. There are a few genes that all cellular life has in common, and these have been used to determine comprehensive trees of life (see page 31).

The age of different cellular life is determined through fossils that can be dated with a good degree of accuracy. Viruses that are related can also be compared, as discussed above, but there are no genes that all viruses have in common, and because of their small size, no virus fossils have been found. This makes dating the oldest viruses impossible with current technology. However, although there are no virus fossils per se, many viral genomes have been incorporated into the DNA of host cells. Once incorporated, these viruses have evolved at the rate of their host, which is generally much slower than the rate at which viruses themselves evolve. Most studies of these viral "fossils" have been of retroviruses (see page 48).

Scientists have proposed three major hypotheses about the origins of viruses:

1 Viruses evolved before cellular life.
2 Viruses were once cells that lost much of their genome because they lived inside other cells and did not need all their original genes.
3 Viruses evolved from small pieces of DNA or RNA that escaped from cells.

There is some evidence to support each of these hypotheses, but nothing that is compelling. The most likely scenario may be a combination. For example, if the basic genes for replication appeared before cellular life, viruses may have acquired other genes from cells over the 3–4 billion years since the first cellular life appeared. Some viruses, such as the giant viruses, seem more likely to have evolved from cells, whereas some of the simplest RNA viruses may have originated before cellular life.

The study of virus evolution has exploded in recent years, leading to the launch in 2015 of a journal dedicated to this topic. The importance of understanding virus evolution has become very apparent since the start of the SARS-CoV-2 pandemic in late 2019. With better understanding of how viruses evolve, scientists hope to be able to predict pandemics before they happen and circumvent the worst of them, but this is not feasible yet.

↓ There are various ideas for the origins of viruses: They may have evolved before cells and led to cellular life; they may have evolved soon after cells first appeared, but before the Last Universal Common Ancestor (LUCA), the cells that gave rise to all of life on Earth; or they may have originated in the different cellular lineages, Eukarya, Bacteria, and Archaea.

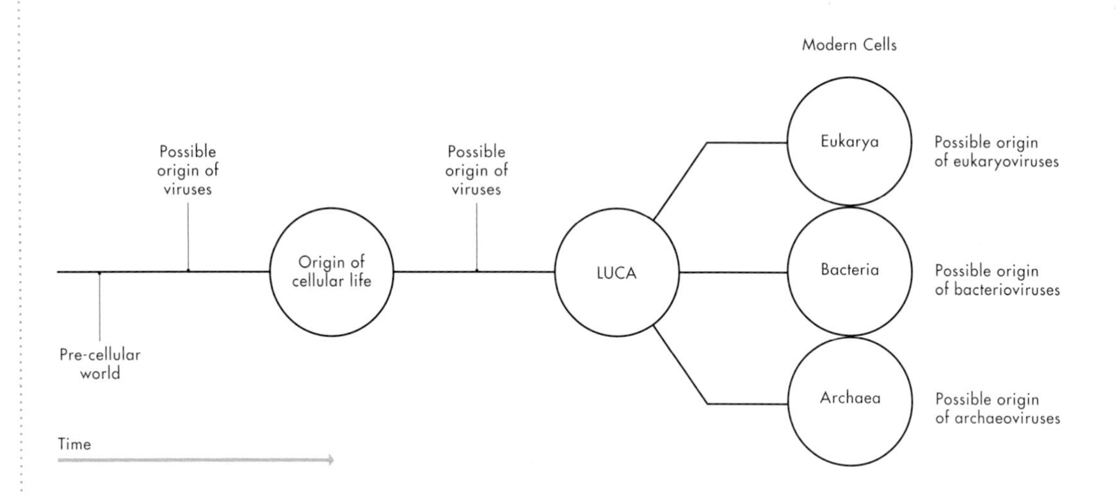

VSV

Indiana vesiculovirus

Classic virus for experimental virology

GROUP	V
FAMILY	Rhabdoviridae
GENUS	Vesiculovirus
GENOME	Linear, single-component, single-stranded RNA comprising about 11,000 nucleotides encoding five proteins
VIRUS PARTICLE	Enveloped, elongated bullet shape, about 75 nm by 180 nm
HOSTS	Sandflies, blackflies, cattle, horses, pigs
ASSOCIATED DISEASES	Mucosal lesions
TRANSMISSION	Insect vector
VACCINE	None

Indiana vesiculovirus, also called vesicular stomatitis virus (VSV), has been a very important player in our understanding of the evolution of minus-sense RNA viruses. An interesting feature of these viruses is that they carry their RNA-dependent RNA polymerase in the virus particle. This allowed scientists to study the enzyme in the laboratory in isolated virus particles—much simpler than trying to sort out the enzyme from the whole cellular milieu.

One of the early studies using VSV showed that the viral enzyme was not capable of removing the last nucleotide that had been added to a growing chain. This is important because this step is crucial for the process of fixing mistakes. If the polymerase inserts the wrong nucleotide, a "proofreading" enzyme will remove it and replace it with the correct one. The enzyme in VSV particles isn't able to carry out this step, however, and so can't proofread. This feature is a major reason why RNA viruses have such high levels of variation.

Many further studies on RNA virus evolution were carried out using VSV, including demonstrating that the addition of chemical mutagens does not significantly change the variation of the virus genome. This means that the virus lives at the edge of evolution and cannot survive any higher mutation frequencies than it already generates during its replication. The biological effects of a highly variable population are that the whole population acts in concert, and this has also been demonstrated with VSV.

Because VSV normally replicates in insects such as sandflies as well as in livestock, it has been studied to examine how its movement between these very different hosts has affected its evolution. This ability of a virus to infect very different hosts requires a lot of plasticity in terms of its adaptability. A number of RNA viruses use this lifestyle of replication in both a primary host and an insect vector. Most DNA viruses don't do this, although the geminiviruses of plants do. Like RNA viruses, these exhibit extremely high levels of variation, which may be a requirement for this type of adaptability.

In recent years VSV has been developed as a vaccine-delivery system. The genetic code for proteins from target viruses is inserted into the genome of a version of VSV that can replicate in people without causing any disease. This is used in the production of the current vaccine for ebolavirus.

→ Colorized transmission electron microscope image of a VSV particle, displaying the typical bullet shape.

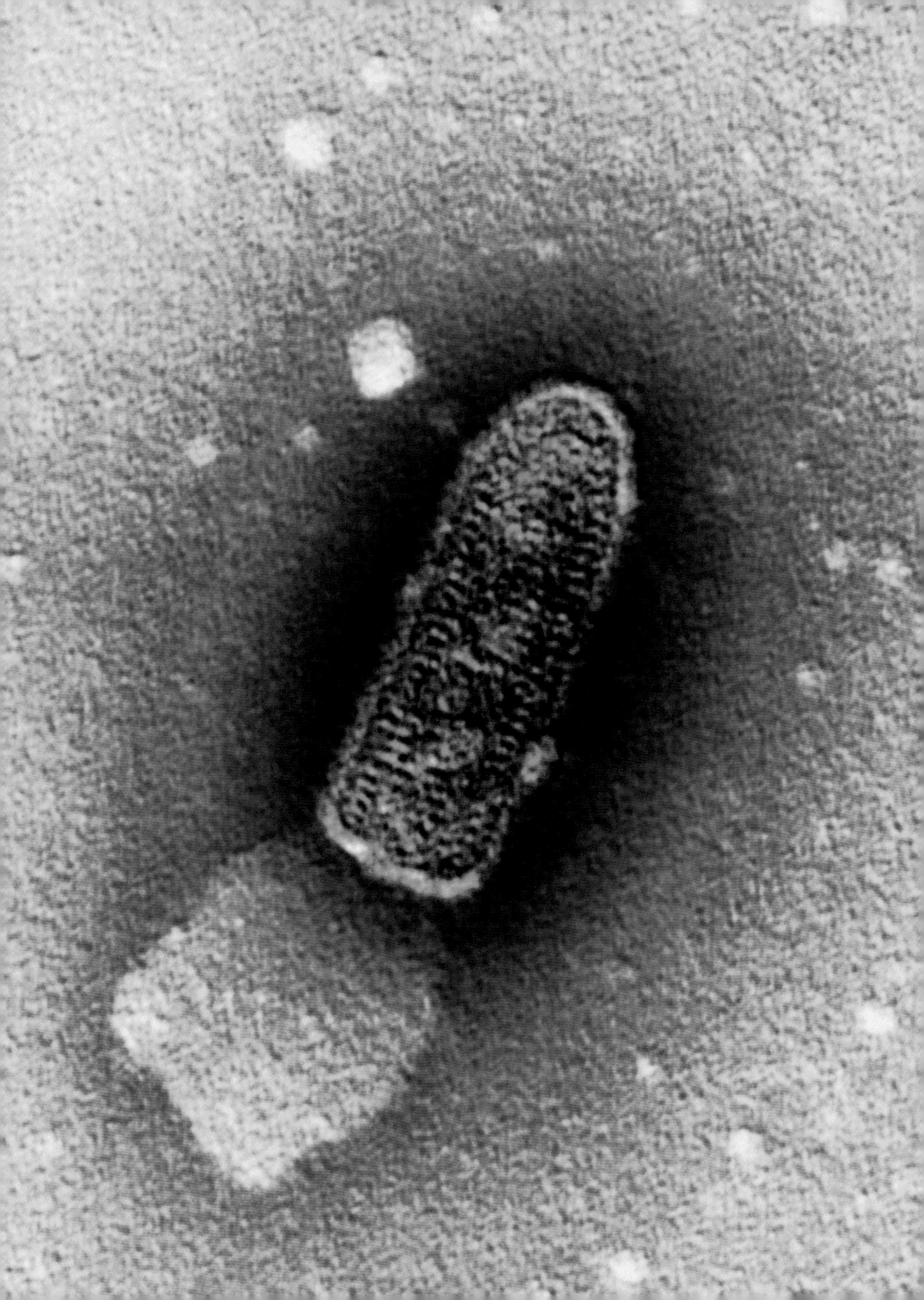

HSV-1

Human alphaherpesvirus 1

Worldwide human virus that sticks around for life

GROUP	I
FAMILY	Herpesviridae
GENUS	Simplexvirus
GENOME	Linear, single-component, double-stranded DNA comprising about 152,000 nucleotides, encoding about 75 proteins
VIRUS PARTICLE	Enveloped, spherical with an icosahedral core
HOSTS	Humans
ASSOCIATED DISEASES	Cold sores, genital sores, meningitis, encephalitis
TRANSMISSION	Direct contact with fluid from lesions
VACCINE	None available

Human alphaherpesvirus 1, also known as herpes simplex virus-1 (HSV-1), is a very common infection in humans, who often acquire it early in life. The infection rate is about 60 percent, as measured by the number of individuals with antibodies against the virus.

HSV-1 lives in neural ganglia and is often dormant for long periods of time. When it activates, it travels down the nerves to cause lesions around junctions of mucous membranes and normal skin. The most common lesions are cold sores, but the virus can also cause genital lesions. At one time it was thought that all genital lesions were caused by the related human alphaherpesvirus 2, but it is now known that both viruses can cause lesions at either site and that the occurrence of HSV-1 in genital lesions is becoming more common. The virus can also infect the eye, which can lead to blindness, and very rarely it can cause brain infections that are severe and can be lethal.

HSV-1 replicates using a virally encoded DNA-dependent DNA polymerase, an enzyme that has a high level of accuracy in copying DNA. For a long time it was assumed that these large DNA viruses would not have very high levels of variation, but in recent years new studies have shown that virus isolated from individuals from around the world is highly diverse. There is variation both between hosts and within the viral population from an individual host. A number of interesting aspects of the HSV-1 genome contribute to this higher than expected variation, including the structure of the DNA, which may promote more errors in the replication process. It also seems possible that some of the variation within individual hosts is due to infection with multiple strains of the virus.

The virus also undergoes extensive recombination events during replication allowing an additional level of variation. This is especially significant when more than one strain infects an individual, so that shuffling of genes occurs. The ever-increasing capabilities of nucleotide sequence analysis are providing virologists with tools to look deeply into viral populations from individuals—studies that have not been undertaken for most large DNA viruses. Such studies will help an understanding of how the virus has remained so stable in its ability to infect humans despite extensive variation. In comparisons with related viruses from Chimpanzees (*Pan troglodytes*) it seems likely that HSV-1 has been infecting humans since our divergence from other primates.

→ Model of the HSV-1 particle, derived from cryogenic electron microscopy data.

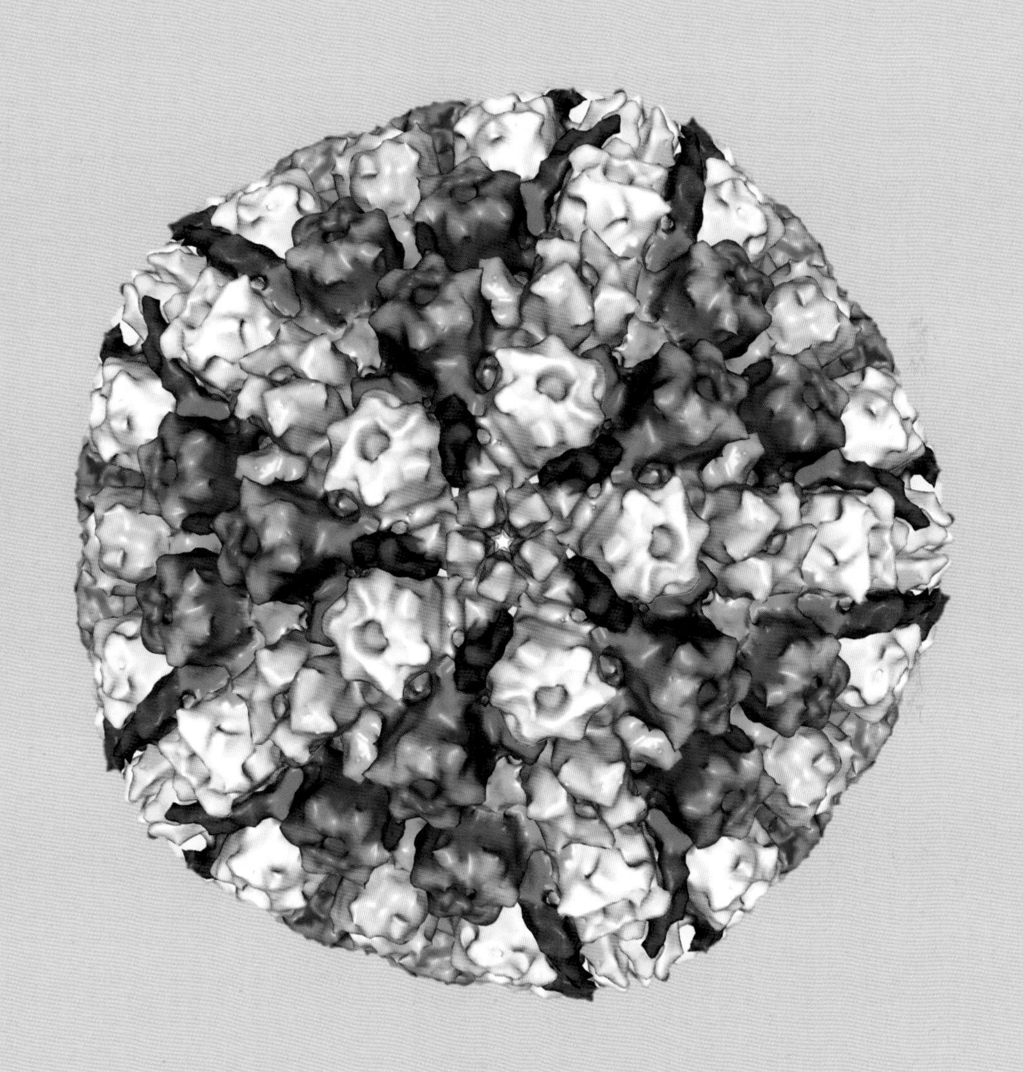

Morbilliviruses

Same virus evolves to infect different hosts

GROUP	V
FAMILY	Paramyxoviridae
GENUS	Morbillivirus
GENOME	Linear, single-component, single-stranded RNA comprising about 16,000 nucleotides, encoding eight proteins
VIRUS PARTICLE	Spherical, enveloped, 100–300 nm
HOSTS	Cattle, humans, and dogs and other carnivores, respectively
ASSOCIATED DISEASES	Cattle plague, measles or rubeola, distemper
TRANSMISSION	Airborne
VACCINE	Live attenuated

The morbilliviruses are some of the most contagious viruses known. Rinderpest (cattle plague) has been eradicated by vaccination. Measles and canine distemper are largely controlled by vaccination, but canine morbillivirus has emerged in wildlife in recent years, posing a threat to wild carnivorous species.

Rinderpest has been reported for hundreds of years, and was one of the most lethal known diseases of cattle. It probably originated in Africa and moved to Europe with the movement of cattle. In the late nineteenth century, 80–90 percent of all cattle in Africa died from a huge epidemic of rinderpest, spurring a lot of research into the disease. Inoculations to help prevent serious disease started in the eighteenth century, and a vaccine made from heat-treated infected tissue was introduced in 1918. In 1957 a vaccine based on an attenuated virus was introduced, and today the disease is considered eradicated as a result of vaccination. It is only the second virus to be eradicated, after smallpox.

The origin of the measles virus is not certain, but epidemics were first reported in the eleventh and twelfth centuries. When measles and rinderpest genomes are compared, it is clear measles evolved from rinderpest around that time. It is thought that close relationships among humans and cattle allowed the rinderpest virus to jump to humans. Measles starts with a cough, fever, and runny nose, followed by a body rash. Although not usually a serious disease, there can be many complications that may lead to long-term effects or even death. Measles is easily prevented by vaccination.

In the early sixteenth century in Central and South America following European colonization, measles epidemics proved devastating for the indigenous people, who had no prior exposure to the virus and hence no immunity. The fatality rate was about 25 percent of infected individuals. Canine distemper was first described in South America in the mid-sixteenth century. It is thought that the measles virus crossed over from humans to dogs to become canine distemper, following the practice of feeding the dead—many of whom had measles—to dogs. The disease was not described in Europe until about 20 years later, supporting the idea that the virus emerged in South America. Canine distemper causes vomiting, diarrhea, and sometimes seizures and death.

→ Artist's drawing of the measles virus based on data from electron micrographs, crystallography, and cryogenic electron microscopy.

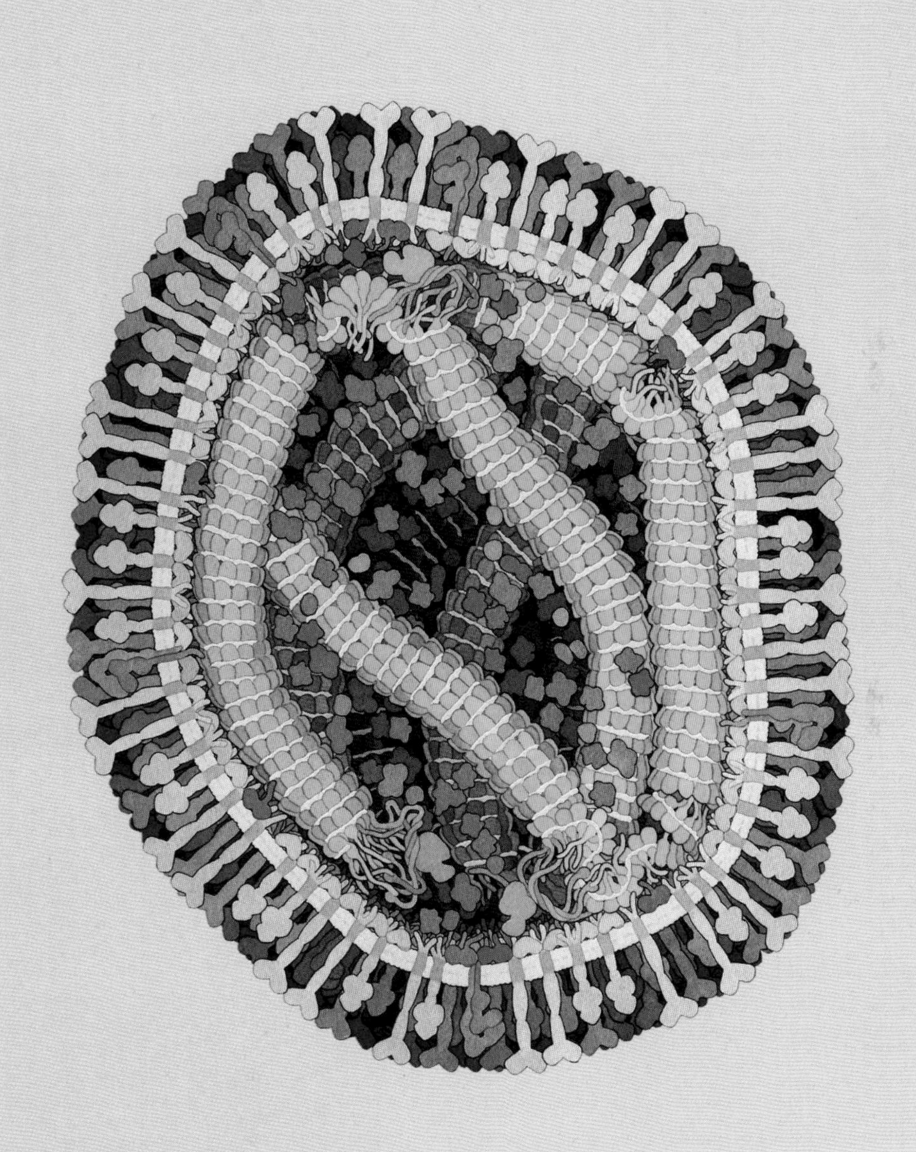

Qubevirus durum

Early model for virus evolution

GROUP	IV
FAMILY	Fiersviridae
GENUS	Qubevirus
GENOME	Linear, single-component, single-stranded RNA comprising about 4,200 nucleotides, encoding four proteins
VIRUS PARTICLE	Non-enveloped, icosahedral, about 26 nm
HOSTS	*Escherichia coli* and related bacteria
ASSOCIATED DISEASES	Cell lysis and death
TRANSMISSION	Diffusion

Qubevirus durum, also known as bacteriophage Qß, was the first virus used for extensive studies of RNA virus evolution. Work with this virus coincided with a theoretical framework for describing RNA virus populations, known as quasispecies. The term was coined from physics rather than biology, and doesn't relate to the biological idea of a species.

The basic idea behind quasispecies is that RNA viruses can generate an enormous amount of variation in a single infection cycle, but this population acts like an individual in terms of selection. This is because all the variants can act together, with different variants providing different functions. Eukarya generally have two copies of each gene (one from each parent), called alleles. In cases where there is a negative mutation in one allele, the other can compensate. This is seen in the mutation that confers cystic fibrosis, where two copies of this "bad" allele must be present for the disease to develop. In a virus population, a similar thing is taking place, with each variant representing an allele, but with the large variable populations there can be hundreds or thousands of alleles.

Qß infects bacteria that have F-pili, hair-like appendages that allow different bacterial cells of the same species to attach to one another and share DNA, or mate. The virus attaches to the side of the F-pilus to enter the bacterial host. Once inside, it replicates until the bacterial cell is filled with progeny virus and bursts open.

Qß was also used in the first studies of the enzyme that copies RNA from an RNA template, called the RNA-dependent RNA polymerase. This enzyme is made up of four proteins, but only one is from the virus; the other three are derived from the host bacteria. This is true for most of these enzymes, which are used by viruses to replicate their genome. At one time it was thought that only viruses used the enzyme, but later an enzyme with a similar function was found in plants, and now the enzyme is known to be widespread. When the nucleotide sequences of the viral enzymes and host enzymes are compared, however, it seems that they are not related to each other but evolved from different origins to have the same function, a process called convergent evolution.

→ Computer-generated model of the structure of Qubevirus durum, using crystallography and cryogenic electron microscopy.

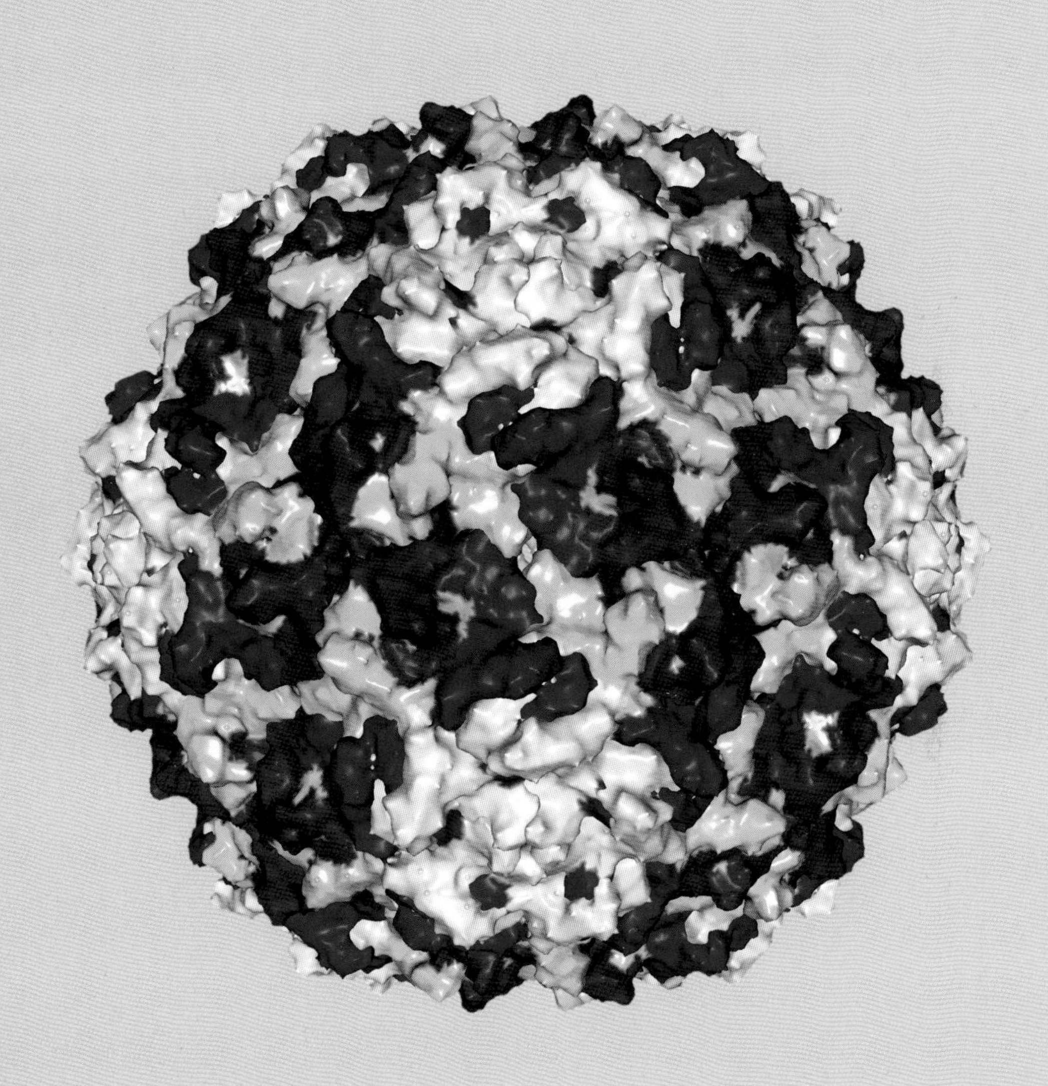

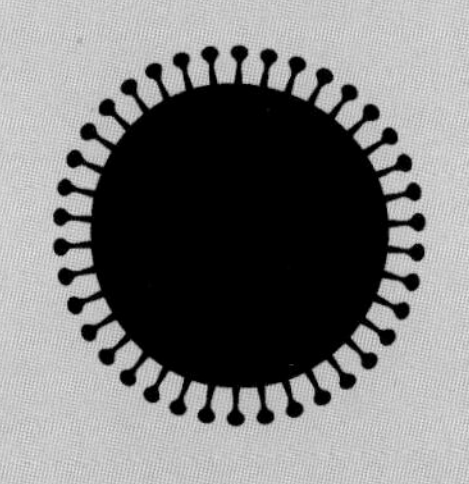

THE BATTLE BETWEEN VIRUSES AND HOSTS

Immunity

Viruses are often thought of as evil agents of disease. While there are certainly plenty of viruses that do cause diseases in everything from bacteria to protists, fungi, plants, and animals (including humans), most don't. We are simply more familiar with those that do because they have been studied much more thoroughly. This chapter focuses on how hosts combat viruses.

There are two major levels of immunity: innate and adaptive. These have long been considered separate, but it is now clear that the two systems interact to some degree through the numerous signaling molecules they elicit. Innate immunity is essential to initiate adaptive immunity, and adaptive immunity relies on innate immunity to get rid of pathogens.

IMMUNE RESPONSES IN DIFFERENT CELLULAR LIFE

	Vertebrate animals	Invertebrate animals	Plants	Fungi	Protists	Bacteria & archaea
ADAPTIVE IMMUNITY	Antibodies	RNA silencing	RNA silencing	RNA silencing	RNA silencing	CRISPR
IMMUNE MEMORY	Yes	Yes	No	No	No	Yes
INNATE IMMUNITY	Physical barriers, white blood cells, defense molecules, cell killing	Physical barriers, white blood cells, defense molecules	Physical barriers, restricted movement, defense molecules, cell killing	Physical barriers, restricted movement	Physical barriers	Physical barriers, restriction enzymes

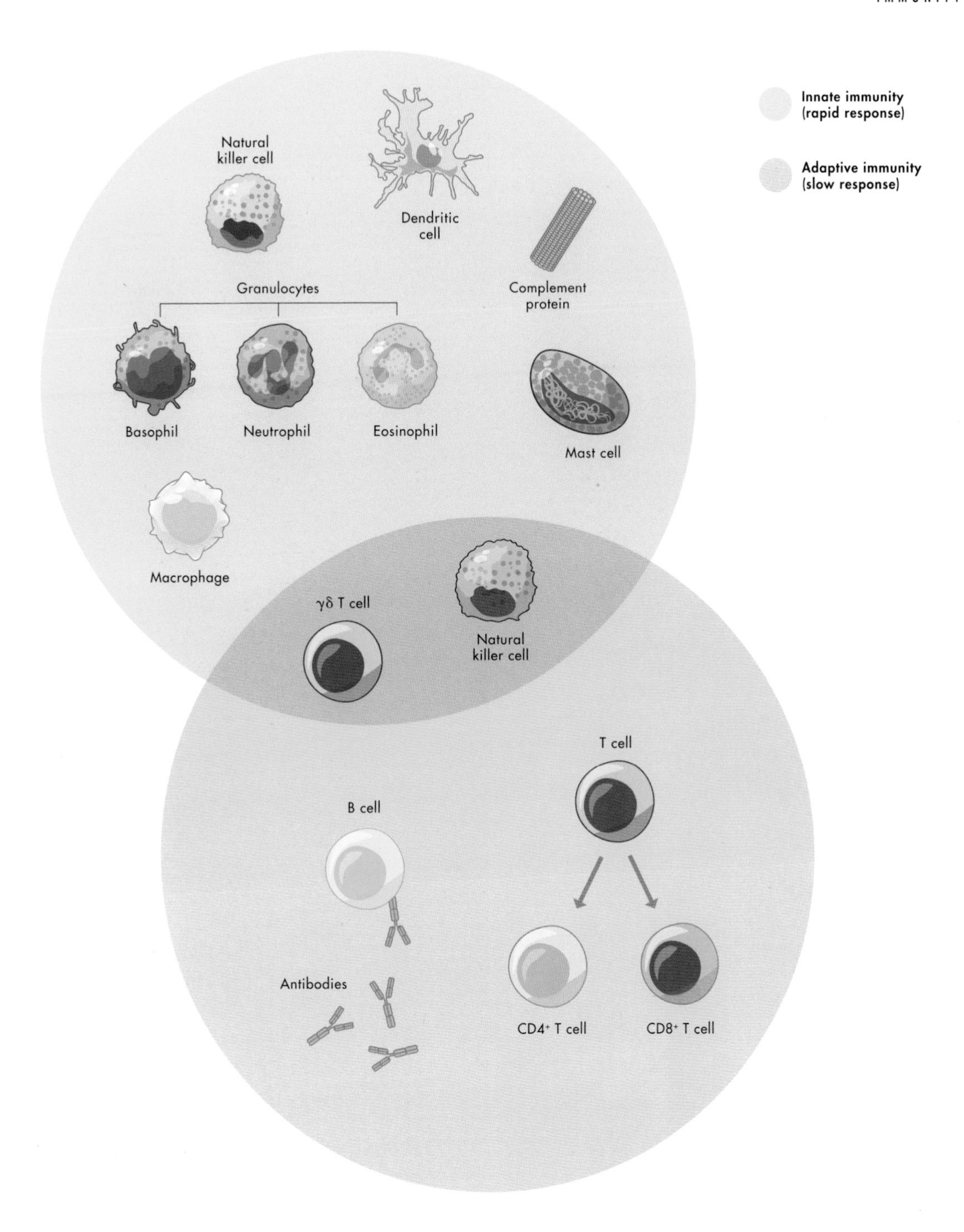

Cells involved in the human immune system

The immune system in vertebrate animals has two major responses to virus infection. Innate immunity occurs rapidly and involves many different kinds of blood cells and lymph cells. The adaptive immune response is slower, and involves B cells and T cells. B cells produce antibodies that are specific to parts of the virus, and these cells can last a long time through memory. T cells can directly kill virus-infected cells.

All cellular life has an immune system to combat pathogens. Being immune to a virus means that you probably won't get infected, or if you do you won't get very sick. Immunity comes in a whole gradient of degrees and the terminology can be confusing, so a few basic terms need to be defined at the outset. These terms may be defined differently in different places, so the explanations below are not absolute, but they are how the terms are used in this book. The definitions apply to all life-forms, from bacteria to plants, fungi, and humans.

Tolerance

Tolerance means that an infection can happen but little or no disease will occur. A tolerant individual can still spread the disease if they are infected, and since they do not have symptoms they may not realize that they are infected. The most famous human example of tolerance was Mary Mallon (1869–1938), a woman infected with a bacteria that causes typhoid fever but who had no symptoms. After emigrating from Ireland to the United States, she worked as a cook, where she exposed many people to typhoid, some of whom died. She was known as Typhoid Mary. Tolerance is common in virus infections.

Resistance

Resistance means that no infection and no disease occur. Resistance can be directly related to the immune system, but it can also just mean that the host can't contract the virus at all for one of many reasons. It is a term often used in agriculture. Many crops are bred for resistance to viral infections, or are developed through recombinant DNA. Sometimes this resistance involves the plant immune system, but at other times it is not very clear how it works.

← European explorers arriving in the Americas brought many diseases with them, including smallpox and influenza, which were lethal to the Indigenous people.

→ Mary Mallon (1869–1938), better known as Typhoid Mary, sits fourth from the right, among a group of inmates quarantined on an island in Long Island Sound. Typhoid is a bacterial disease, but Mary's story is a great example of tolerance.

Susceptibility

Susceptibility and immunity often occur on a gradient. A host never before exposed to a virus will be completely susceptible to it and is likely to get infected. Previous exposure of any kind, even if it didn't result in disease, will usually confer some degree of immunity. There are also some instances where partial immunity may look like tolerance. For example, multiple exposures to a virus will slowly build immunity, so that an individual may seem to be tolerant to infection. Viruses that are endemic in an environment result in this type of immunity in the population. A virus is often most virulent when it begins to infect a naive host population. That is why the indigenous people of the Americas often died from influenza or measles, or even the common cold viruses, after the arrival of the Europeans colonizers.

Innate immunity

The first line of defense against a foreign agent is innate immunity. It is not specific to the virus, but rather a generalized response. Innate immunity starts with physical barriers, including skin, mucous membranes, plant cuticles, and cell walls of bacteria and archaea, all of which prevent foreign entities such as viruses from entering a host. In some animals, mucosal surfaces such as the respiratory tract, gastrointestinal tract, and urogenital tract also contain numerous harmless microbial communities, collectively termed the microbiota, that play a major role in preventing or outcompeting invading pathogens.

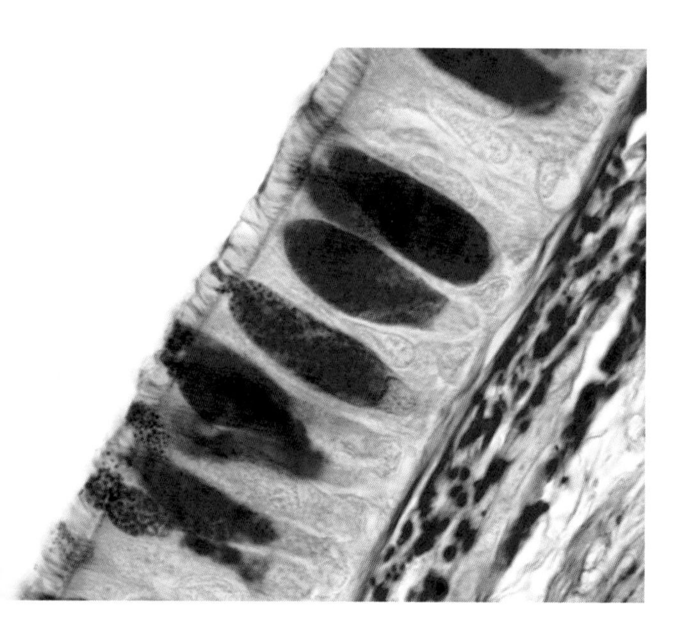

↑ Epithelial cells from the respiratory tract showing the hair-like cilia, and the mucus-secreting goblet cells.

Once the physical barriers are breached, the next level of innate immunity kicks in. Secretions such as tears, stomach acids, and mucus are common in animal innate immunity, and contain several antimicrobial substances that can kill invading pathogens. In addition, the cells lining the respiratory tract of mammals have cilia, hair-like extensions, which move to exclude foreign matter, including pathogens. When a foreign agent is detected, chemical signals such as histamine bring blood to the area, starting the process of inflammation. This involves several kinds of white blood cells, which are important in regulating the response, removing unwanted entities such as viruses or bacteria, and releasing other chemicals. An elevated temperature, either locally at the site of a wound or as a whole-body fever, is part of inflammation. Viruses of warm-blooded animals usually have a very narrow temperature tolerance, and increasing the temperature in their environment slows or stops their replication. The fever might make the host feel miserable, but it is playing an important role in combating virus infection.

Viruses induce vertebrate animal hosts to make specific small molecules that help fight the virus, such as interferons. These messenger molecules are involved in communication among different responses and cell types to help orchestrate the proper response. Some gut bacteria are also important in helping ensure proper levels of interferon are produced in response to a virus infection.

Another layer of innate immunity involves the nonspecific recognition of molecular signatures that are not part of the host. These are called microbe-associated molecular patterns (MAMPs), and can be nucleic acids or proteins. With RNA viruses, MAMPs are often types of RNA that the virus makes but cells do not, such as double-stranded RNA, or RNA that is modified in a different way than cellular RNA. MAMPs trigger a cascade of chemical signals that target these foreign agents. This type of innate immunity is active in plants and animals, although the specific cell receptors and effectors differ. Alternatively, the viruses cause damage to the cells, and the damaged and leaked cellular debris alerts the innate immunity. This is a particularly important response in mammals to prevent both infection and serious damage caused by the pathogens.

Cells and factors involved in innate immunity

The innate immune system in vertebrate animals involves recognition of microbe-associated molecular patterns (MAMPs), which go on to trigger a cascade of events. Some of these cells can also be activated by stress or cancer. In plants, a similar recognition of MAMPs occurs, but the downstream pathways are different. Viruses often evolve ways to avoid these systems to circumvent immune responses.

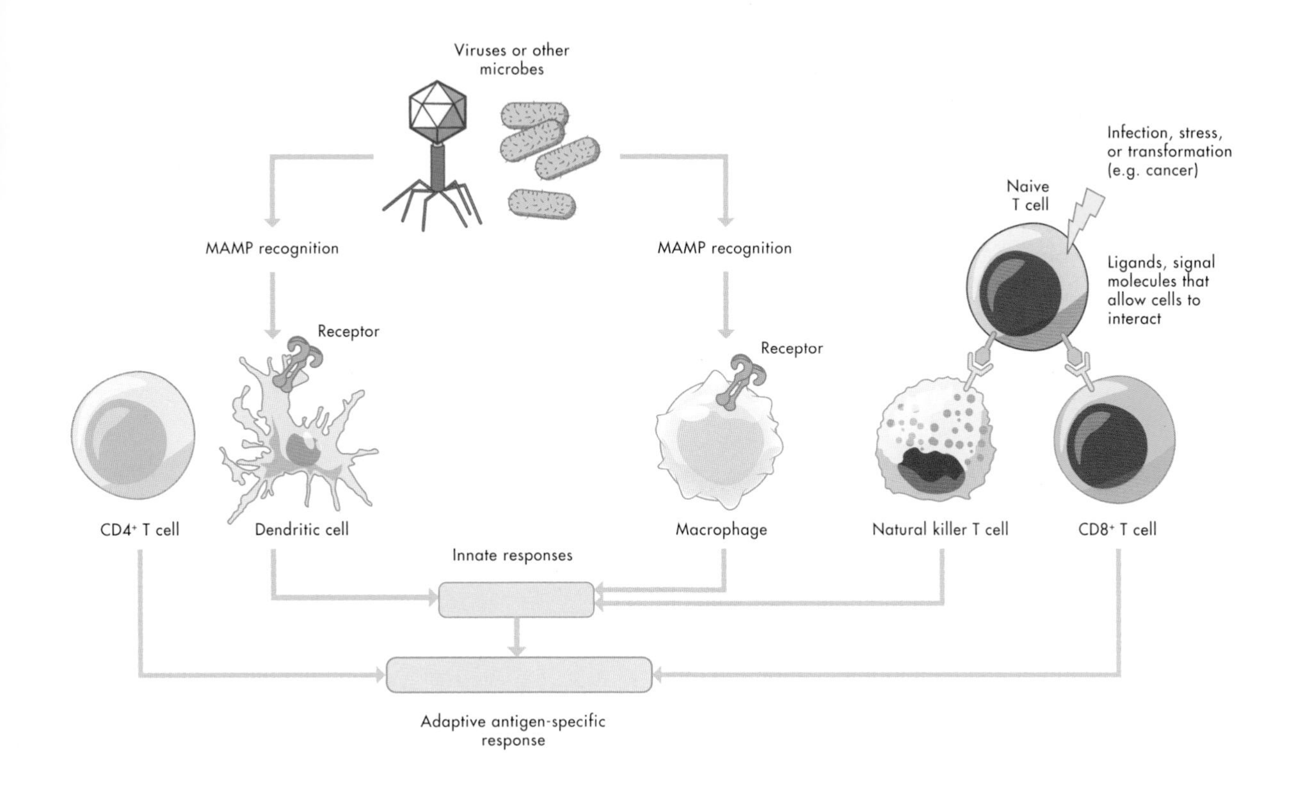

Once the host recognizes that a cell is infected by a virus, it may kill it to prevent further replication and spread of the virus. In plants this can often be seen on leaves as small spots of necrotic tissue, known as local lesions. Other aspects of innate immunity in plants involve physical barriers or resistance to insect predation, and resistance to movement throughout the plant.

In most cases, once the innate immune response is activated it can respond quickly to any foreign invaders. This means that a virus infection can prime a host to go to battle against another virus. If the first virus is rather benign, such as a cold virus, this can be an advantage when the host is confronted with a more severe virus. This type of priming works across different microbes as well, so that a bacterial infection can prime the innate immune system to be ready to combat a viral infection. In plants this system is called systemic acquired resistance and involves plant hormones, including salicylic acid, which is the basis for the common drug aspirin.

← Innate immunity response to virus infection on a leaf of Quinoa (*Chenopodium quinoa*). This plant has a very well adapted innate immune response to viruses, and will produce these necrotic spots, called local lesions, when infected by various plant viruses.

→ A white blood cell, surrounded by red blood cells. There are several different types of white blood cells, including B cells and T cells.

Adaptive immunity

All cellular life—including animals, plants, fungi, bacteria, and archaea—has some kind of adaptive immune system (see table page 162). The system in animals has been studied in most detail and uses two lines of defense: cellular immunity and humoral immunity.

Cellular immunity involves T cells, which rapidly kill virus-infected cells to prevent further spread. Humoral immunity involves B cells, which are made in the bone marrow and produce antibodies that specifically recognize and bind to components of the virus. This binding triggers a series of effects. The antibody may directly inactivate the virus, bind to the virus and prevent it from entering cells, or label the virus so that other cells can remove it (see the diagram on page 163). The T and B cell systems cooperate with each other and the components of innate immunity to completely rid the body of the pathogen.

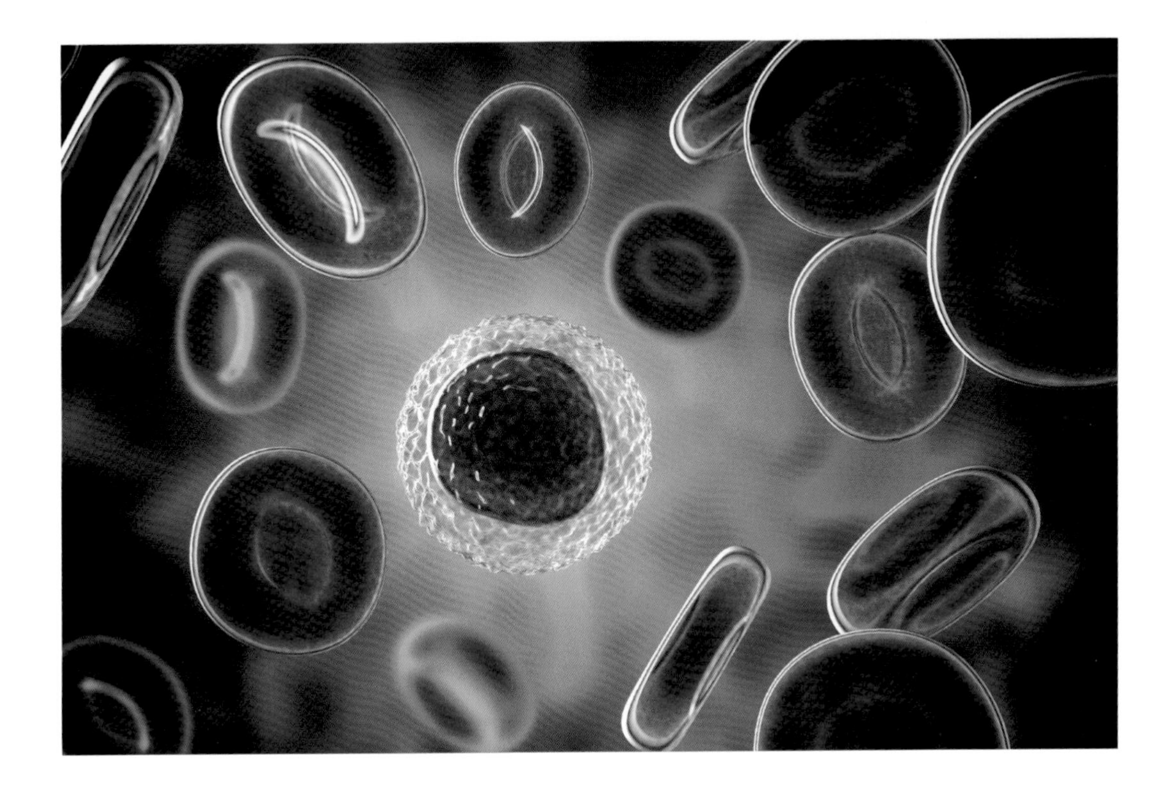

← Artist's rendition of a B cell releasing newly produced antibodies.

One very important feature of B cells is that they have memory. Once a B cell has recognized a virus, that recognition remains in the population of B cells even after the virus infection has cleared. If the cells encounter the same virus again, they can therefore very quickly make lots of antibodies. Similarly, T cells also have memory. If they encounter a virus again, they rapidly replicate and make more T cells that are capable of killing virus-infected cells to prevent the spread of infection. The life of memory cells is variable, lasting for the host's lifetime in the case of some viruses, including smallpox, or for only a year or two for other viruses, including rhinoviruses. It isn't clear what dictates the lifespan of a specific memory B or T cell.

Plants, invertebrate animals, fungi, and protists have a very different kind of adaptive immune system, known as RNA silencing or RNA interference. This was first discovered in plants and then later in a nematode. It involves the production of small interfering RNAs (siRNAs) that recognize the viral genome and target it for degradation. In a plant, siRNAs move through the plant ahead of the virus infection, so they are poised to stop the virus when it moves out of the cell that was initially infected. Unlike in B cell- or T cell-mediated immunity, there is no known memory in RNA silencing immunity.

RNA SILENCING

RNA silencing is an adaptive antiviral response in plants (where it was first discovered), invertebrates, fungi, and protists. Replication of RNA viruses, or transcription of some DNA viruses, results in double-stranded RNA. Cells don't make any double-stranded RNA aside from very short pieces of it. These RNAs trigger the RNA silencing pathway. An enzyme in the cell called DICER chops the double-stranded RNA pieces into very short segments of 21 or 22 nucleotides, called small interfering RNAs (siRNAs), that have the same sequence as one strand of the double-stranded RNA. These are replicated by a host enzyme, and then enter a complex with a protein called argonaut. The siRNAs then bind to the viral RNA, which they recognize by sequence identity, and the viral RNA is cut into small pieces. The pathway is similar in invertebrates, fungi, and protists, and comparable pathways are found in many different cellular life-forms that are involved in regulating RNA production.

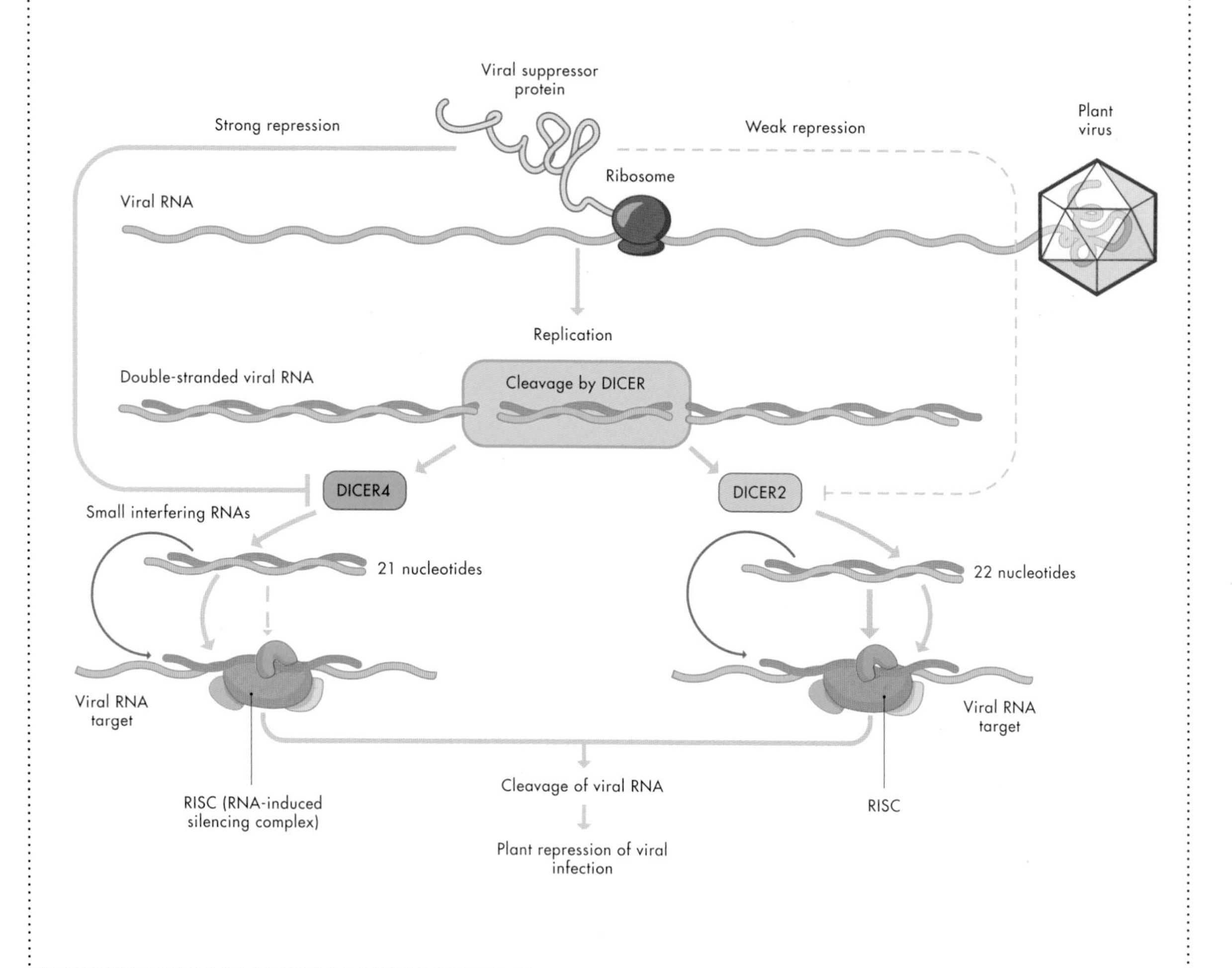

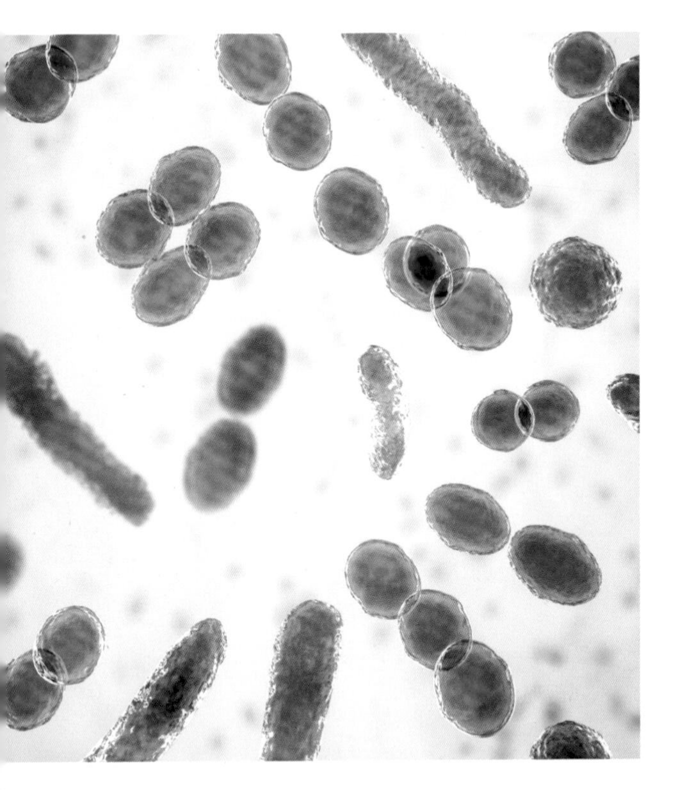

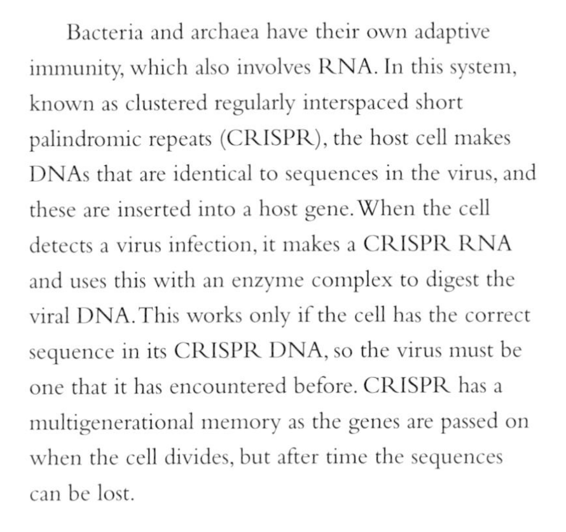

Bacteria and archaea have their own adaptive immunity, which also involves RNA. In this system, known as clustered regularly interspaced short palindromic repeats (CRISPR), the host cell makes DNAs that are identical to sequences in the virus, and these are inserted into a host gene. When the cell detects a virus infection, it makes a CRISPR RNA and uses this with an enzyme complex to digest the viral DNA. This works only if the cell has the correct sequence in its CRISPR DNA, so the virus must be one that it has encountered before. CRISPR has a multigenerational memory as the genes are passed on when the cell divides, but after time the sequences can be lost.

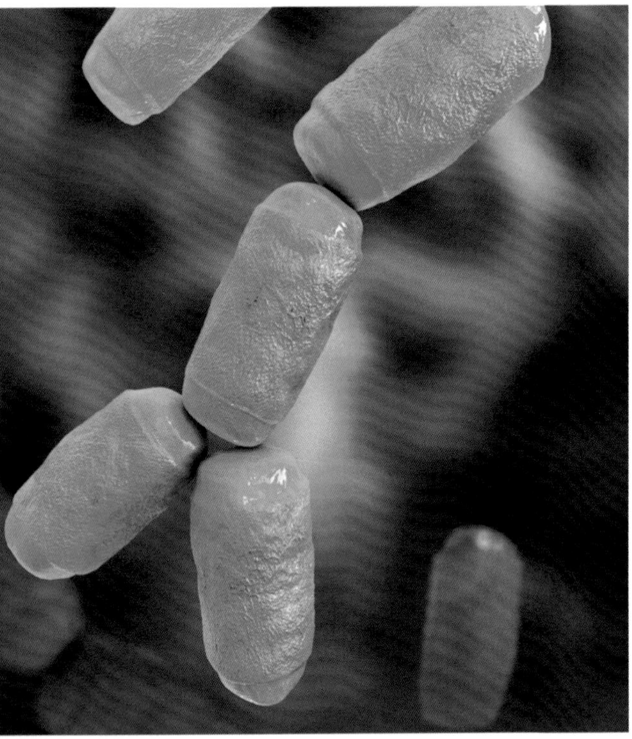

↖ Illustration of various shapes of common bacteria, including rods and cocci.

← Three-dimensional illustration of a common archaea normally resident in the human gut.

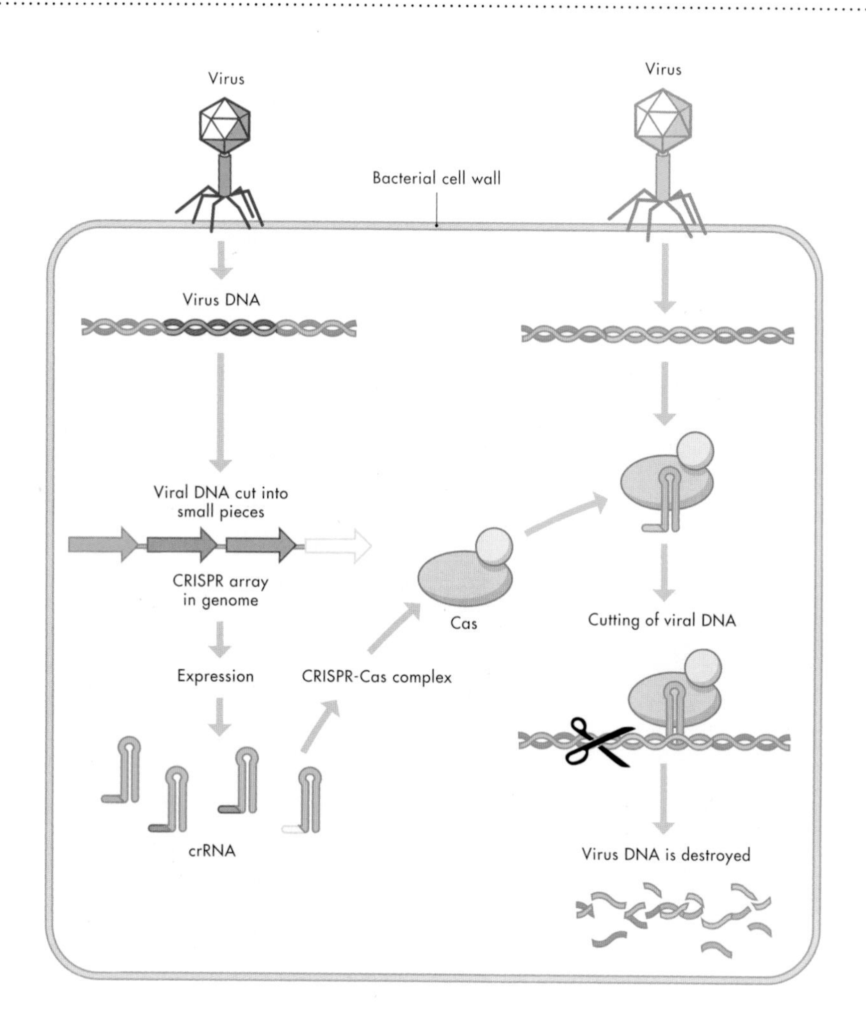

ADAPTIVE IMMUNITY WITH CRISPR

The adaptive immune system in bacteria and archaea is called clustered regularly interspaced short palindromic repeats (CRISPR). When a virus enters the bacterial cell, the virus genome is cut into small pieces by the host enzymes. These are then inserted into CRISPR genes in the bacterial genome, between short palindromes of nucleotide sequences. When a new virus infects the bacteria, the CRISPR gene is transcribed and cut up at the end of each palindrome, leaving RNAs that will enter an enzyme complex (Cas). If sequences in the Cas complex match sequences in the incoming virus, the virus genome is chopped to pieces. This provides a memory-based immunity. CRISPR and the Cas complex have been developed into a tool for modifying genomes of eukarya, and also used to engineer plants and even humans—in at least one case in China, the technique was used to create embryos resistant to HIV.

Vaccination

A virus infection usually elicits a robust immune response that is remembered for years or even the host's entire lifetime. A vaccination can involve a whole variety of things: a related virus, a killed virus, a live attenuated virus, a part of a virus expressed by another virus, DNA from a virus, or, most recently, an RNA molecule that can direct the synthesis of a viral protein once it gets inside a cell. All of these methods have been very successful against different viruses.

RELATED VIRUS VACCINE

Smallpox was a scourge for human populations around the world for centuries, and began to move globally as early as the sixth century CE. The fatality rate was about 30 percent, and those who recovered were usually scarred for life. The practice of variolation began in China long before anyone even knew what a virus was. In this, people who had not had smallpox were inoculated with some of the fluid from a smallpox sore, either by scratching their skin or by inhalation. Variolation was dangerous, but not as dangerous as a full-blown infection with the virus. This practice was later used in Europe and among European settlers of the Americas.

← Etching of a vaccination clinic for the poor in New York City in 1872, before the viral nature of smallpox was known.

At the end of the eighteenth century, an English doctor named Edward Jenner (1749–1828) noticed that milkmaids were rarely infected with smallpox, and indeed milkmaids were often reported to be the most beautiful women in a village, probably because they didn't have smallpox scars. Jenner hypothesized that infection with the benign cowpox (which the milkmaids caught from cows) prevented infection by smallpox. He tested this idea on the young son of his gardener, inoculating the lad with cowpox and later challenging him several times with smallpox. The boy never developed the disease. Jenner called this process vaccination, after the disease agent of cowpox, vaccinia.

Vaccination successfully replaced variolation as a means to prevent smallpox and gained widespread use. A global smallpox eradication program was started in the 1950s, and in 1980 the World Health Organization declared the process complete. Smallpox remains the only human virus disease to have been eradicated by vaccination. For successful eradication, a virus cannot have any wild hosts that could constitute a reservoir.

Progression of smallpox eradication

The timeline of smallpox eradication. The dates with each continent show when the disease was eradicated. The last known case of smallpox was in 1977.

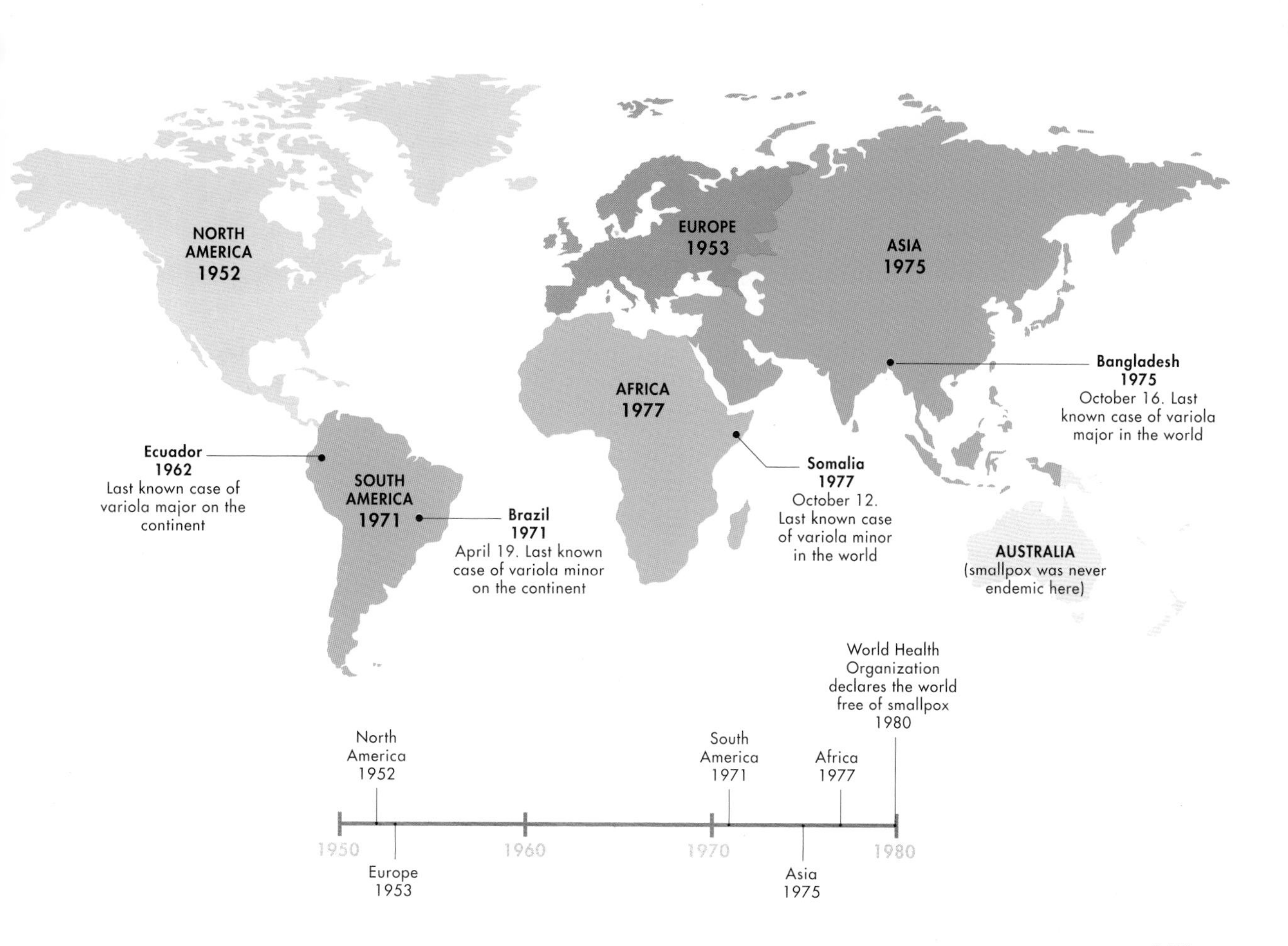

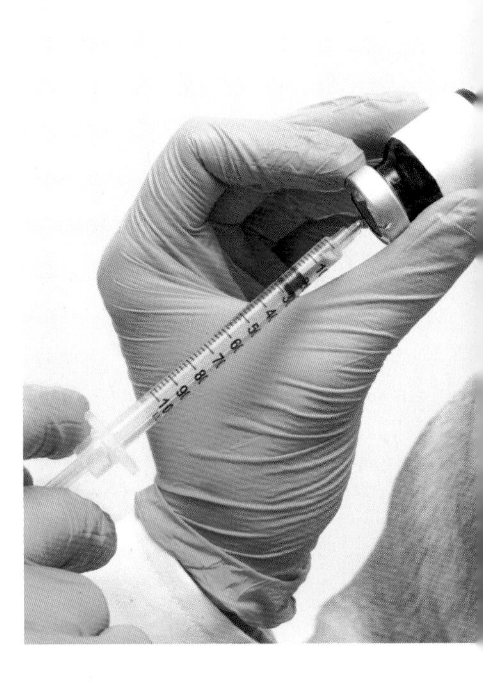

KILLED VIRUS VACCINES

In the late nineteenth century French microbiologist Louis Pasteur (1822–1895) developed a killed vaccine for rabies from an infected rabbit and was successful in preventing the disease in a person who had been bitten by a rabid animal. Although there were problems with the initial vaccine, it led to the eventual development of one that is very effective, and in many parts of the world rabies has become a very rare disease because most pets are vaccinated against it. Usually, vaccinations are administered before exposure to a virus, but the rabies virus grows so slowly that vaccination following a bite can be effective.

↖ Painting of Louis Pasteur (1822–1895) by Albert Edelfelt, oil on canvas, 1886. Pasteur did not know that rabies was a virus when he developed his vaccine for the disease; the discovery of viruses occurred a few decades later.

↑ Dogs are generally vaccinated for rabies as puppies, and then receive a booster every year. This has largely eradicated the disease in humans in some parts of the world.

↗ A polio vaccination clinic distributing the sugar-cube live attenuated virus.

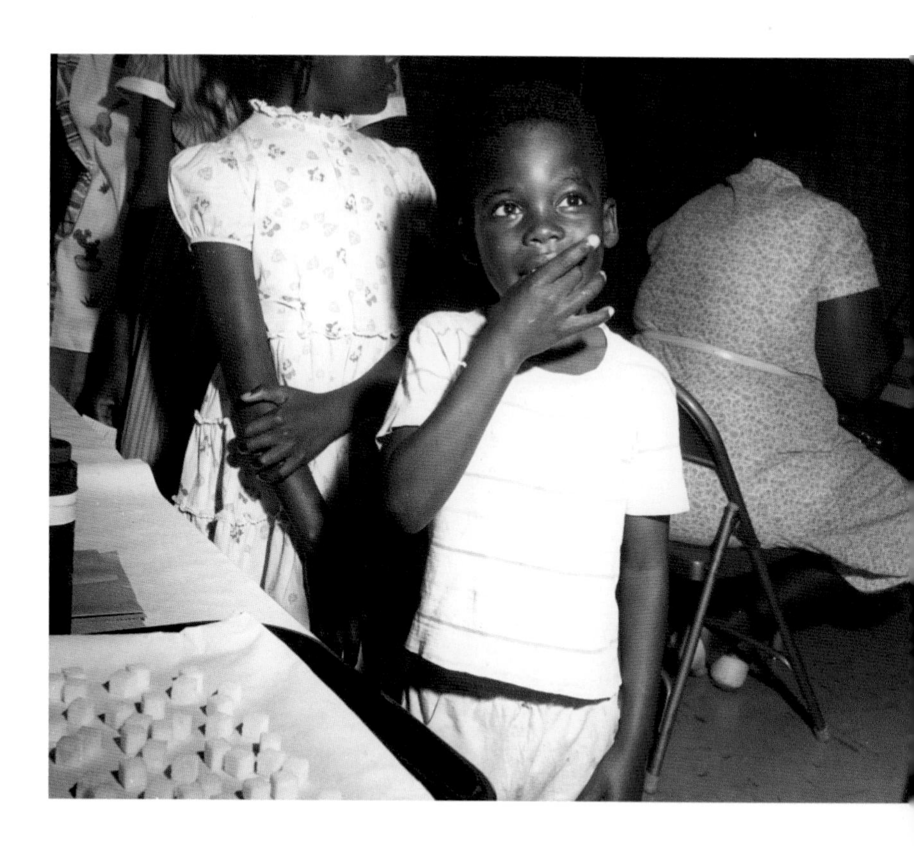

LIVE ATTENUATED VIRUS VACCINES

A vaccination can also be created from a whole virus that has been developed in the laboratory to cause no disease, known as a live attenuated virus. In this, the virus is grown in a laboratory in an atypical host cell type, and over time it evolves to lose its pathogen characteristics. This type of vaccine was very common throughout the last half of the twentieth century, after success with using the method for yellow fever virus. The widely used polio vaccine that has been administered from the 1950s by a sugar cube is a live attenuated vaccine. Polio is also prevented by vaccinating with a heat-killed virus, a form of vaccine that was developed before the live attenuated vaccine. Although this is usually used today in developed countries, the logistics are easier and compliance is much higher with the sugar cube method, and so the live attenuated vaccine is still used in much of the world. One problem with live attenuated virus vaccines is that the virus can very rarely revert to a more lethal form. This is the main reason why polio has not yet been eradicated.

RNA-BASED VACCINES

How well a vaccine will work is difficult to predict, because there are so many variations in the immune system and these are not all understood. Often, scientists must learn through trial and error. In 2019 a vaccine was developed using an RNA molecule that could express the spike protein of severe acute respiratory syndrome coronavirus 2 (SARS-CoV-2), the virus responsible for the COVID-19 pandemic. This vaccine proved to be very successful, and early analysis showed that it elicited an even more robust immunity than natural infection. However, this was not really a new idea. DNA vaccines were studied throughout the 1990s and used in animal systems, and RNA-based methods were being studied as treatment regimes for cancer.

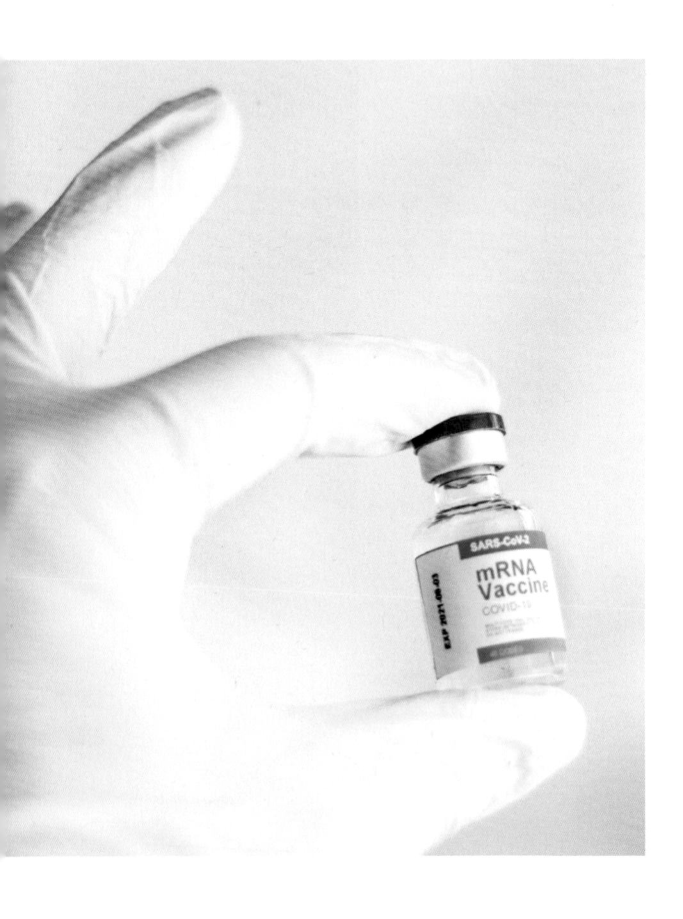

VIRUSES FIGHT BACK

Viruses have many ways to avoid or fight back against the immune response of the host. Some simply hide—for example, RNA viruses hide in complexes in the cell where they replicate. Double-stranded RNA viruses never fully uncoat themselves, so they only expose their single-stranded RNA to the host cell and complete their replication process inside the virus particles. Some viruses shut off the protein synthesis of their hosts, or encode enzymes that destroy host proteins, thus suppressing other parts of the immune response.

Viruses evolve rapidly and can change the proteins that are targeted by a host's immune system to reduce or completely circumvent the immune response. This is especially important in influenza virus (see page 252), where new, slightly different variants tend to emerge every season. Other viruses, including most rhinoviruses, do not elicit long-term memory B cells. It's not clear what controls the longevity of immune memory; many virus infections or vaccinations provide life-long immunity.

In plants, viruses have evolved different ways to avoid being targeted by RNA silencing. Some viruses block the host enzymes required for silencing in the cell, while others prevent the small RNAs from moving out of the initially infected cell. In mixed infection with different unrelated viruses, that ability to block RNA silencing will affect both viruses, and can result in a much more severe infection.

T CELL VACCINES

All the vaccines discussed so far elicit a strong B cell response in the body, resulting in the production of antibodies. They can also induce a T cell response, but this is often only mild. In recent years there has been more research into a different kind of vaccine, one that elicits a strong T cell response. T cells that are primed with this type of vaccine kill viruses and virus-infected cells when they encounter them. One reason for the development of this type of vaccine is that in some viruses—most notably dengue and Zika viruses—antibodies don't always inactivate the virus but instead assist it in entering cells. T cell vaccines have worked well in studies with Zika virus.

This idea is also being explored for influenza to see if a different approach could provide a vaccine that has a longer immune memory. When a person is infected with influenza, their immunity usually lasts for up to 10 years, but current vaccines haven't been able to replicate this type of immune memory. Part of the problem with the flu vaccine is that it uses a very limited target of the virus, which can change from year to year, whereas with an actual infection antibodies are made to many parts of the virus, some of which don't change significantly over time.

↑ Plants can be vaccinated with mild strains of a virus, and they are protected from severe strains. Here a mild strain of cucumber mosaic virus (CMV) was used to vaccinate tobacco plants. After they were inoculated later with a severe strain they did not display any significant symptoms. (A) Uninfected, (B) Severe CMV after vaccination with Mild CMV, (C) Severe CMV, (D) Mild CMV.

← During the COVID-19 pandemic, the first mRNA-based vaccines were introduced. Although many people were afraid of them because they were new, they are, in fact, probably the safest vaccines ever made. Unfortunately, SARS-CoV-2 does not elicit a very long antibody memory, whether from a vaccine or natural infection.

PLANT VACCINATIONS

Plants have been vaccinated on an experimental basis, but not in a broad way to combat viruses in the field. It has long been known that plants infected with a mild strain of a virus will be immune to a more severe strain, and in fact this attribute was once used to determine whether a virus was a different strain of a known virus or a new virus altogether. Plants would be inoculated with a battery of mild isolates, and then tested with a new unknown virus. If the plants were immune to the new virus, then it was a strain of the virus already infecting the plant. It is not always clear how this works in a plant, but in some cases it is probably due to the activation of the RNA silencing pathway.

Immunity and disease

Having an immune system is a critical requirement of living in a world filled with microbes. Many microbes—including viruses—do not pose a threat to other life and may even be beneficial (see the following chapters), but others are pathogenic and protection from these is of utmost importance.

The dark twist of the immune system, however, is that it also makes the host sick. In fact, many symptoms induced by viruses are actually caused by an overactive immune response. Fever and inflammation, common in human virus infections, are caused by the innate immune system trying to get rid of the virus. Interferons, immune defense molecules specifically induced by viruses, can contribute to inflammation and muscle aches. As the disease progresses, these responses are carefully downregulated, and symptoms subside after the pathogen has been removed from the body. In some cases, however, the downregulation of immunity is disrupted and leads to severe disease, as is often observed in severe cases of COVID-19.

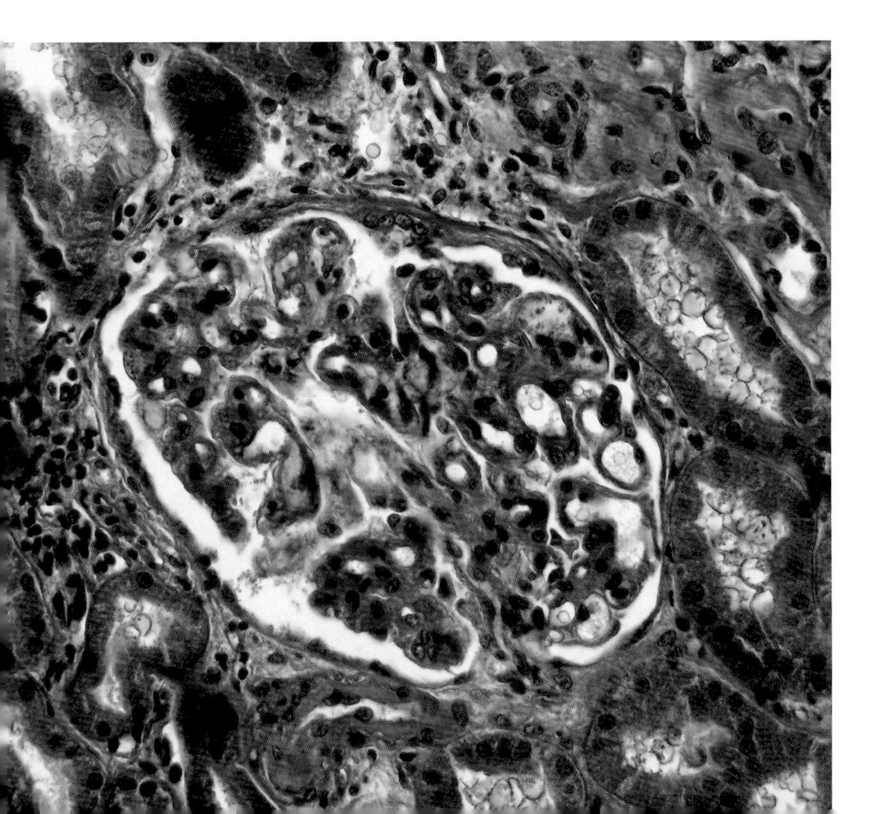

← A light micrograph of stained kidney tissue affected by glomerulonephritis, an inflammation of the small blood vessels in the kidneys that is often caused by virus infections, such as hepatitis B virus.

↗ Autoimmune diseases occur when the immune system attacks healthy tissue. This artist's rendition shows nerve cells being attacked by antibodies.

In plants, defense molecules elicited by viruses can also cause severe problems. For example, reactive oxygen species are produced in response to a number of different plant pathogens. These molecules are almost always damaging to host membranes, and other hormones produced as part of the plant defense response also damage plant cells.

Pathology based on antibodies and T cells is also an important part of the disease course in virus infection. With some viruses, high levels of virus–antibody complexes can cause kidney disease. T cells that are designed to kill infected cells can take the killing too far and cause tissue damage.

In plants and likely other life-forms that use RNA silencing as an adaptive immune response, the siRNAs involved in degrading the viral RNA can be similar to the RNA from plant genes and may destroy the messenger RNA meant to produce plant proteins. Research has shown that a number of symptoms induced by plant viruses have actually been caused by this accidental targeting of plant genes. The antiviral siRNAs can also interrupt cell regulation, another cause for disease in virus-infected plants.

The animal adaptive immune system includes a phenomenon known as immune tolerance, which is the ability of the immune system to recognize proteins that belong to the body. This is very important, as it prevents the immune system from attacking the body. Immune tolerance develops early in life and is fully established shortly after birth. A breakdown of this system results in the development of an autoimmune disease, where the body makes antibodies that attack its own proteins. Some autoimmune diseases have been correlated with virus infections, but the connection is not yet fully understood. Sometimes, a pathogen may have a protein that looks similar to the host protein and the host may start attacking the body after the infection has been removed.

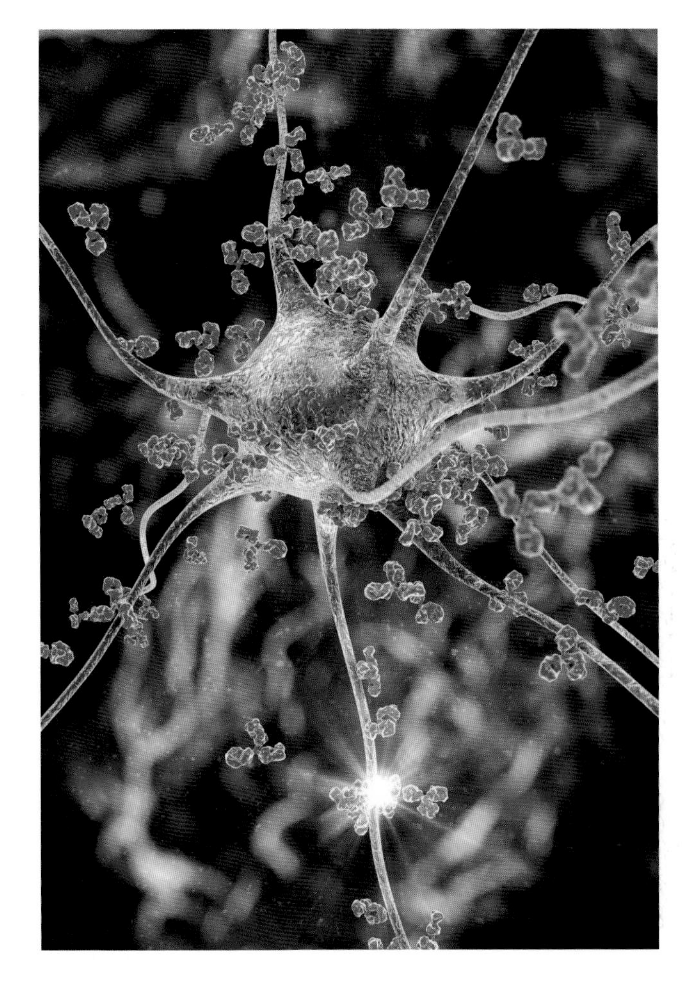

The ongoing battle between pathogenic viruses and hosts is sometimes referred to as an arms race. One side makes a change, then the other side compensates with its own change. The evolutionary capacity for rapid change in viruses far outweighs that of the host, so the arms race is not a fair fight. In the end, the virus usually wins. However, winning does not mean making the host sick; to win the race, the virus just has to be able to replicate efficiently. If that makes the host sick, too bad for the host, but disease is not a selective force in virus evolution.

Antiviral drugs

The discovery of antibiotics in the early twentieth century revolutionized the course of infectious diseases, but these drugs have no effect on viruses. It is complicated to target viruses with drugs because they use the host metabolism for all aspects of their life cycle, and so a drug that targets a virus is often toxic to the host.

Nucleotide analogs were an early class of antiviral drugs. These are molecules that look a lot like a nucleotide, so the viruses incorporate them into their genomes when they copy themselves, but then they do not allow further replication. While nucleotide analogs can be effective, they are also mutagens to the host. In general, the host can tolerate mutations much better than the virus, so if the virus is severe then the risk associated with administering nucleotide analogs can be worthwhile. Drugs that target the interaction between viruses and hosts, such as cell receptors for the virus, are also often toxic. The host receptor did not evolve to let the virus in and always has an important function for the host, so targeting it will negatively affect its normal function.

Other drugs target enzymes specific to a virus. Protease inhibitors prevent a virus from processing its proteins into functional units, while reverse transcriptase inhibitors target the enzyme that copies the genome of Type VI viruses. These enzymes are not found in the host, so these drugs are less toxic.

Bacteria evolve to become resistant to antibiotics quite quickly, but in viruses this happens even more quickly because they have an enormous capacity for evolution (see page 138). Although a number of antiflu drugs have been developed, most have had limited value because the virus rapidly evolves to overcome them. A strategy that has been successful to get around this problem is to use several different drugs at once. This has allowed people with human immunodeficiency virus (HIV) to be treated and live a relatively normal life, and has also led to efficient treatments for the hepatitis C virus. Experimental treatments are also using the small interfering RNA systems of plants and protists to combat human viruses.

Immune antisera have long been used as a primary treatment against the rabies virus. In these, antibodies are raised in other animals, including sheep and horses, and then used to treat people after they have been bitten by a rabid animal. Vaccines for rabies have, however, made this treatment less common. Human immune sera were used for prevention of hepatitis A virus infection, especially for travelers, before a vaccine was available for this virus. Immune sera have also been used successfully in some cases of COVID-19. Human sera require very careful screening to be certain other pathogens are not introduced. The use of animal antisera is limited because sera from a specific animal can be used only once in a person's lifetime. Once someone has received an animal antiserum, their immune system recognizes it as a foreign substance and will raise an immune response against it. If an antiserum from the same species of animal is administered again, it will therefore be destroyed by the immune system before it can be effective.

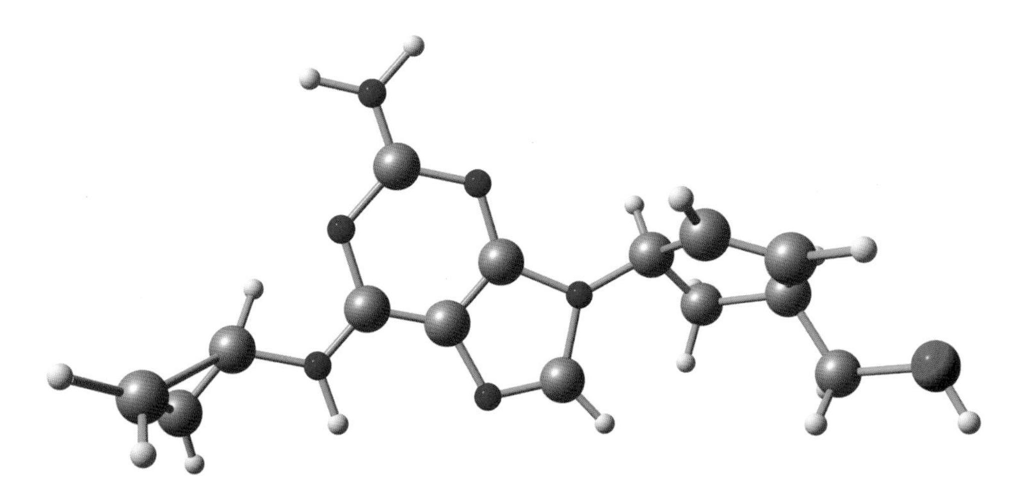

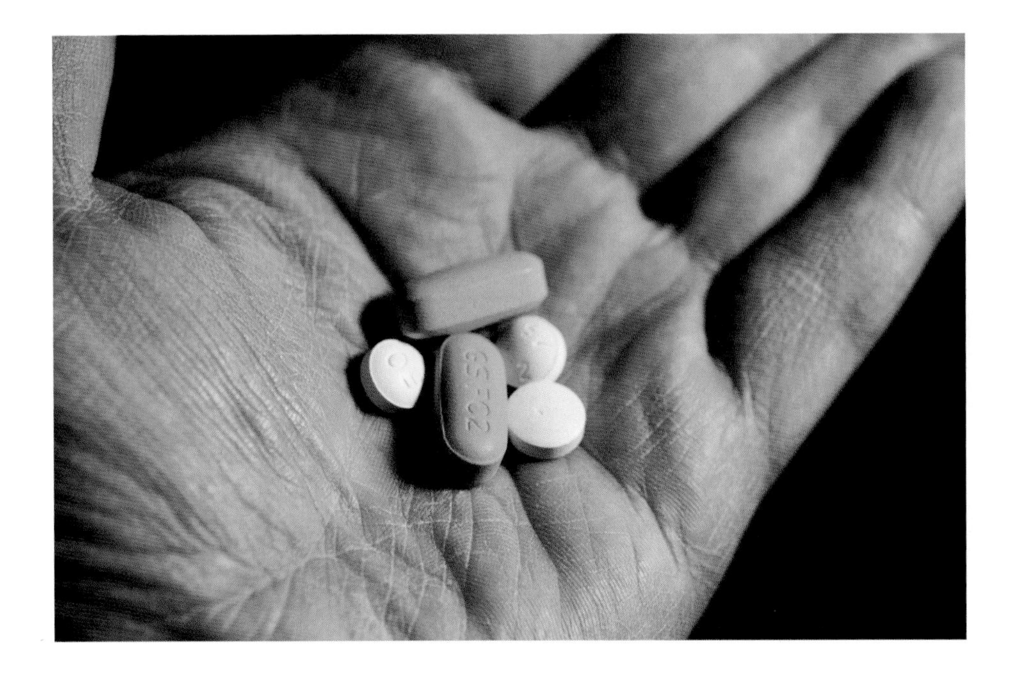

The immune system provides an amazing defense against viruses, and scientists are continually studying new ways to assist the immune response. One thing is certain: there will be more viruses in the future, and they will continue to find ways around immune responses. The battle continues.

↑ The chemical structure of abacavir, an anti-HIV drug (top). This drug inhibits the reverse transcriptase enzyme by mimicking a nucleotide. Most people with HIV take a mixture of different antiviral drugs every day (above).

Vaccinia virus

Virus that cured the world of smallpox

GROUP	I
FAMILY	Poxviridae
GENUS	Orthopoxvirus
GENOME	Linear, single-component, double-stranded DNA comprising about 190 kb, encoding about 260 proteins
VIRUS PARTICLE	Enveloped, brick-shaped, about 250 nm long and 200 nm wide
HOSTS	Cows, horses; can infect many other mammals in experimental conditions
ASSOCIATED DISEASES	Cowpox, horsepox
TRANSMISSION	Contact with lesions
VACCINE	None

There aren't many examples of viruses that are used as vaccines against other viruses, but vaccinia virus is one example of this method and was the first vaccine ever developed. In fact, the term "vaccine" comes from the name of this virus.

Vaccinia virus probably originated from cowpox virus, but it was propagated in laboratory settings for decades and as such its precise origin is obscure. It is also extremely similar to horsepox virus, and it is possible that this was the original virus rather than cowpox. However, Edward Jenner chose to use cowpox in the late eighteenth century to vaccinate his gardener's nine-year-old son against smallpox. The experiment was successful, and vaccines were born.

Recently, a related virus, monkeypox virus, has been spreading among some human populations and causing concern. Monkeypox is generally not life-threatening, and a good vaccine is available. In addition, because of the similarity to vaccinia virus, it is likely that smallpox vaccinations confer some protection. In the early days of smallpox vaccination there was no reliable way to store the vaccinia virus, so it was often passed from human to human. Initial inoculation was achieved by spreading material from a pustule on the skin of a person infected with cowpox to a scratch on the skin of the patient's arm. Once a pustule formed here, this was then used to transfer the virus to others, and so on. Of course, other pathogens were sometimes inadvertently transferred too. Luckily, vaccine safety has come a long way since those early days. Since smallpox is now deemed to have been eradicated, vaccination against the disease is no longer routinely carried out.

Vaccinia virus is considered a giant virus, because it is large enough to be seen through a light microscope (most viruses are visible only through electron microscopy). It and other members of the Poxviridae family are unique among large DNA viruses in that they replicate in the cytoplasm of the cell rather than the nucleus. This means that the virus has to encode all of its own proteins for the replication process, unlike other large DNA viruses, which use the host enzymes for this.

→ Colorized transmission electron microscope image of vaccinia virus.

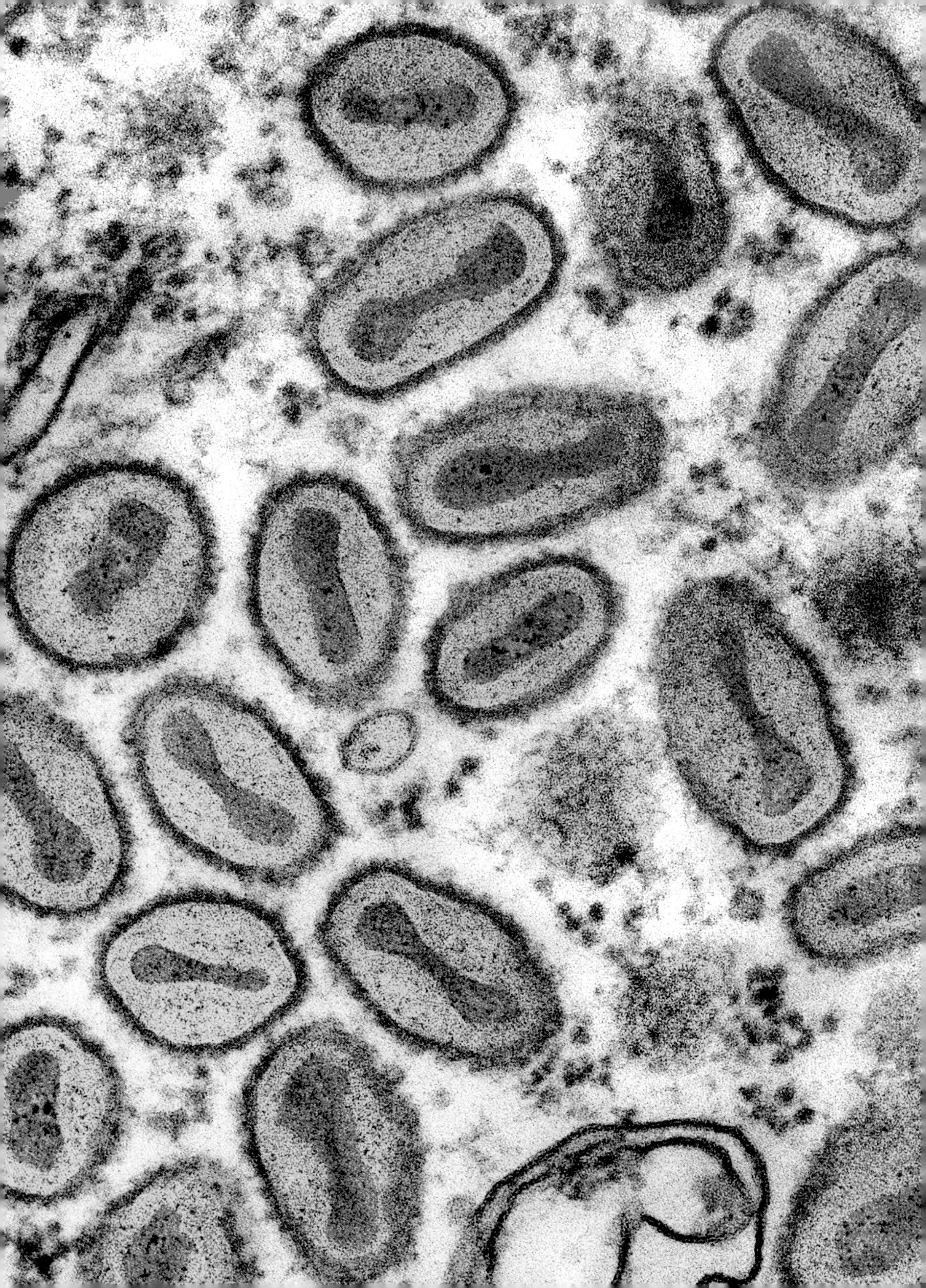

RSV

Human orthopneumovirus

Virus that has learned to evade the immune response

GROUP	V
FAMILY	Pneumoviridae
GENUS	Orthopneumovirus
GENOME	Linear, single-component, single-stranded RNA comprising about 11 kilobases, encoding 10 proteins
VIRUS PARTICLE	Enveloped, spherical, about 150 nm
HOSTS	Humans; related viruses infect other mammals
ASSOCIATED DISEASES	Bronchitis, cold, pneumonia
TRANSMISSION	Airborne
VACCINE	None available

Human orthopneumovirus is often better known by its older name, respiratory syncytial virus (RSV). It is the most common cause of viral respiratory disease in infants, usually resulting in bronchitis, but it can infect people of all ages.

In adults, RSV usually manifests as a common cold, but in older adults it can cause pneumonia. There is no vaccine, but a monoclonal antibody-based drug can be administered in severe cases. Monoclonal antibodies are made in the laboratory and are usually mouse or human in origin. They mimic the antibodies the body makes in response to viral infection.

RSV is found in droplets released by infected individuals, which are then either inhaled or picked up on the hands by touching contaminated surfaces, followed by touching the face. The virus enters through the nose or the eyes. The most important prevention protocols are therefore good handwashing techniques and mask wearing.

The virus is very clever at avoiding detection by the innate immune response. A primary trigger for innate immunity against RNA viruses is the unique type of RNA viruses make but cells do not. RSV, along with many other respiratory RNA viruses, hides its unique RNA inside cells by inducing them to make membrane-bound structures where the virus replicates. In addition, the virus can destroy the host mRNA that is used to make some of the proteins required for the innate immune response.

Young children tend to be more dependent on the innate immune response to combat virus infections, because their adaptive immune response has not yet matured enough to recognize viruses that they have not been exposed to previously. The ability of respiratory viruses such as RSV to evade the innate immune response may be one reason why young children are so prone to respiratory infections.

→ Artist's rendition of the structure of RSV showing the inner and outer structures. The RNA genome is shown in yellow and red in the interior.

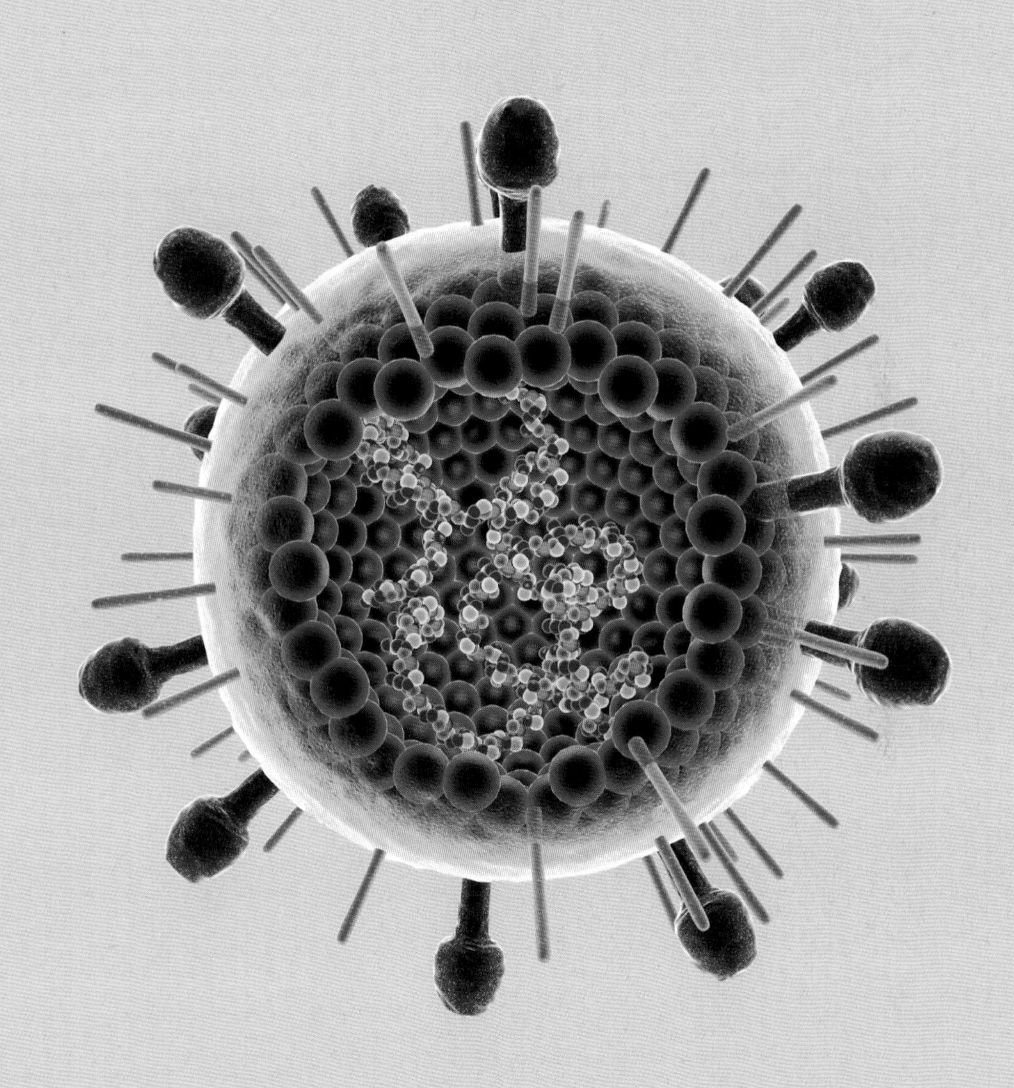

TEV

Tobacco etch virus

Virus that led to the discovery of RNA silencing

GROUP	IV
FAMILY	Potyviridae
GENUS	Potyvirus
GENOME	Linear, single-component, single-stranded RNA comprising about 9,500 nucleotides, encoding 11 proteins
VIRUS PARTICLE	Long filamentous rod, about 730 nm in length
HOSTS	Nightshades and other perennial weeds
ASSOCIATED DISEASES	Leaf etching, stunting, mottling, vein clearing
TRANSMISSION	Aphids, dodder

Tobacco etch virus (TEV) was first described in 1921 in Jimson Weed (*Datura stramonium*). It was first thought to be a genetic anomaly of the plant, until it was shown to be transmitted to other plants by grafting. The virus was found several years later in tobacco.

In 1980, when the first genetic engineering experiments in plants became possible, portions of virus genes were put into plants to see if they could act like a vaccine, just as mild strains of viruses could protect plants from more severe infection. The first virus studied in this way was tobacco mosaic virus, but soon a number of other viruses were used too, and plants immune to TEV were generated in the early 1990s. In an effort to understand how this type of immunity worked, scientists engineered a plant to make just the RNA from the virus, but no protein. These plants turned out to be fully immune to the virus, leading to the discovery of RNA silencing. This type of adaptive immunity was later found in a number of other organisms, including nematodes, insects, and fungi.

TEV has proven to be a valuable tool for other fundamental studies. Like all viruses in the Potyviridae family, most of its proteins are synthesized as a large polyprotein that is then cut up into functional proteins by a virus-encoded enzyme called a protease. Foreign proteins can be inserted into the viral genome and then specifically cut out of the viral polyprotein by the viral protease. This provides a way of testing a protein in a plant to see if it works as expected, before going through the more complicated process of making a transgenic plant.

→ Computer-generated model of a cutaway of the related and similarly structured Potato Virus Y showing part of the helical capsid structure with the viral genome (in orange).

Dengue virus

A challenge for vaccine development

GROUP	IV
FAMILY	Flaviviridae
GENUS	Flavivirus
GENOME	Linear, single-component, single-stranded RNA comprising about 11 kb, encoding 10 proteins
VIRUS PARTICLE	Enveloped, spherical, about 50 nm
HOSTS	Humans and other primates, mosquitoes
ASSOCIATED DISEASES	Dengue fever (breakbone fever), hemorrhagic fever
TRANSMISSION	Mosquitoes
VACCINE	None approved

Dengue fever (originally called breakbone fever) was first described in the late eighteenth century in several locations at the same time, in Asia, Africa, and the Americas. Although many people who are infected have only mild or no symptoms, the disease can be very painful—as its original name implies. Infected people usually recover completely.

Dengue virus infection numbers have increased around the world since the Second World War. This is likely because of an increase in numbers of people migrating from rural to urban environments, where the vector, Yellow Fever Mosquito (*Aedes aegypti*), replicates very well in standing water in old tires and flowerpots. There are now four strains of the virus, which probably originated in rural areas where mosquitoes bite wild primate hosts and people.

In about 1 percent of cases, a more severe form of the disease occurs, a hemorrhagic fever with a fatality rate of about 25 percent. There is evidence that hemorrhagic fever results when someone is infected for a second time with a different strain of the virus. This is thought to be because the individual has antibodies that cross-react but are not specific to that strain, and rather than inactivating the virus, they can actually help the virus enter host cells. This has made it nearly impossible to develop a safe vaccine. However, it has also prompted research into new ways of making vaccines that would elicit a cellular immune response rather than a B cell response. The cellular response involves T cells, which rapidly kill virus-infected cells. A DNA vaccine against the closely related Zika virus that elicits a T cell-only response has been very effective in a mouse model, but has not yet reached the human vaccine trial stage.

The Yellow Fever Mosquito, the vector for dengue and other severe viruses such as Zika, chikungunya, and yellow fever, is currently limited to tropical and subtropical climates. However, as the climate warms, the range of the mosquito is increasing, and dengue has been reported in the southern United States and other parts of the world where it was not previously found.

→ Computer-generated model of dengue virus, drawn from data from crystallography and cryogenic electron microscopy.

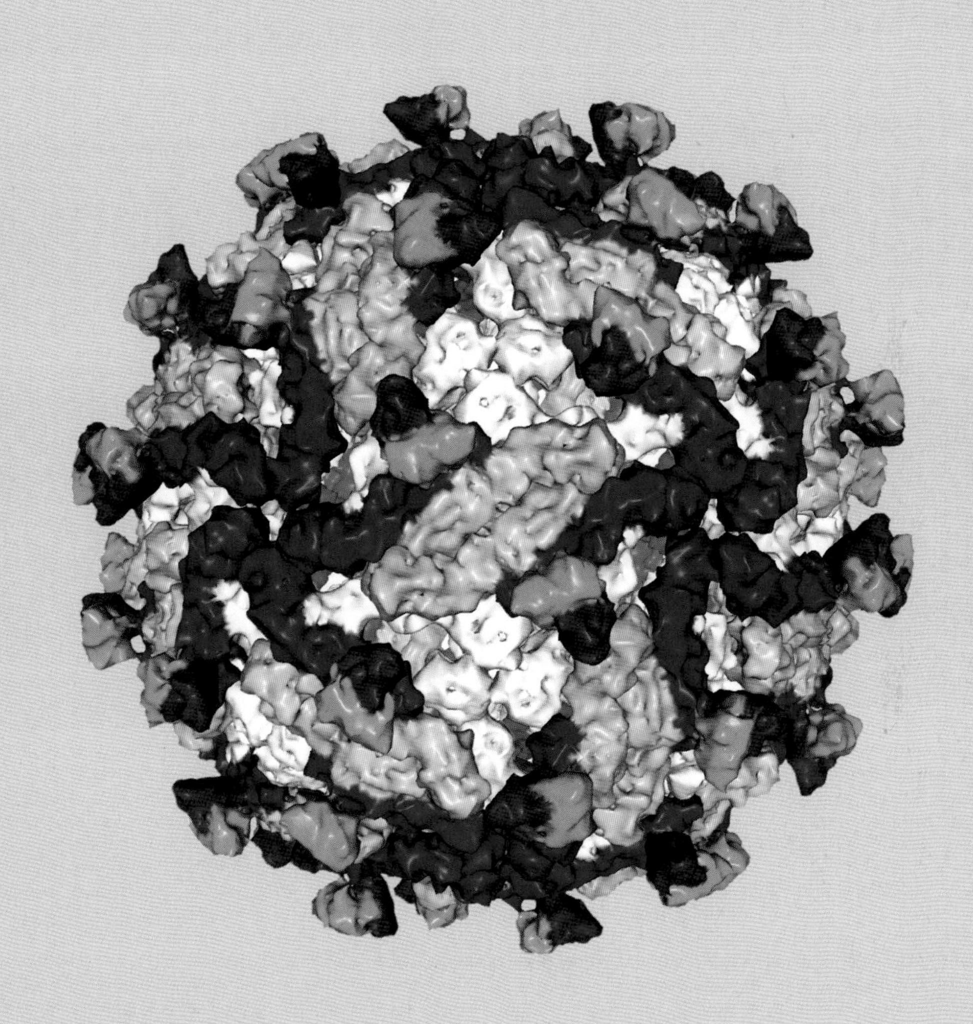

Escherichia virus T7

A virus that fights back against its host's immune system

GROUP	I
FAMILY	Autographiviridae
GENUS	Teseptimavirus
GENOME	Linear, single-component, double-stranded DNA of about 40,000 nucleotides, encoding about 55 proteins
VIRUS PARTICLE	Icosahedral head of about 60 nm, with a short contractile tail
HOSTS	*Escherichia coli* and related bacteria
ASSOCIATED DISEASES	Cell lysis and death
TRANSMISSION	Diffusion

Escherichia virus T7, often called T7 phage, has been used as a model system for many studies into phage biology and molecular biology. It is thought that it prompted the discovery of "bacteria-eating" viruses, or bacteriophages, by French microbiologist Félix d'Herelle (1873–1949) in the 1920s. It was also studied extensively by German scientist Max Delbrück (1906–1981) in his research into virus replication, work that garnered him a joint Nobel Prize in 1969.

T7 phage was one of the first complete genome sequences ever determined, in 1983. When grown at 37 °C (98.6 °F), the standard growth temperature of Escherichia coli, it has a rapid life cycle of just 17 minutes, from infection to lysis of the cell, but this can lengthen to 30 minutes when the temperature is decreased to 30 °C (86 °F). The virus is very easy to purify in large quantities, which is one reason why it has been popular for basic studies into viruses.

Bacteria use numerous strategies to defend against viruses, many in the innate immunity category. One strategy, called the restriction-modification system (R-M), uses enzymes to cut the viral genome into pieces as it enters a cell. R-M enzymes are one of the most important and fundamental tools in the molecular biology tool chest, because they cut DNA at specific nucleotides. They are used in almost all recombinant DNA and cloning experiments. The host bacteria add methyl groups to their own genomes to prevent being degraded by the R-M system. T7 phage makes a protein very early in infection that sequesters the R-M enzymes to prevent them from cutting up its own genome.

→ Computer-generated model of T7 phage from cryogenic electron microscopy data, showing the landing gear that it uses to attach itself to a bacterial cell.

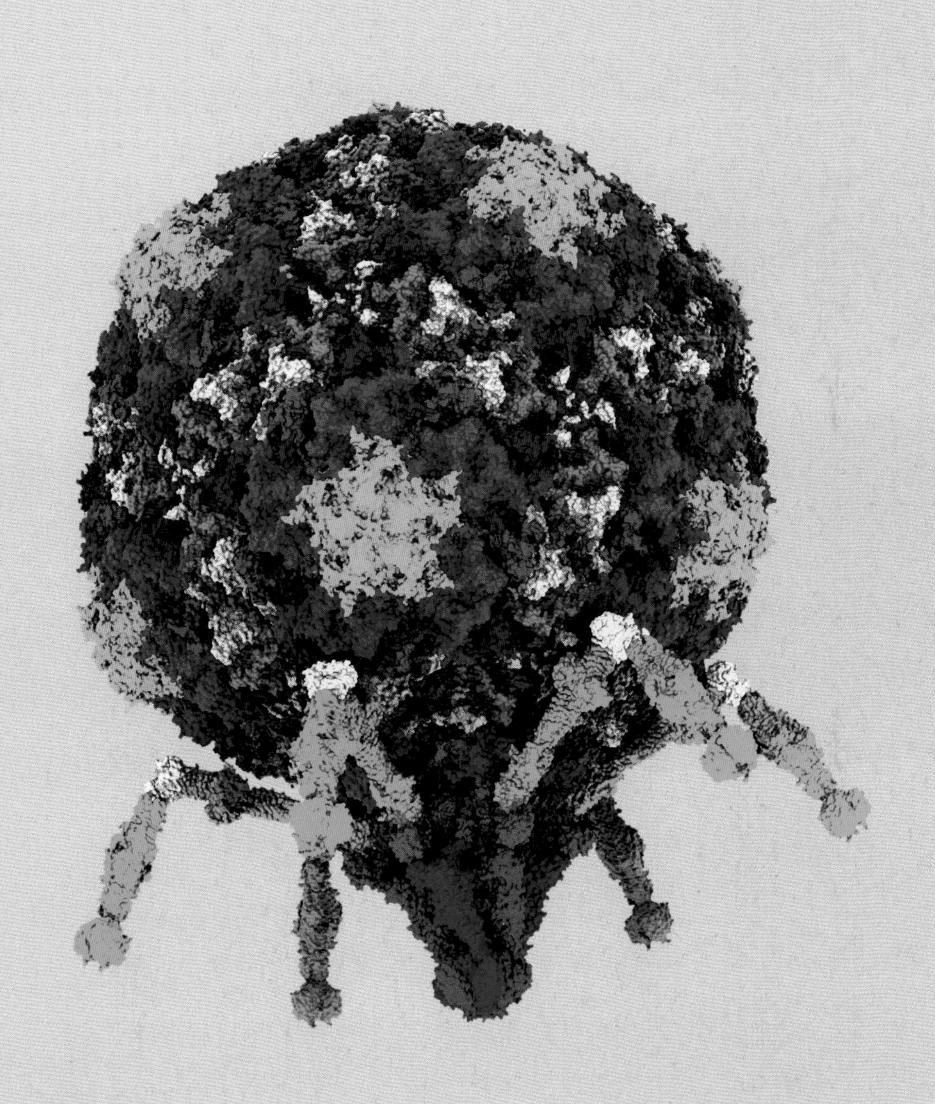

VIRUSES IN ECOSYSTEM BALANCE

Viruses in the sea

In the late 1980s scientists estimated the number of viruses in the sea by adding a milliliter of filtered seawater to a plate of a lab strain of the bacteria *Escherichia coli*. Bacterial viruses often kill their hosts, and when a virus infects *E. coli* on a petri dish it makes a little hole in the bacterial lawn where the cells have died. Each hole represents a single infectious virus. Using this method, the researchers calculated that there are about a million viruses that can infect *E. coli* in a milliliter of seawater.

This didn't include any of the viruses that don't infect *E. coli* or those that don't kill their hosts. Later estimates were made using electron microscopy or fluorescent imaging (see page 36), which revealed all the viruses present in a sample of seawater, regardless of their hosts. These put the figure of marine viruses at 10 million virus particles per milliliter of seawater. What are all these viruses doing? Which organisms are they infecting? These are complex questions and we certainly don't know all the answers yet, but we do know the genome sequences for many marine viruses and we also know that most of them infect microbes. Understanding the details relies on sophisticated computer analyses and comparisons, part of the important field of bioinformatics.

↖ A plaque assay is an analysis for viruses that can infect bacteria. The sample is filtered to remove bacteria and then added to a bacterial lawn. When a virus infects a bacterial cell, it kills the cell and spreads to nearby cells, which are also killed. This leaves a small hole, or plaque, in the bacterial lawn, each of which represents a single virus.

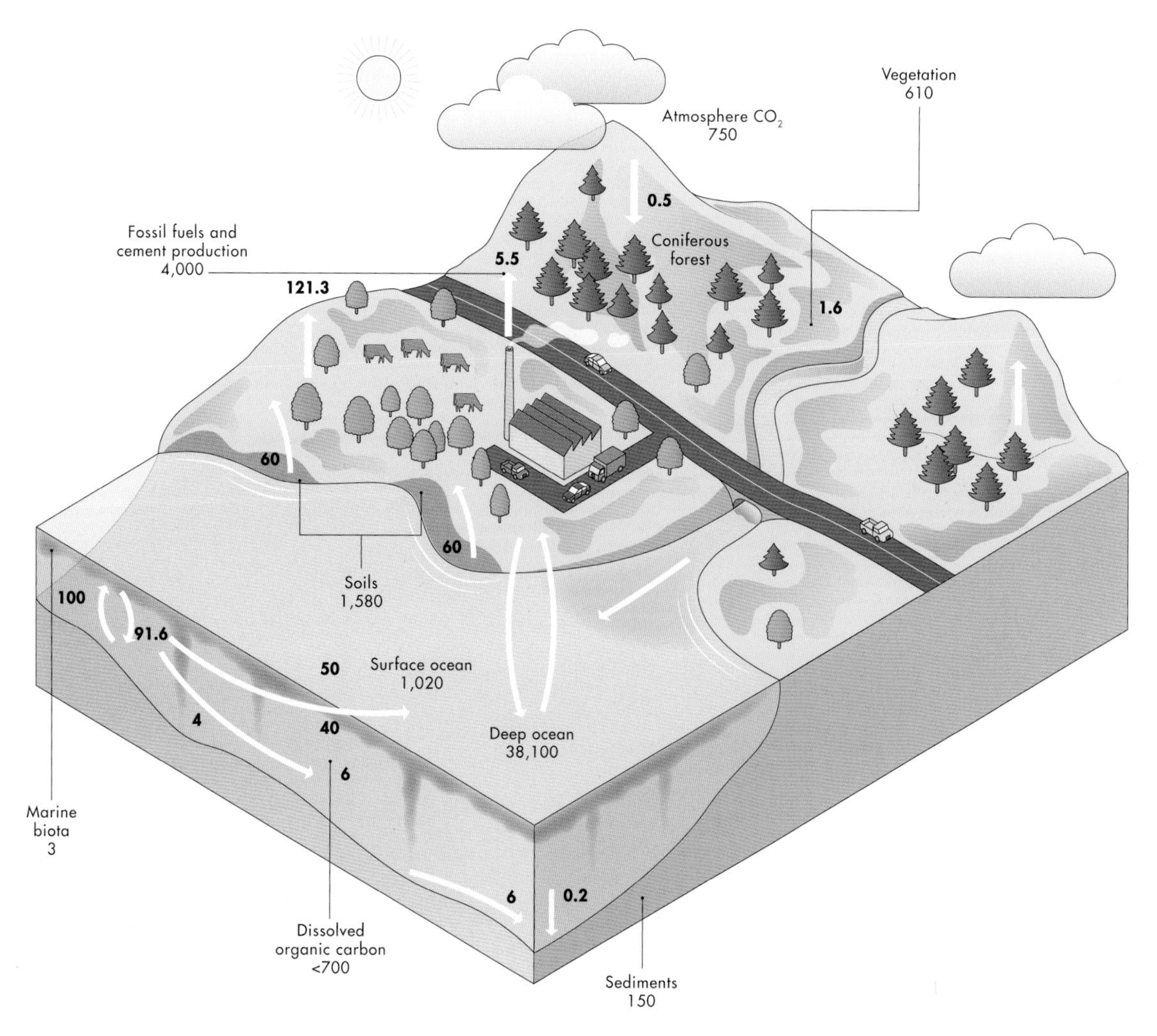

Carbon balances

The global carbon cycle relies heavily on the oceans. The lysis of a large portion of marine microbes by viruses every day contributes a majority of carbon release from the oceans.

Plain numbers represent storage in gigatonnes of carbon

Bold numbers represent fluxes in gigatonnes of carbon per year

Microbes are the main contributors to the biomass of the oceans, and most are bacteria. In terms of sheer numbers, however, viruses exceed other microbes by at least tenfold. Some 20–40 percent of marine microbes are killed every day by viruses that rupture their cells through a process called lysis. This is critical to the oxygen and nitrogen cycles of the oceans, and for retaining nutrients in the biotrophic layers of the oceans. If a microbe dies without being lysed, it will sink to the bottom of the ocean and its nutrients will be lost. However, if it is lysed then it becomes dissolved organic matter that stays in the biotrophic layer and is available for use by other life-forms. In addition, viruses transfer genes among microbes—it has been estimated that as many as 10^{29} gene transfer events occur in the world's oceans every day.

Viruses also have an impact on the biochemistry of microbes in the sea. For example, the metabolisms of marine photosynthetic cyanobacteria are controlled in part by the viruses that infect them. The viruses carry genes for enzymes that affect carbon cycling, photosynthesis, and nutrient cycling. Sometimes these controls favor the viruses, such as when a virus shuts down metabolism in a bacterium to induce starvation and trigger the synthesis of nucleotides that the virus needs for its replication.

Viruses keep the oceans healthy

Without viruses to release the majority of nutrients in the biotrophic portions of the oceans, most of these nutrients would sink to the depths and be removed from the pool of available nutrients.

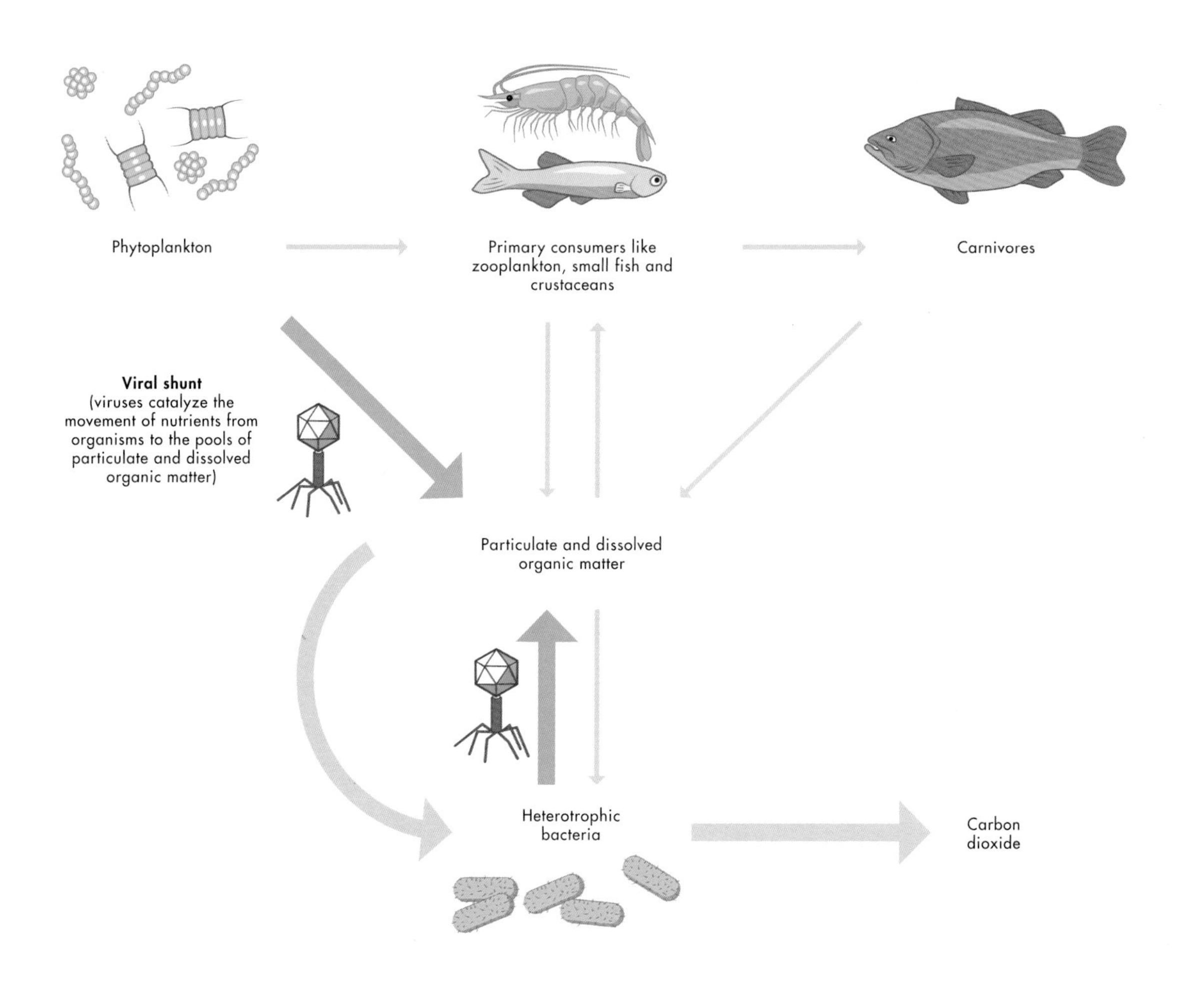

Viruses, insects, and plants

Plants and insects have been interacting for eons, probably since they both appeared on land about 480 million years ago. Viruses of plants largely rely on insects to get around (see pages 117–19), and the relationships among them are fascinating examples of the intricacies of ecology and evolution.

Viruses not only manipulate plants to express volatile compounds that attract insects, but they can also do this with precision. In some cases, if an aphid is already carrying a plant virus then it will be less attracted to a virus-infected plant than an uninfected aphid. This implies that the virus manipulates both plant and insect to enhance the overall spread of the virus. In another example, aphids feeding on some virus-infected plants are more likely to develop wings than those feeding on virus-free plants, thus aiding dispersal of the virus to other plants.

Thrips are very small insects that can be infected by, and transmit, some plant viruses. The insects do a lot of damage to plants and induce them to make antifeeding compounds. Juvenile insects don't thrive on these damaged plants unless they are also infected by a virus. If they feed on plants that have been infected experimentally rather than by thrips and therefore don't have thrip damage, they do even better. Male thrips feed more heavily on virus-infected plants than on uninfected plants. Female feeding isn't affected by the virus, but males are more likely to transmit the virus.

Some of these plant–insect–virus interactions spill over to other viruses too. For example, if a plant is infected by more than one virus, both viruses can benefit from changes in volatile compounds that attract insects, even if only one of the viruses is responsible for those changes.

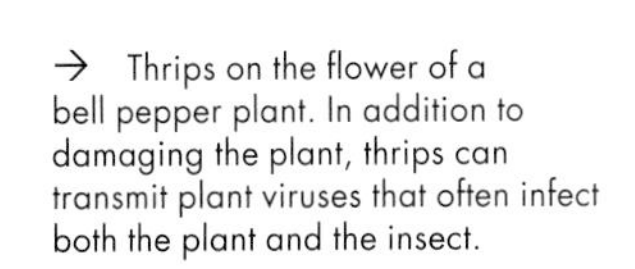

→ Thrips on the flower of a bell pepper plant. In addition to damaging the plant, thrips can transmit plant viruses that often infect both the plant and the insect.

Viruses of plants and fungi

Fungi preceded plants in the move from the oceans to life on land, and it seems likely that plants needed to establish relationships with fungi in order to colonize the land. Today, wild plants are almost always colonized by fungi. These perform many important functions for the plants, including improving uptake of nutrients and tolerance to drought, salt, and high temperatures (see page 240). Fungi also act as a communication network among plants growing in forests. Because fungi can also be pathogens of domesticated plants, they are often excluded from crops, and so the relationships between plants and fungi have only been fully appreciated relatively recently.

Many virus families infect both plants and fungi. As more and more genome sequence analyses are completed for viruses, these relationships are becoming increasingly obvious. Even though animals also interact with fungi, and fungi are evolutionarily closer to animals than to plants (see the diagram on page 31), there are few examples of virus families that infect both animals and fungi. The relationships between crop plants and fungi in particular have been studied in detail, and researchers have found that some fungi can grow through plant cells and exchange small molecules. These interactions provide ample opportunities for the transmission of viruses, even though this has not been well documented. Much of our understanding about how plants and fungi have shared their viruses comes from comparison of the nucleotide sequences of the viruses, but in at least one case a plant virus was found infecting a fungus and could be moved into plants by colonizing them with the infected fungus.

One interesting virus family that is shared by plants and fungi is the Narnaviridae. Some of these viruses infect mitochondria, the organelles common in all Eukarya and derived from ancient bacteria. The polymerase (the enzyme that copies the virus genome) of these viruses looks most like a bacterial virus. This is not surprising, because the mitochondria still function like bacteria in many ways. There is also a plant virus that has this type of polymerase, but the rest of its genes are derived from another family of viruses that infect plants.

← The velvet mushroom (*Flammulina velutipes*) is found in a brown version and a white version. The difference is because the brown ones are infected by a virus.

Population control by viruses

Ecological studies of forest insect populations have been ongoing since the mid-nineteenth century. Early on, researchers noticed that these populations would rise and fall in regular cycles over several years. However, it wasn't until a century later that the role of viruses in these cycles became apparent.

↓ The Indian Meal Moth (*Plodia interpunctella*) is largely a pest of the grain industry, and often feeds on cornmeal. As a larva it can chew its way into many containers and is hard to control.

In many cases, a virus infection can kill off a large percentage of an insect population when it becomes too dense. The insect population then slowly builds up again, until it reaches a critical density and the cycle starts again. These types of cycles occur in many different insect species and with a variety of viruses, although members of the Baculoviridae family are most commonly involved.

Some of these insect–virus relationships are complex. For example, populations of the Indian Meal Moth (*Plodia interpunctella*) undergo cyclic patterns of boom and bust even if they are not infected with a virus, although populations that are infected have longer cycles. The population crashes are caused by pressure on food resources. As the population increases, food becomes scarcer and eventually many members of the population die from starvation. Virus-infected insects are smaller and have lower food requirements, so these populations take longer to run out of food.

Sometimes other players are involved in insect population fluctuations. Gypsy Moth (*Lymantria dispar*) populations are primarily controlled by mice that feed on them. In winter, the mice eat acorns, but when acorn production is low, mice numbers decline and during the next season the Gypsy Moth population explodes—until it is brought down by virus infection (see page 218).

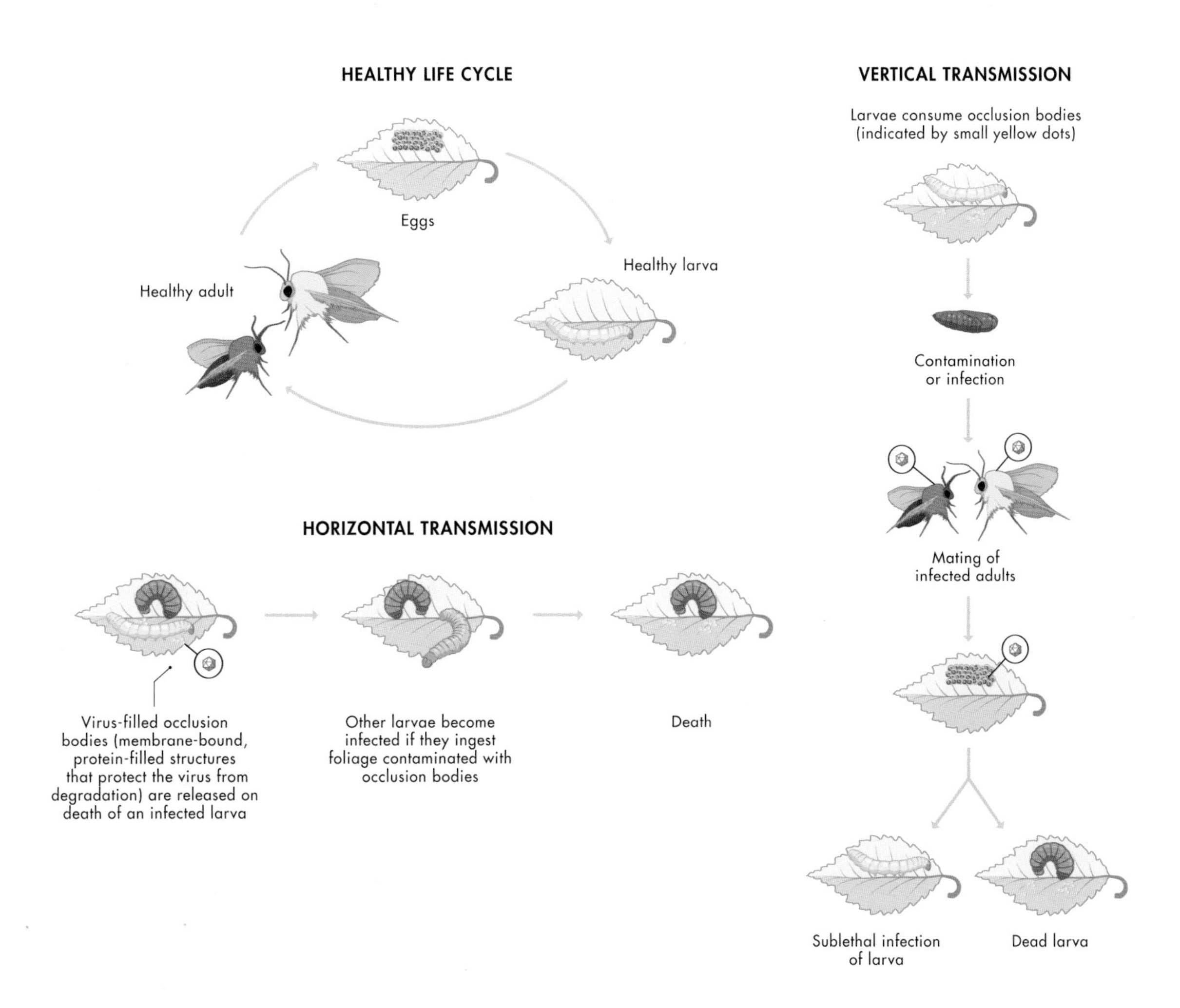

Virus transmission in two different ways

Transmission of baculoviruses can be horizontal among individuals through environmental contamination, or vertical from parents to offspring. Vertical transmission can occur if larvae consume occlusion bodies but pupate before death. Adults that are sublethally infected as larvae might then transmit virus to their offspring, either on or in the eggs. This can lead to an active infection, killing the offspring, or a covert infection that is passed on to offspring that will survive. These alternate cycles allow the viruses to be maintained in the insect population, awaiting the time when insect numbers become unsustainable.

Some species of sea slug also have cyclic population patterns. The Emerald Green Sea Slug (*Elysia chlorotica*) is a remarkable animal that incorporates the chloroplasts from the algae it eats, turning it green and also photosynthetic. The adult population of Emerald Green Sea Slugs dies annually, coinciding with a large rise in the concentration of a retrovirus that infects them. Although it is not completely clear that the virus is responsible for the annual die-off, this does seem likely.

The marine phytoplankton *Emiliania huxleyi* is responsible for large algal blooms that can be seen on satellite imagery. The algae draw carbon from the atmosphere to form their outer calcite shell, and are responsible for large fluctuations of carbon in the oceans and the atmosphere. These huge populations are extinguished by infection with a virus in the Phycodnaviridae family. However, the algae also exist in a different form, one that doesn't create the calcite shells and is not susceptible to virus infection. The type that forms shells is diploid (it has two copies of its genome), while the one that doesn't and is resistant to the virus is haploid (it has only one copy). When two haploid cells fuse, they become diploid. The diploid cells reproduce asexually until the population becomes huge, at which point they are killed off by the virus and only the haploids survive. These cycles have a seasonal nature, with the diploids growing when the surface temperatures of the water warm in the summer.

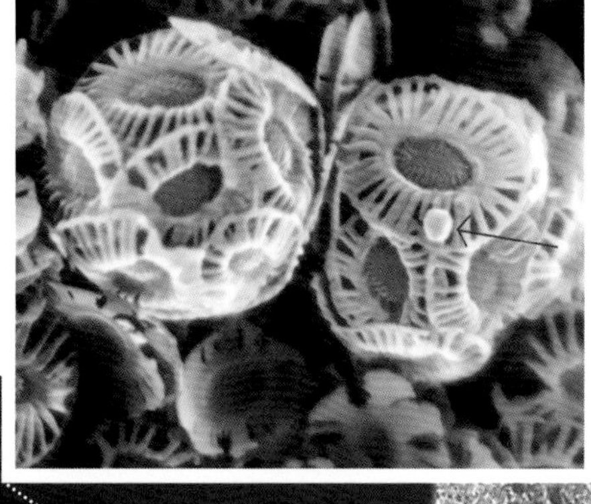

←↓ Blooms of the phytoplankton *Emiliania huxleyi* can be seen on satellite images, as in this bloom off the southwest coast of England. The algae have a hard outer shell made of calcite that reflects light.

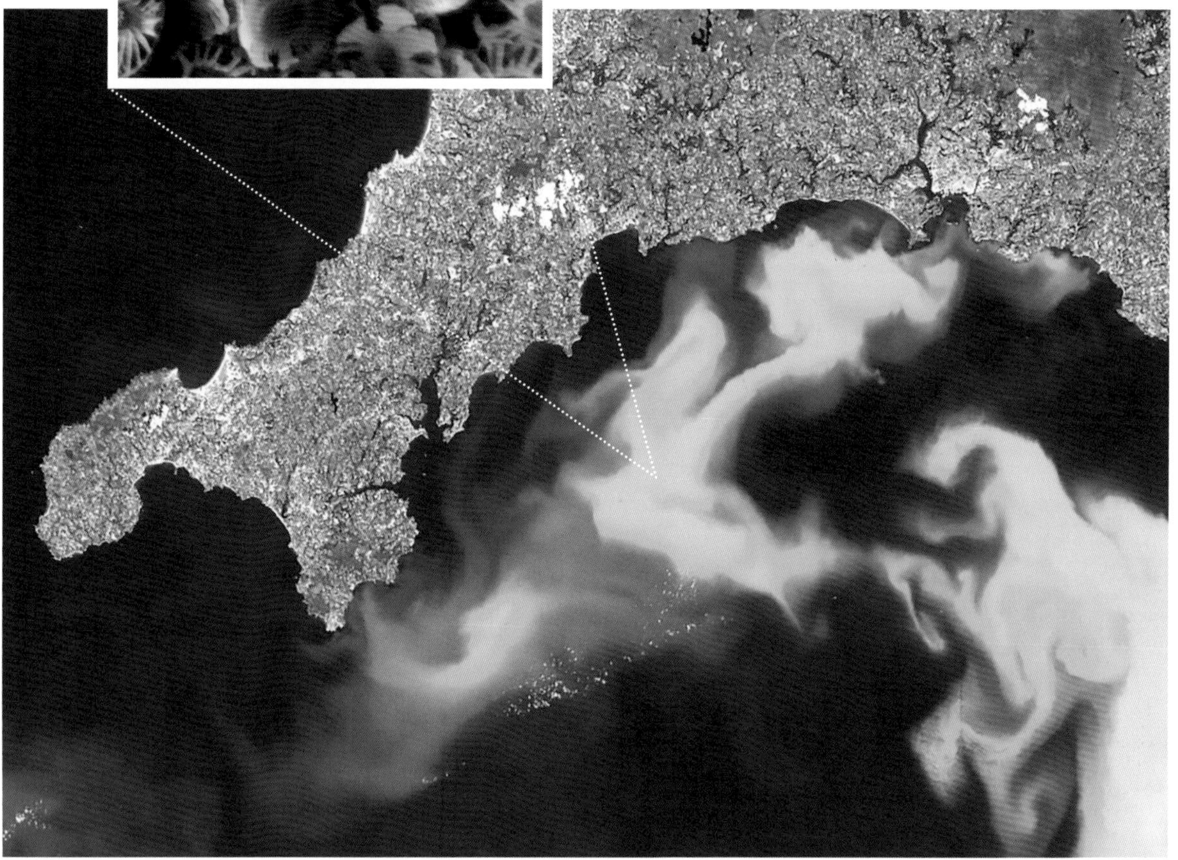

Effects of viral infection on host behavior

Many viruses influence the behavior of their hosts, and this can usually be related to increased virus transmission. The bornaviruses are one example, with infection increasing aggression in a number of rodent species and the virus being transmitted during the associated biting behavior.

↑ Iridoviruses are the only viruses that have a natural color. The color is not a pigment, but is derived from the reflective properties of the virus particle. Insects that are infected with iridoviruses can display the color of the virus, like this blue roly-poly, or woodlouse. In crickets, infection with an iridovirus increases mating behavior.

Rabies virus also increases aggression in many host species, as well as a condition known as hydrophobia, or fear of water. If the host doesn't drink water, the concentration of virus in its saliva increases. Simian immunodeficiency virus (the virus from which HIV evolved) infects many wild primates and similarly increases aggression in these animals.

Some of the most interesting impacts of viruses on behavior occur in insects. The human pathogens dengue virus and La Crosse encephalitis virus are transmitted by mosquitoes. The viruses increase the feeding rate of the hosts, and those infected with the La Crosse virus have much higher rates of feeding on new hosts. This increases transmission of the virus to new human hosts.

There are a number of other examples of manipulation of insect behavior by viruses, including increased mating in crickets infected with an iridovirus (which incidentally turns them blue). Another example is seen in parasitoid wasps, which lay their eggs inside an insect. The egg develops in the insect and the hatched larva eventually kills it by eating its internal organs, but this is a slow process. The parasitoids can generally tell if their insect host has already been parasitized by another wasp, and will avoid laying eggs in these as the chances of the second egg surviving are very low. However, when the wasp is infected with a virus it will preferentially lay its eggs in insects that are already parasitized. The second egg will not develop, but the wasp will transmit the virus to the other larva.

Effects of host ecology changes on virus outcomes

Poliovirus has been infecting humans for millennia and can cause the severe neurological disease poliomyelitis, commonly known as polio. However, this disease was extremely rare until the twentieth century, when it became an epidemic, causing paralysis, deformities, and often death, especially in children. So what changed? The answer is that people changed.

Poliovirus is a waterborne virus, transmitted by the fecal–oral route. Prior to the twentieth century sources of drinking water were full of the virus and almost everyone was infected with it as an infant around the age of weaning, when maternal antibodies are no longer available. In infants, polio is very mild, but this infection provides a lifelong immunity. In the early twentieth century the role of contaminated water as a source of cholera was recognized, and efforts were made to clean up water supplies. Filtration was used initially, but after the First World War most water was sanitized with chlorine. This removed poliovirus from drinking water supplies and infants no longer contracted polio. There was still plenty of poliovirus around, however, because the modernization of sewage systems didn't start until the 1960s and 1970s. As a result, older children and adults were exposed to poliovirus in the environment, often while swimming. When the virus infects anyone past the infant stage, the dreaded paralytic disease poliomyelitis is much more common.

↗ The iron lung was used during polio outbreaks in the mid-twentieth century to help patients breathe. The illness causes paralysis in many parts of the body, and when it affects the diaphragm patients die because they can't breathe. The iron lung, a negative-pressure ventilator, saved many lives, with patients usually having to spend at least two weeks encased in the machine.

→ A ten-year-old polio victim from Bangalore, India, in 2006. The last case of polio in India was reported in 2011 and it has since been free of the disease.

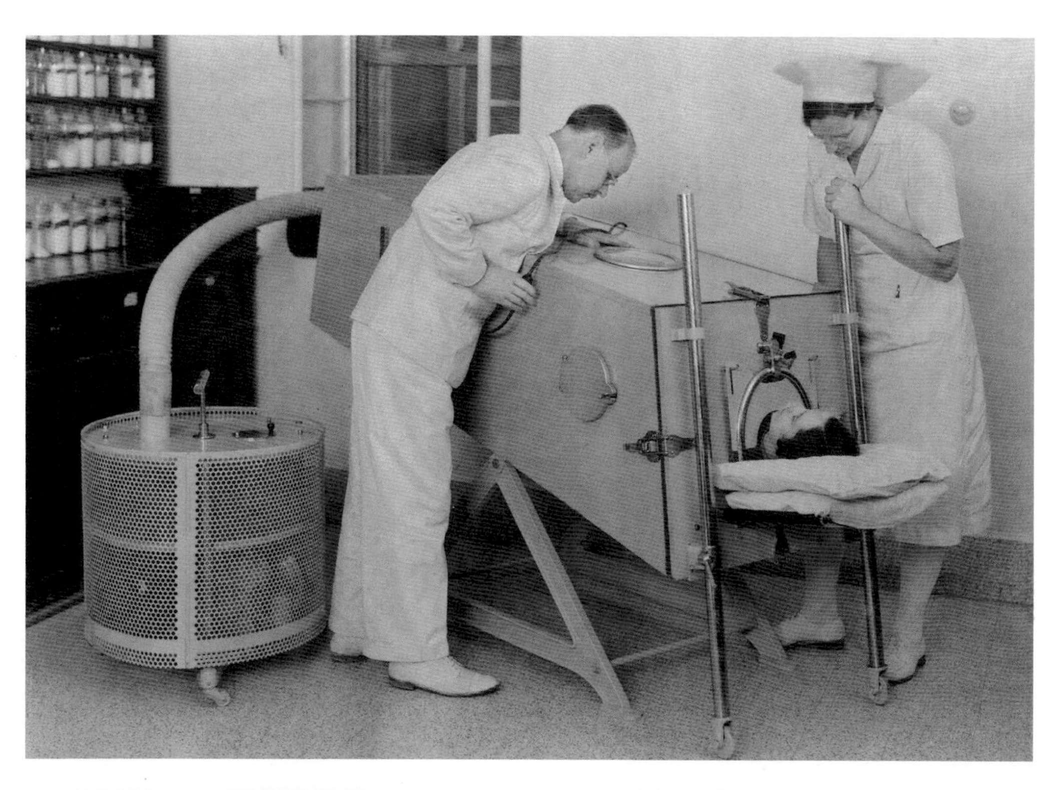

Other changes in human ecology have had an impact on viral diseases such as yellow fever, dengue, and chikungunya. Yellow fever was spread worldwide with the movement of people, and dengue and chikungunya have been spread in recent decades by the migration of people from rural areas to urban centers. Deforestation and increased human populations have added to these issues, and climate change is having an impact too, especially in the case of insect-vectored viruses because the range of the insects is increasing.

Humans have also unwittingly moved viruses of plants and domestic animals around the world with them. Many plant viruses have become epidemic as plants are cultivated in new places, and viruses of native plants have jumped into crops.

Spread of a plant virus around the world

The Tomato (*Solanum lycopersicum*) is native to South America but is cultivated around the world. Plants introduced to the Middle East encountered a local virus that causes tomato yellow leaf curl, a serious disease. The virus then moved around the world, infecting Tomato plants in many locations.

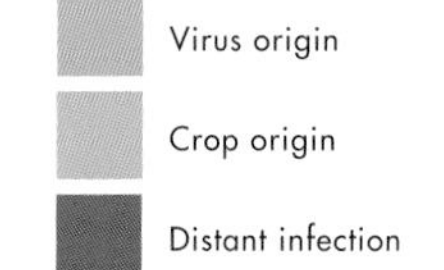

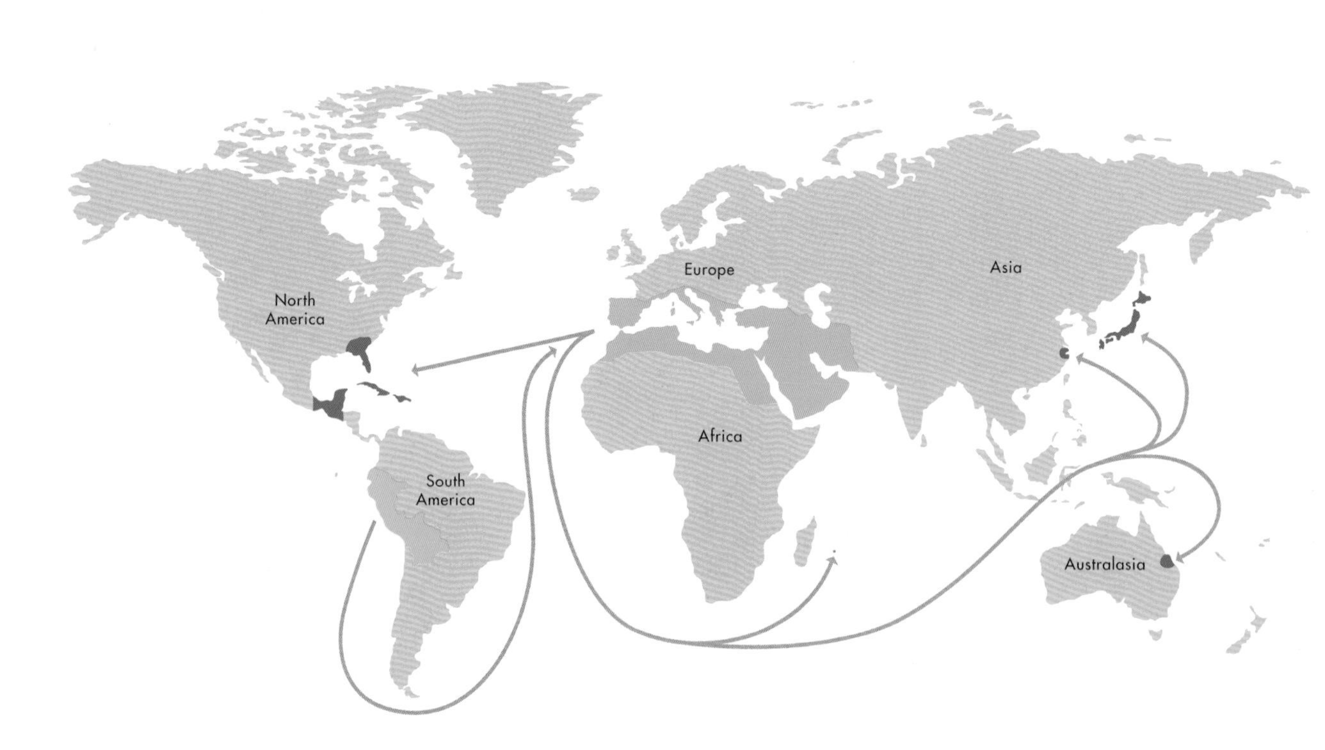

VIRUSES AND INVASIVE SPECIES

As humans have moved around the globe, they have taken plants, animals, and microbes with them, sometimes intentionally and sometimes accidentally. These newcomers can become invasive when they take over a niche occupied by a native species, and viruses can help in this process. For example, in the grasslands of California the invasive Wild Oat (*Avena fatua*) attracts large numbers of aphids that vector a virus. The virus is most damaging to native bunchgrass, and so has contributed to the success of the exotic oat.

When Europeans arrived in the Americas, they brought their diseases with them, including several viruses that shaped much of the future of the indigenous people. Smallpox may have been the most devastating, killing entire communities who had never been exposed to the virus before and had no tolerance to it. Even normally less severe viruses, including influenza, measles, and rhinoviruses (common cold viruses), were often lethal owing to a lack of immunity among indigenous populations (see page 232).

Humans have used viruses to try to combat invasive species. One example is the control program for the European Rabbit (*Oryctolagus cuniculus*) in Australia in the 1950s. Just 24 rabbits were originally released in 1859 but, freed from their natural predators, they bred, well, like rabbits, and by the middle of the twentieth century there were an estimated 600 million. Myxoma virus, which infects the Brazilian Rabbit (*Sylvilagus brasiliensis*) but causes little disease, was found to cause myxomatosis, a disease that killed the European Rabbit. This virus was introduced to Australia in 1950, and by 1952 the rabbit population had decreased to 100 million. The program wasn't entirely successful, however, because the deaths stopped at this point. It turned out that the virus had adapted to its new host and evolved to become less virulent.

↓ Rabbits at a water source in the Australian outback in the 1950s.

Enterovirus C

Virus whose disease course changed when humans cleaned up their drinking water

GROUP	IV
FAMILY	Picornaviridae
GENUS	Enterovirus
GENOME	Single-component, single-stranded RNA comprising about 7,500 nucleotides, encoding 11 proteins
VIRUS PARTICLE	Non-enveloped, icosahedral, about 30 nm
HOSTS	Humans
ASSOCIATED DISEASES	Poliomyelitis
TRANSMISSION	Waterborne
VACCINE	Live attenuated, or heat-killed mix of three serotypes

Enterovirus C, also known as poliovirus, has been notorious for many reasons, but most people associate it with the severe disease poliomyelitis (polio) or infantile paralysis, which spread around the globe in the twentieth century.

Today, poliovirus is close to being eradicated thanks to vaccination, but there are still a few hundred cases of polio worldwide every year. This is largely because the live attenuated vaccine can very occasionally revert to a more virulent form (see page 177). Recently vaccine escapes have been found in sewer water in New York, Jerusalem, and London, indicating some circulation of the virus, and there has been one case of paralysis in New York. Attenuated vaccine is easy to administer because it is given in the form of a sugar cube, rather than the injection that is required for the heat-killed vaccine. In some war-torn parts of the Middle East, a few cases of wild polio, the original virus, have been found in recent years as well. Additionally, the COVID-19 pandemic has relaxed some vaccination campaigns, and there is a serious threat of polio across much of central Africa, and in Palestine and Ukraine. All of Africa remains in the status of "not yet eradicated."

There is evidence of poliomyelitis in at least one Egyptian mummy, indicating that the virus has been with humans for a long time. However, the disease was very rare until the virus was removed from drinking water by chlorination.

Enterovirus C was the first human virus to be cloned in such a way that the clone could be used to infect cells. Many viruses have been cloned now, and this is an extremely useful tool in studying how viruses interact with their hosts. In 2002 the entire genome of enterovirus C was synthesized artificially in a laboratory from the known sequence of the nucleotides. The synthetic genome could infect cells, making it the first example of life created artificially.

Experiments with enterovirus C have demonstrated that antibiotics used to fight bacteria can also have an effect on a virus infection. In general, antibiotics are not active against viruses, and in fact they can even increase disease in some viruses like influenza. However, viruses that are taken up in the gut can use the gut bacteria to enhance their infection, so eliminating these bacteria can decrease virus infection, even though it is generally deleterious to overall health.

→ The structure of enterovirus C, generated by cryogenic electron microscopy.

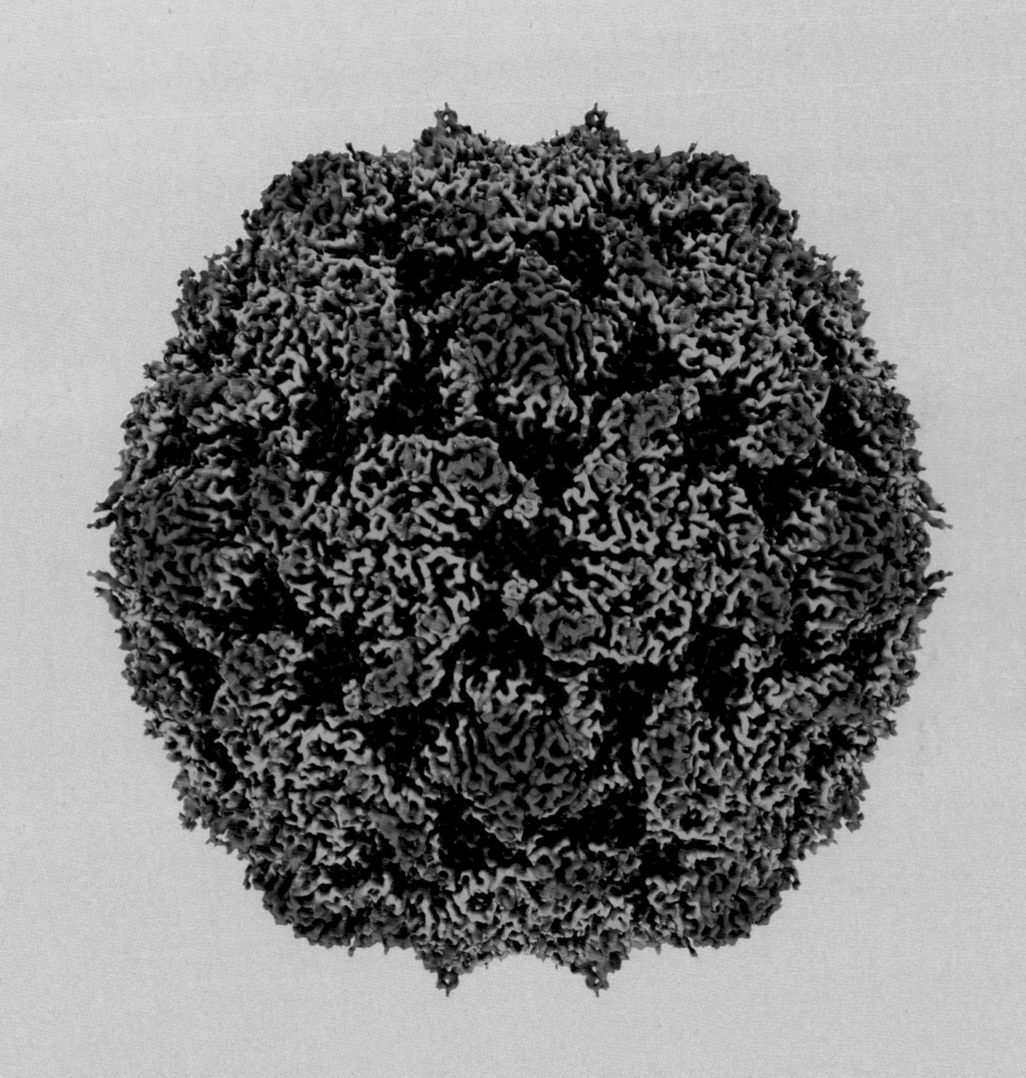

TYLCCNV

Tomato yellow leaf curl China virus

Virus that helps an invasive insect

GROUP	II
FAMILY	Geminiviridae
GENUS	Begomovirus
GENOME	Circular, single-component, single-stranded DNA comprising about 2,700 nucleotides, encoding six proteins
VIRUS PARTICLE	Non-enveloped, twin-icosahedral, about 22 nm by 38 nm
HOSTS	Tomato (*Solanum lycopersicum*), tobacco (*Nicotiana* species)
ASSOCIATED DISEASES	Yellowing, leaf curling
TRANSMISSION	Whiteflies

Tomato yellow leaf curl China virus is one of 13 viruses that probably originated from a single virus, first isolated in the Middle East. Each virus species has diverged rapidly in its location, and is designated by the country where it was isolated.

Many plant viruses are sensitive to heat, and heat treatment of the growing tip of plants is an old method used to get rid of viruses in plants that are grown from cuttings rather than seeds. In the Middle East the normal temperatures for cultivating Tomato plants (*Solanum lycopersicum*) frequently reach 40 °C (104 °F). However, tomato yellow leaf curl virus replicates better at these high temperatures, and also confers heat tolerance to its Tomato host.

Most viruses in the Geminiviridae family are transmitted by Silverleaf Whiteflies (*Bemisia tabaci*). These insects have been moved around the globe and are invasive in many places. They feed on the underside of plants and are especially prevalent in greenhouses, where populations can increase rapidly. They don't cause significant damage to the plant by themselves, but they have been found to transmit about 60 different plant viruses and as such pose a huge threat to agriculture.

In China, the tomato yellow leaf curl China virus is vectored by two different whitefly biotypes (a biotype is a subgroup of a species that shares the same genotype), a native biotype and an invasive biotype. The invasive biotype lives longer and has more offspring when it colonizes virus-infected plants, whereas the native biotype is not affected by the virus. This has allowed the invasive biotype to displace the native one, enhancing the spread of the virus. The virus is therefore an indirect mutualist of the invasive whitefly.

→ The structure of Ageratum yellow vein virus, a close relative of TYLCCNV, as determined by cryo-EM.

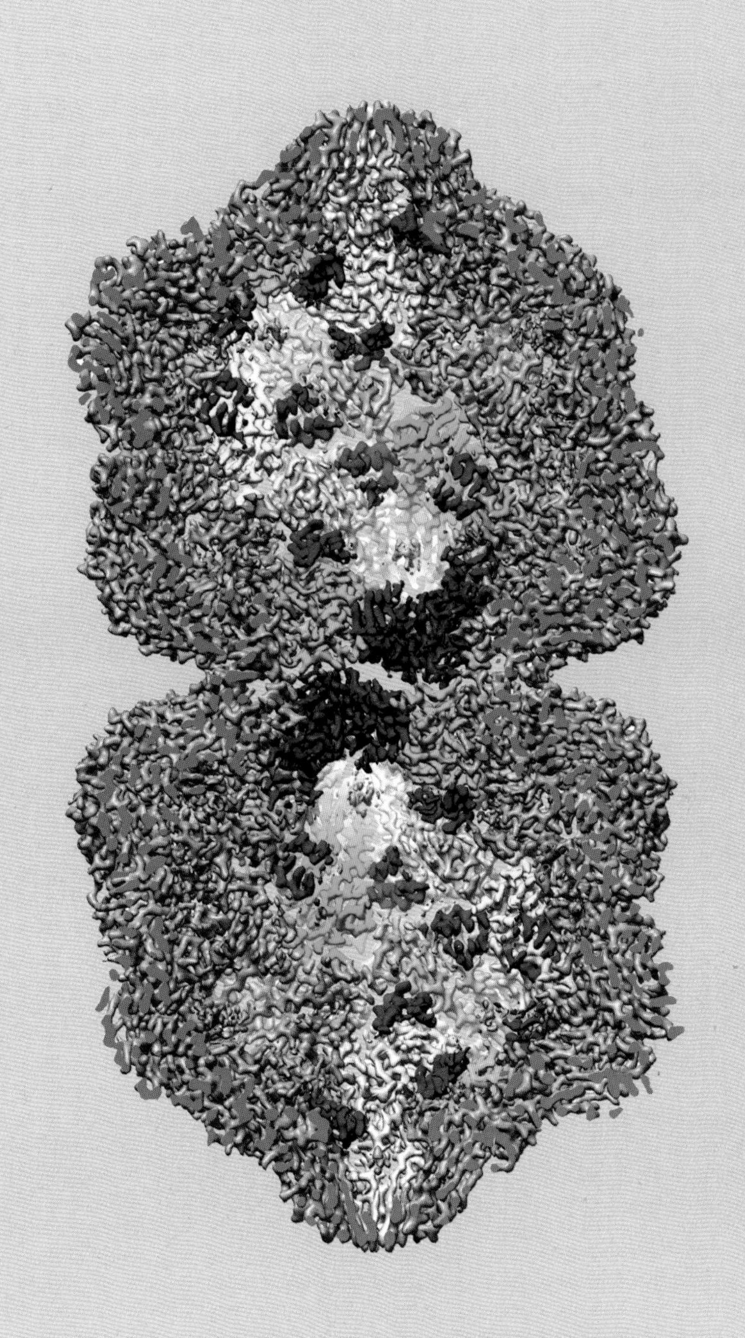

CMV

Cucumber mosaic virus

Virus that can infect both plants and fungi

GROUP	IV
FAMILY	Bromoviridae
GENUS	Cucumovirus
GENOME	Linear, three-component, single-stranded RNA comprising about 8,500 nucleotides, encoding five proteins
VIRUS PARTICLE	Non-enveloped, icosahedral, about 28 nm
HOSTS	More than 1,200 plant species
ASSOCIATED DISEASES	Mosaic, yellowing, stunting, leaf distortions
TRANSMISSION	Aphids

Cucumber mosaic virus (CMV) was first discovered infecting cucumbers, where it causes mosaic symptoms on the leaves and necrosis on the fruits. The virus has since been described in 1,200 plant species—more hosts than any other known virus, and there are probably more that have not yet been described. Interestingly, almost all cucumber plants grown today are resistant to the virus.

CMV is one of the most well studied RNA viruses of plants. Its divided genome makes it a good model virus for genetic studies, and it was one of the earliest plant viruses to be cloned in a way that could make infectious viruses. This also made it an excellent model for studies in experimental evolution.

One of the most remarkable recent discoveries with CMV is that it can also infect fungi. The virus was found in a fungal pathogen of Potato plants (*Solanum tuberosum*) that causes tuber rot, and this strain was nearly identical to the CMV strains found in plants. In laboratory experiments researchers found that the fungal virus could infect plants and induce typical symptoms. Another strain of the virus, originally isolated from melon plants, could also infect the fungus, and the virus could be transmitted between closely related strains of the fungus. When researchers used infected fungus tissue to colonize plants, the virus was transmitted to the plants. The reverse experiment also worked: when virus-free fungus was used to colonize virus-infected plants, the virus was transferred to the fungus. A few other fungi could also be experimentally infected with CMV.

Although there are a number of examples of viruses that can infect insects and plants, other cross-kingdom viruses are rare. This has important implications about the origins of viruses. If the same virus can infect different kingdoms, researchers can't know for sure what the original host is, but it is possible that a virus like CMV originated as a fungal virus and became an epidemic plant virus after jumping into plants.

→ The structure of cucumber mosaic virus, derived from cryogenic electron microscopy data.

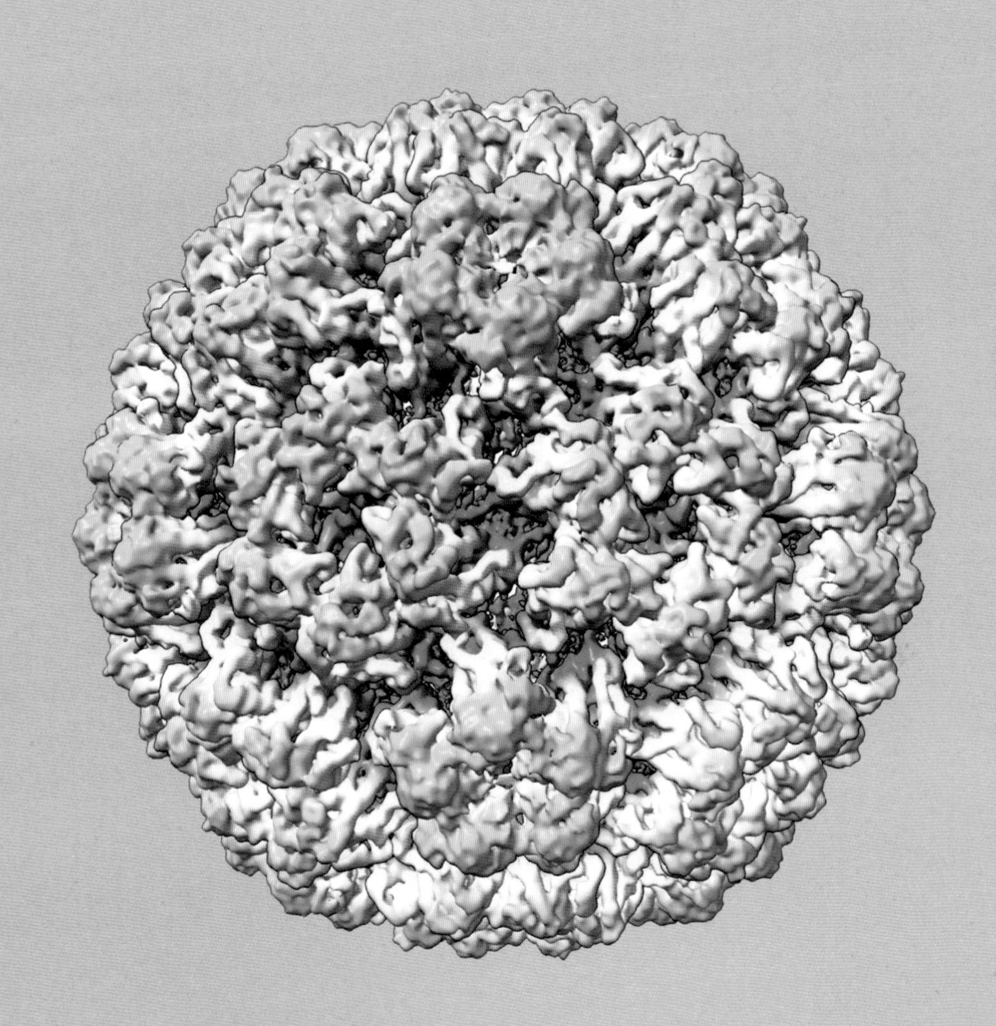

SVSYN5

Synechococcus virus Syn5

Virus critical for nutrient flow in the oceans

GROUP	I
FAMILY	Autographiviridae
GENUS	Voetvirus
GENOME	Linear, single-component, double-stranded DNA comprising about 46,000 nucleotides, encoding 61 proteins
VIRUS PARTICLE	Non-enveloped, icosahedral with a short tail and tail fibers
HOSTS	*Synechococcus* species
ASSOCIATED DISEASES	Cell lysis
TRANSMISSION	Waterborne

Cyanobacteria are photosynthetic bacteria that live in the oceans, and include the genera *Synechococcus* and *Prochlorococcus*. They are responsible for much of the photosynthesis that occurs there and are major producers of oxygen.

The numbers of marine cyanobacteria are controlled by viruses such as synechococcus virus Syn5 (SVSyn5), which infects and kills the microbes. If the microbes are not killed this way, they sink to the bottom of the sea when they die, taking all of their nutrients with them. When the viruses lyse the bacteria (see page 197), their contents remain in the upper levels of the oceans and are available for other organisms to use in the continuous cycle of life.

There are billions of different virus species that infect marine cyanobacteria and all are involved in this massive recycling, which goes on every day. Without them, the oceans would die and so too all of life on Earth. While these viruses could be considered pathogens of the bacteria, they can also aid their hosts by providing important genes in the metabolism of the bacteria, called auxiliary metabolic genes. These genes are especially important when the bacteria encounter extreme environments, such as deep-sea vents. While scientific study is discovering more and more viruses in the oceans all the time, we still know very little about which viruses infect which hosts. Virologists are developing better tools to uncover these mysteries, but it will take time to sort out all the relationships among viruses and cyanobacteria.

→ Structural model of the synechococcus virus Syn5.

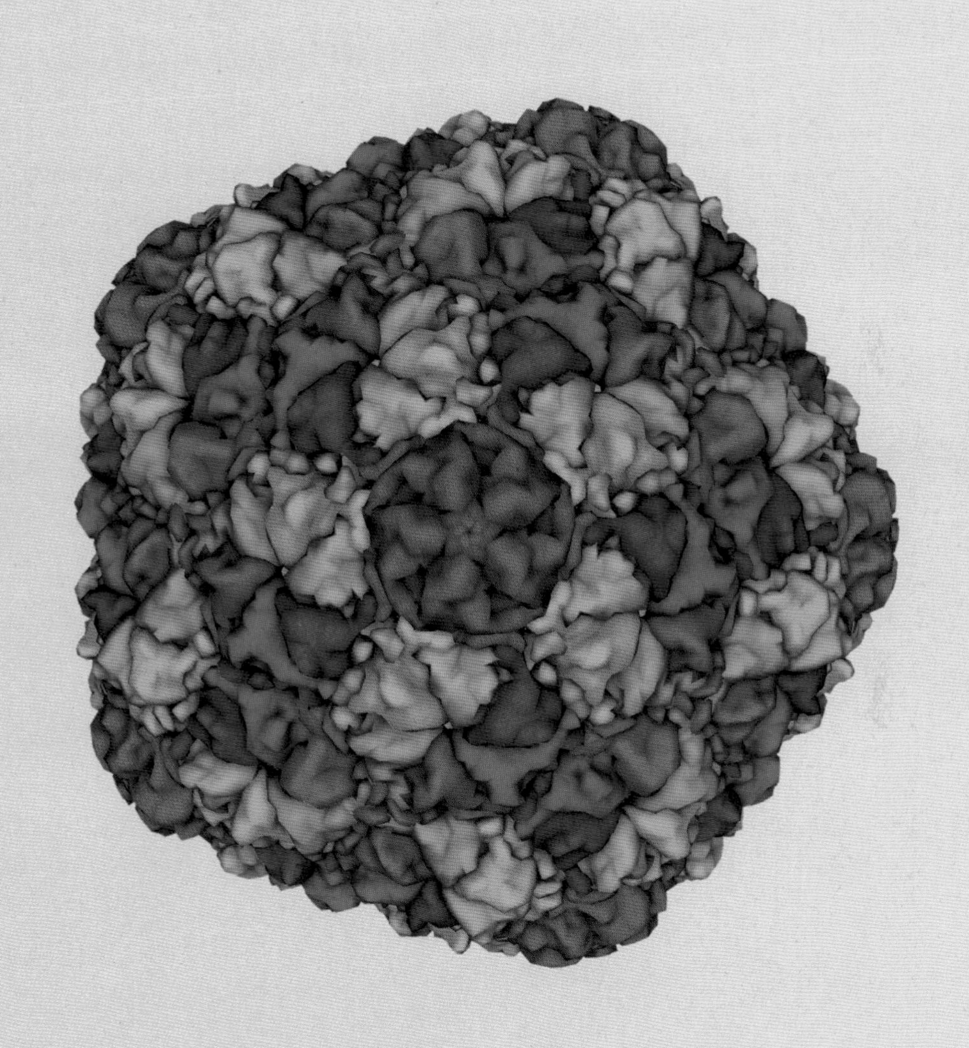

LdMNPV

Lymantria dispar multiple nucleopolyhedrosis virus

Insect virus important for population control

GROUP	I
FAMILY	Baculoviridae
GENUS	Alphabaculovirus
GENOME	Circular, double-stranded DNA comprising about 167,000 nucleotides, encoding about 165 proteins
VIRUS PARTICLE	Enveloped, rocket-shaped core
HOSTS	Gypsy Moth (*Lymantria dispar*)
ASSOCIATED DISEASES	Treetop disease
TRANSMISSION	Ingestion

→ Artist's rendition of the Lymantria dispar multiple nucleopolyhedrosis virus, showing a cutaway of the inner core of the particle.

Population boom-and-bust cycles in Gypsy Moths (*Lymantria dispar*) were reported in Europe as early as the mid-nineteenth century, although the cause of these population fluctuations was unknown. In the 1860s the Gypsy Moth was introduced to the United States, where it caused a lot of damage to northeastern forests and has spread to the south and west.

Lymantria dispar multiple nucleopolyhedrosis virus (LdMNPV) is a natural enemy of the Gypsy Moth, infecting the larva of the insect and causing the boom-and-bust population cycles. The virus can be transmitted vertically (see page 112), when it causes a nonlethal infection. This often happens when the insects are at low density. When the insects reach high density, however, the virus is more likely to be transmitted horizontally, in which case the disease becomes lethal.

The virus also changes the moth's behavior in two ways. First, it delays molting, so the insect spends more time feeding in the leafy canopy. Second, instead of hiding during the day, the insects feed continuously and climb to the tops of the trees. The increased biomass from the delay in molting and the continuous feeding together provide more material for the formation of occlusion bodies, the infectious form of the virus. Upon the death of the infected moths in the treetops, their virus-filled bodies liquefy and millions of occlusion bodies rain down through the canopy and onto the forest floor. These are then ingested by newly hatched larva, starting the cycle over.

LdMNPV is used in a commercially available form known as Gypchek by the United States Department of Agriculture to control the Gypsy Moth. This method of biocontrol works well because the virus is very specific to the Gypsy Moth and doesn't infect any other insects (although related viruses in the Baculoviridae family do infect other insects).

THE GOOD VIRUSES

Symbiosis and symbiogenesis

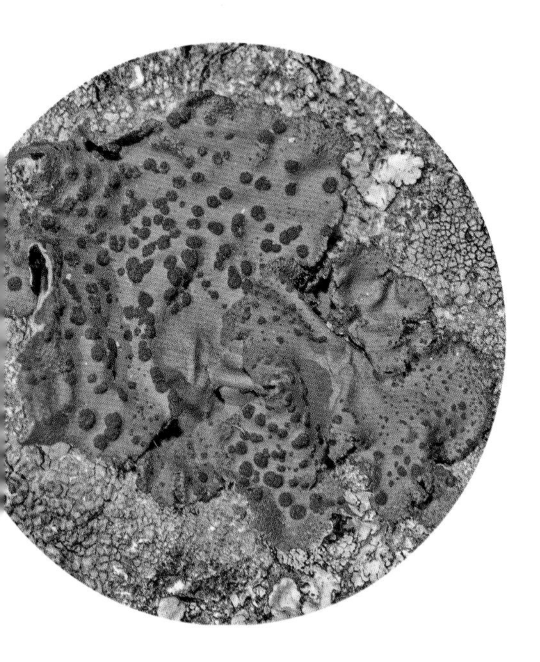

Can viruses really be good for us? In fact, although the news is usually about bad viruses, most don't cause any disease in their hosts, and some are clearly beneficial or even required for their hosts' survival. The previous chapter explored some of the ways viruses are necessary to keep the planet's ecosystems in balance. This chapter will cover some of the more direct ways that viruses benefit their hosts.

The term symbiosis was coined in the nineteenth century to describe lichen as a mixture of fungal and bacterial cells living together. It is often confused with mutualism, but the two are distinct. Symbionts are different entities that live in an intimate relationship, which can be beneficial to both parties (mutualistic), but can also be neutral or antagonistic, as with pathogens. All viruses are symbionts; most are probably neutral, some are mutualists, and a few are pathogens. Symbiogenesis happens when symbionts fuse to form a new entity, such as the emergence of eukaryotic cells after a prototype cell fused with a bacteria that became the cell mitochondria. Viruses also can be symbiogenic—there are numerous examples of virus genes fusing with the genome of their host. The easily recognized retroviruses alone represent about 8 percent of the human genome, or about five times more than the amount of the genome coding for proteins.

Human endogenous retrovirus K (see page 56) is a symbiogenic virus in the human genome that is required for the formation of the placenta. Other endogenous retroviruses are also essential. In some cases the virus carries a gene for a critical protein, while in others it affects different genes that are found nearby in the genome. For example, amylase is an enzyme required for the digestion of starches. It is found in the gut, but in humans it is also in saliva thanks to an endogenous retrovirus that allowed it to be made in the salivary glands. In some cases an endogenous retrovirus can protect the host from other related viruses. This occurs in mammals, plants, insects, and fungi.

↖ Lichen were the first entities described as symbionts, a fungus and an algae growing together. This example, *Umbilicaria phaea*, was used in a study to find viruses, and several new viruses were identified related to plant and bacterial viruses.

Mammalian placenta

Most mammals develop a placenta during gestation of the fetus by the fusion of multiple maternal cells. This structure, the syncytiotrophoblast, provides nourishment to the developing fetus and a barrier to transmission of most infectious agents. The critical protein for cell fusion is syncytin, which is encoded by an endogenous retrovirus. The villi provide extended cell surfaces for the exchanges between maternal and fetal tissue.

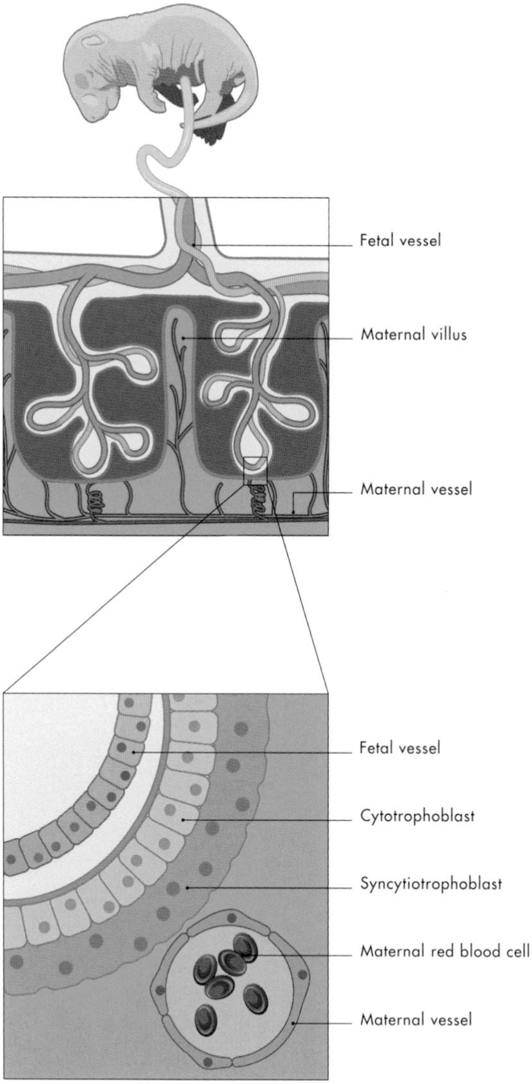

Fetal vessel
Maternal villus
Maternal vessel

Fetal vessel
Cytotrophoblast
Syncytiotrophoblast
Maternal red blood cell
Maternal vessel

Family tree of placental mammals

All placental mammals have an endogenous retrovirus that provides syncytin, but they don't all have the same one. There are four clear lineages (boxed) of the syncytin-encoding viruses, which indicates the possibility that placental mammals evolved more than once, or that an ancestral endogenous retrovirus was replaced in some lineages.

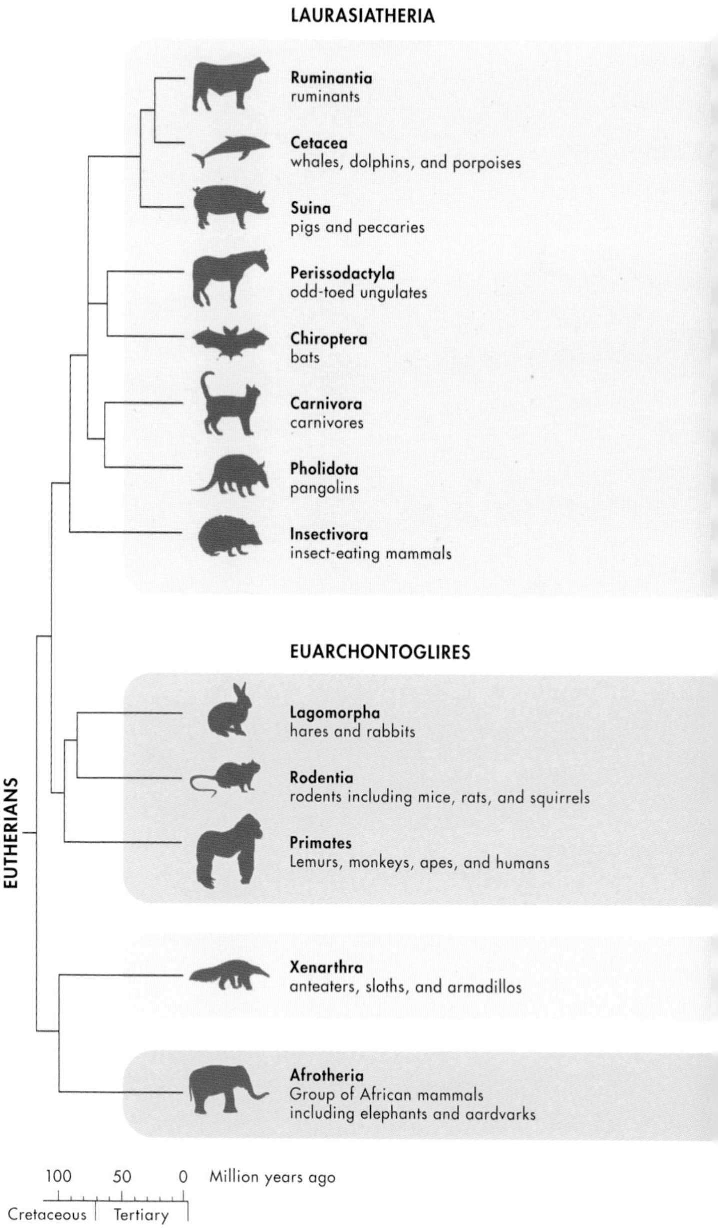

LAURASIATHERIA

Ruminantia
ruminants

Cetacea
whales, dolphins, and porpoises

Suina
pigs and peccaries

Perissodactyla
odd-toed ungulates

Chiroptera
bats

Carnivora
carnivores

Pholidota
pangolins

Insectivora
insect-eating mammals

EUARCHONTOGLIRES

Lagomorpha
hares and rabbits

Rodentia
rodents including mice, rats, and squirrels

Primates
Lemurs, monkeys, apes, and humans

Xenarthra
anteaters, sloths, and armadillos

Afrotheria
Group of African mammals including elephants and aardvarks

EUTHERIANS

100 50 0 Million years ago
Cretaceous | Tertiary

It isn't surprising that so many retroviruses are found in genomes, since these viruses have to integrate into a genome as part of their life cycle (see pages 84–85). However, sequences from all the classes of viruses can be found in genomes. In most cases no one knows how they got there or whether they are active, but in a few cases they may protect the host from related viruses.

A mutualistic virus infecting parasitoid wasps that use caterpillars to raise their offspring is in the process of becoming symbiogenic. The genes for most of the normal virus proteins, such as the coat and replication proteins, have moved to the genome of the wasp. The wasp uses the virus particles as a delivery system for its own genes, which suppress the immune system of the caterpillar and are deposited when the wasp lays its egg in the caterpillar. Without being able to suppress the caterpillar immune system, the wasp egg would be expelled by the caterpillar's body and therefore unable to develop. There are thousands of these mutualistic viruses, which have been associated with their wasp hosts for a very long time.

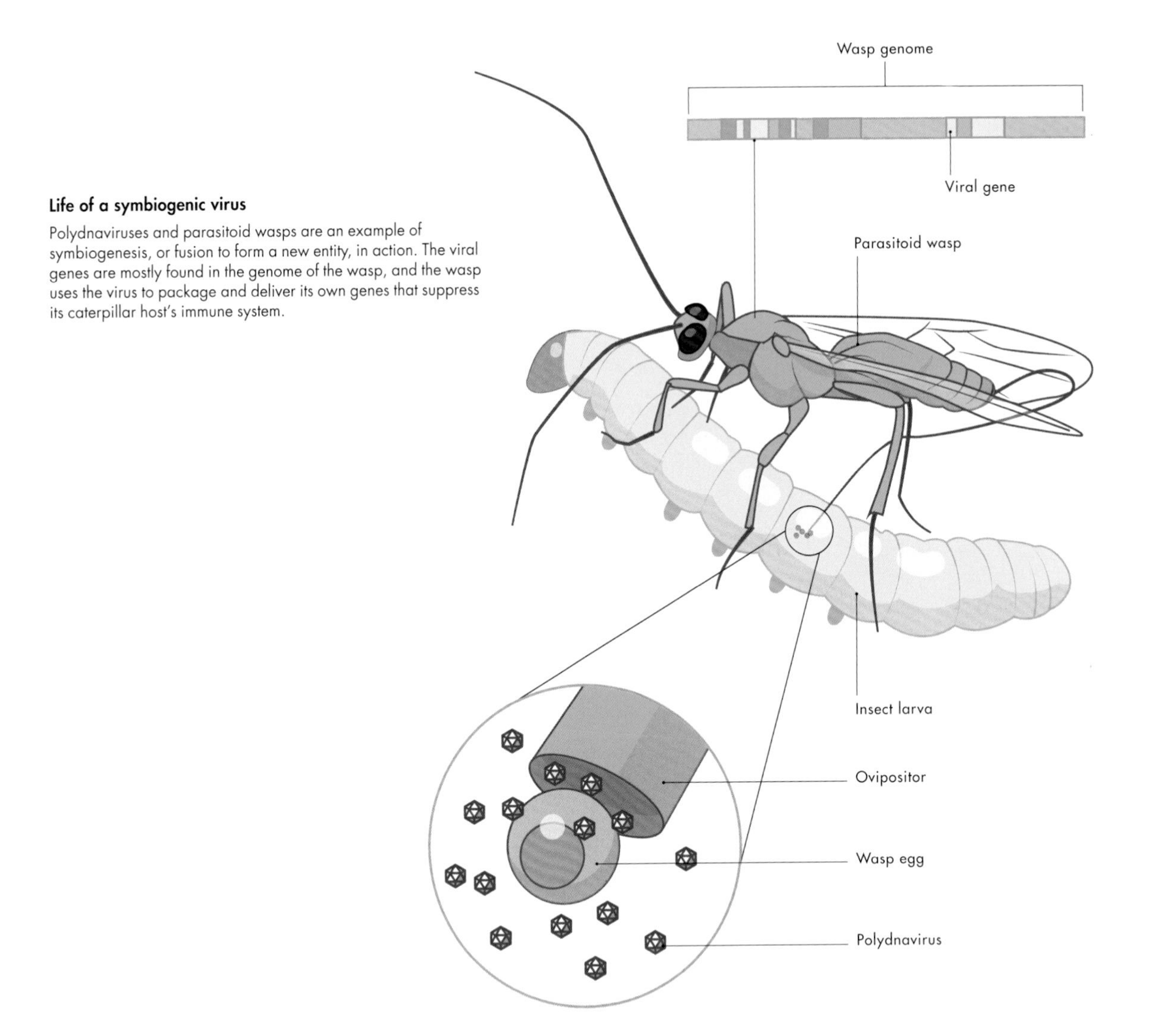

Life of a symbiogenic virus

Polydnaviruses and parasitoid wasps are an example of symbiogenesis, or fusion to form a new entity, in action. The viral genes are mostly found in the genome of the wasp, and the wasp uses the virus to package and deliver its own genes that suppress its caterpillar host's immune system.

↑ Cocoons of braconid parasitic wasps on the caterpillar of a parasitized hawk-moth.

Viruses in human and animal health

In the 1980s a diagnosis of human immunodeficiency virus (HIV) infection was considered a death sentence, because it caused a catastrophic effect on the immune system, leading to acquired immunodeficiency syndrome (AIDS). However, some people contracted the virus but the infection never progressed to AIDS. For some this may have been because they were also infected with pegavirus C (formerly known as hepatitis G virus). This doesn't cause any disease in humans, but it can delay the onset of AIDS.

→ The waters of Chesapeake Bay contain high levels of *Vibrio cholerae*, but the bacteria do not cause cholera because the toxin-producing viruses are not present.

↓ This miniature from the Swiss Toggenburg Bible (1411) shows people suffering from what is believed to be bubonic plague. Mice are protected from bubonic plague by infection with a herpesvirus, and it is possible that a similar virus protects modern humans.

Other human viruses can also suppress the disease caused by different viruses, including cytomegalovirus, a type of herpesvirus that can suppress infection with HIV and influenza. There are also some viruses in mice, used as a model for human diseases, that suppress diseases. For example, a mouse herpesvirus that is related to human viruses can suppress the bacteria that causes bubonic plague—the dreaded Black Death that decimated populations in the Middle Ages.

The human virome includes many viruses that are made by the microbes that live in the gut. In many animals, bacteriophages from the normal microbiome are staged at the mucous membrane entry points to the body. These phages depress the ability of bacteria to attach to the membranes, effectively preventing infection by pathogenic bacteria.

Bacteriophages are very important mutualists of many bacteria. Although they don't benefit the human hosts of the bacteria, they allow their bacterial hosts to invade the human body. Although rare now due to vaccination, diphtheria was once a disease that spread terror, especially in crowded communities. Not all cases were severe, with the severity depending on the production of a toxin that allows the bacteria to invade the host tissue in the respiratory tract. However, it isn't the bacteria themselves that actually encode the toxin, but a mutualistic bacteriophage. A similar situation occurs with cholera (see page 244). The cholera bacterium has to be infected with two mutualistic viruses to produce the toxin that allows it to invade the human gut tissue. Phages also encode many other toxins—for example, the dreaded Escherichia coli contaminants of food are really just the normal bacteria that live in the human gut that have been infected with a phage carrying a toxin gene from Shigella bacteria.

The importance of the microbiome in proper human gut function is well known now, but most people once considered all bacteria to be evil germs. Diversity is an important part of the gut microbiome, and this occurs in the developing infant gut thanks to phages that kill off the most abundant bacteria, allowing others to develop. While the gut bacteria are critical for many aspects of digestion and metabolism, an experiment in mice found that a norovirus related to the virus that causes gastroenteritis outbreaks on cruise ships could replace the functions of the bacteria in establishing normal gut activity.

Saving a host from stress

A relationship between two different organisms that is beneficial only under certain circumstances is called conditional mutualism. A number of different plant viruses may be pathogens under normal conditions, but when infected plants experience drought, they survive longer without water than their noninfected counterparts.

↓ Hot Springs Panic Grass (*Dichanthelium lanuginosum*) grows in the geothermal soils of Yellowstone National Park in Wyoming. However, the grass cannot tolerate the high soil temperatures unless it is colonized by a fungus that is, in turn, infected with a virus. The fungus is found in other plants in nonthermal soils in Yellowstone, but it is not infected with the virus in these sites.

Viruses can also protect plants from cold stress, allowing them to survive light frosts that would otherwise kill them. In Yellowstone National Park in Wyoming, a panic grass grows in the geothermal soils, which often experience temperatures as high as 55 °C (131 °F). The plant is able to tolerate these high temperatures only because it is colonized by a virus-infected fungus (see page 240).

Conditional mutualism in a plant–virus symbiosis

Plants that are infected with a virus (symbiotic) may exhibit symptoms under normal conditions; however, when drought stress occurs, the virus-infected plants do better than the virus-free plants (nonsymbiotic).

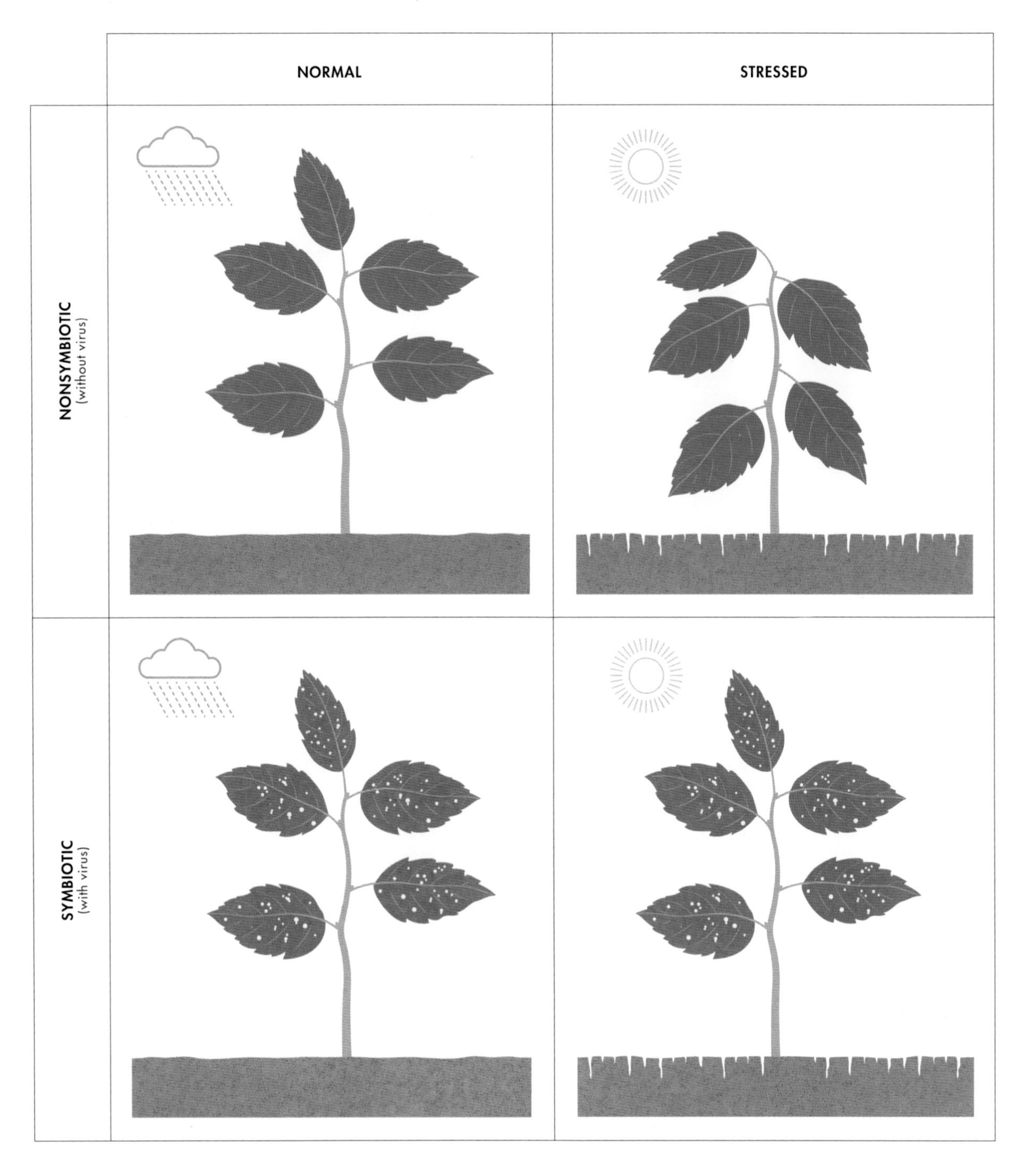

Aphids are small insects that feed on plants. They are damaging to the plants in two ways: they feed on the plant tissue, taking up the sugars that are meant to nourish the plant; and they carry viruses and other microbes that can cause disease. Peppers (*Capsicum annuum*) are infected by a persistent virus that is transmitted only vertically (see page 238). Plants that carry the virus aren't as attractive to aphids as virus-free plants, and the aphids feeding on them don't replicate as well as those feeding on uninfected plants. The virus doesn't cause any disease in the plant, so this is a purely mutualistic relationship.

The Rosy Apple Aphid (*Dysaphis plantaginea*) is affected by a virus that causes the insect to grow wings. The virus-infected winged aphid is smaller than its virus-free counterpart and doesn't reproduce as well, so in general the virus is a pathogen. However, when an aphid colony on a plant becomes very large, there is an advantage to growing wings as these allow the insects to move more easily to fresh plants. When a winged aphid lands on a plant, it deposits some of the virus into the plant tissue. The virus doesn't replicate in the plant and is not passed on to the aphid's offspring, but when the aphid population on the plant increases, the odds increase that a new aphid nymph will acquire the virus from the plant tissue. This causes it to grow wings so that it can move off to a new food source, and so the cycle starts again.

↖ A population of Rosy Apple Aphids (*Dysaphis plantaginea*) with both winged and non-winged morphs. The winged (virus-infected) aphids become more abundant in the fall, when the colonies move from apples to overwinter on plantains.

↑ The Pea Aphid (*Acyrthosiphon pisum*) has a mutualistic relationship with a bacteriophage.

Aphids can have other complex mutualistic relationships with viruses. The Pea Aphid (*Acyrthosiphon pisum*) carries a gut bacterium that produces a toxin. This toxin protects the aphid from a parasitoid wasp that lays its eggs in the aphid by killing the wasp larva. However, it isn't actually the bacterium that makes the toxin, but a bacteriophage that infects it.

VIRUSES IN NATURAL GERM WARFARE

Bacterial viruses are often integrated into the genome of the host, and bacteria with these integrated viruses are immune to getting infected by the same or related viruses. These viruses sometimes move out of the genome to replicate and kill the host, such as when the bacteria encounter competitors in their environment. The virus remains in the integrated state in most bacteria in a population, but a few will excise the virus from their genome to produce hundreds of copies of it, which can then infect and kill competitors. This allows the host to invade new territories. Archaea use a similar strategy to get rid of competitors.

Yeasts are also affected by killer viruses, but these don't directly infect competing yeasts. Instead, the viruses make a lethal toxin. The yeast that is infected with the virus is immune to the toxin, but uninfected yeasts are rapidly killed, leaving all the nutrients in the environment available for the infected yeast (see page 246).

Viruses have been involved in the expansion of human populations into new environments too. When Europeans colonized other continents, including Australia and the Americas, they took their viruses with them. The indigenous populations in these places were completely susceptible to these viruses because they had never encountered them before and had no immunity to them. The common cold virus, measles, influenza, and other viruses proved lethal to the local populations. It is estimated that 90 percent of the indigenous peoples of the Americas died within a decade of the arrival of the Spanish, due to warfare and disease.

↓ Indigenous Americans died in huge numbers from smallpox after the arrival of Europeans.

A killer bacterial virus

Killer phages help their bacterial hosts invade new territories. Bacteria may have the genome of a virus hidden in their own genome. If the bacteria are invaded by other bacteria, a small number will release the virus in their genome to kill off the invaders. Most of the original bacteria survive, but the invaders are killed.

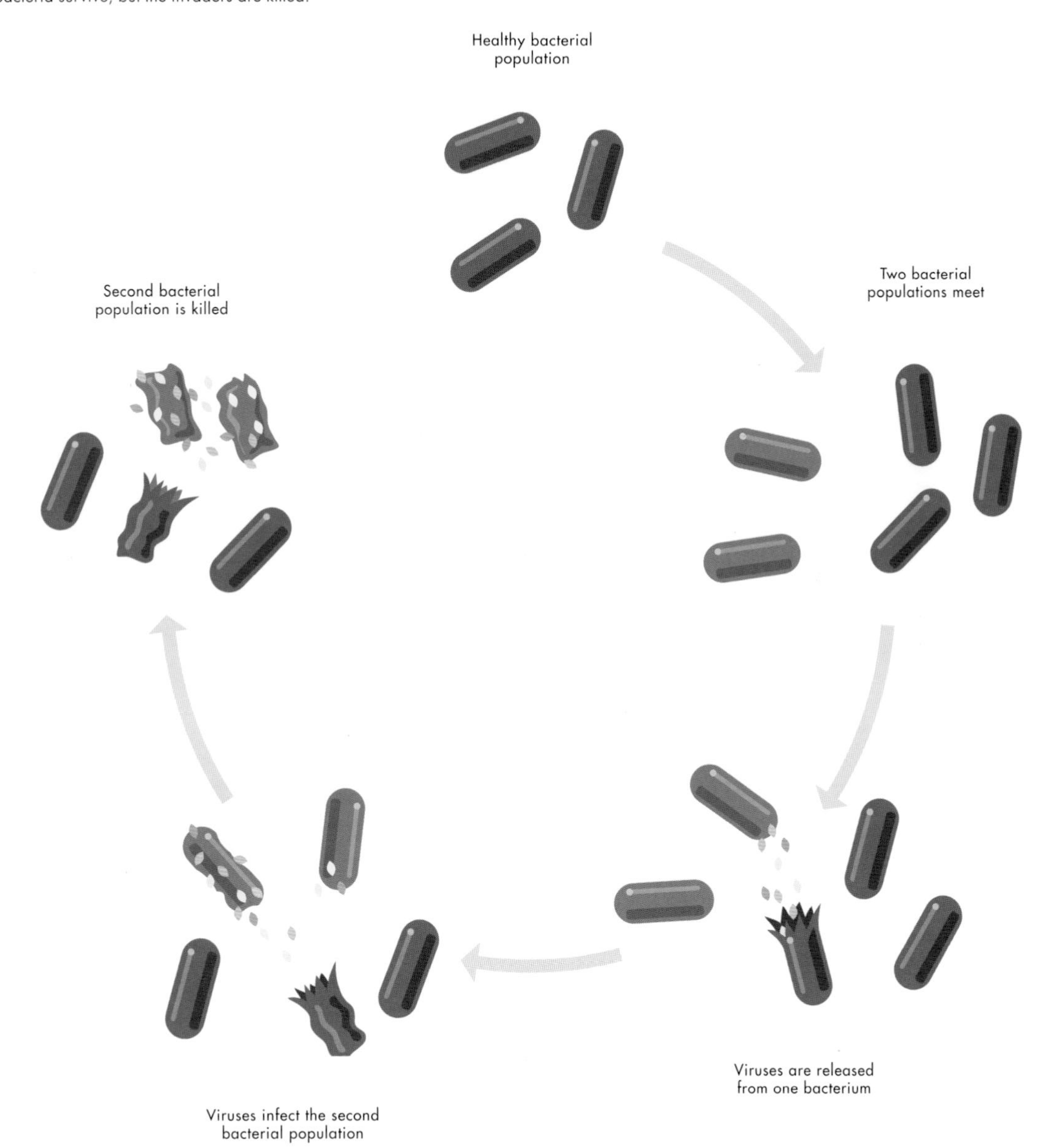

Exploiting viruses to control pathogens

The eastern portion of the United States was once covered with forests dominated by American Chestnuts (*Castanea dentata*). These huge, stately trees were not only beautiful, but they were also important in providing wood for construction. Chestnut wood is extremely durable and can be found in most houses in the region built toward the end of the nineteenth century.

VIRUSES AND CHESTNUT TREES

Around 1904, a fungal pathogen accidentally imported on an Asian chestnut tree was found in an American Chestnut at the New York Zoological Garden. Within a few years the mighty American Chestnut forests began to die off, and by 1950 most were gone, done in by the deadly chestnut blight. The disease was introduced to Europe in the 1930s, but by the 1960s there were reports of European Chestnuts (*Castanea sativa*) in Italy that were recovering from the disease. The recovering trees were still infected by the fungus, but it wasn't killing them. What was the difference? It turned out that the European fungus was infected with a virus. This virus could be transferred to uninfected fungus in culture, provided they were closely related, and the experimentally infected fungus didn't kill the trees. This virus–fungus pair was released in forests in Europe, and as a result chestnut blight is now mostly controlled on the continent.

Back in the United States, scientists have conducted many experiments into the virus and fungus in the hope of restoring the American Chestnut forests here, but so far they have not been successful. They have been able to save one tree at a time, by isolating the fungus from the infected tree, growing it in a lab and introducing the virus, and then returning the virus-infected fungus to the tree. The virus then spreads through the fungus on the tree, and the tree is saved. The catch is that, unlike in Europe where the fungus is all quite similar, there are many different strains of the fungus in the United States and the virus can't be transmitted among these in forests. There is still hope that an engineered version of the virus that can be more easily transmitted could restore the American Chestnut, but this hasn't yet been widely deployed in forests. As Asian chestnuts are resistant to the disease, botanists have also developed breeding programs to introduce the resistance to the American Chestnut, but trees have a long generation time and breeding programs are very slow.

↗ The chestnut blight fungus (*Cryphonectria parasitica*) killed off huge chestnut forests in the eastern United States. The devastation is seen in this 1930 photo of a forest in Chattahoochee National Forest in Georgia.

→ A healthy American Chestnut tree in Tennessee in 1915, before the blight hit.

BACTERIOPHAGES

In the early twentieth century two independent scientists studying bacteria discovered that their cultures sometimes developed holes in the bacterial lawn growing on a petri dish (see page 196). If they removed fluid from the holes and added this to other bacteria, they found that this produced the same types of holes. It turned out that they had discovered bacteriophages, the viruses of bacteria. Viruses don't actually eat bacteria (the literal meaning of bacteriophage), but they can kill them, as discussed above. Scientists soon recognized that using viruses that could kill the bacteria was a potential way of combating bacterial infections. They introduced viruses to a boy to cure him of dysentery and, later, to treat cholera, bubonic plague, and other bacterial diseases. However, politics and the discovery of penicillin eclipsed the idea of using phages to cure bacterial disease, although the work remained ongoing in the Soviet Union. Recently, this idea has been revived, fueled by the increasing occurrence of antibiotic-resistant bacteria. Use of phage therapy in agriculture could help to control plant pathogens such as bacterial wilt or animal infections such as Salmonella in poultry. There have been numerous experimental studies on using phages to combat plant pathogens, and there has also been successful experimental work with phage therapy in humans who are infected with lethal antibiotic-resistant bacteria. In the future this could become a standard treatment for bacterial infections.

→ *Rhizoctonia solani* is a fungal pathogen of grains, shown here on rice (left), while fire blight (*Erwinia amylovora*) is a bacterial disease in apple trees (right). Each of these diseases may possibly be controlled by a virus.

↓ Artist's rendition of a bacteriophage infecting a bacterial cell. The virus lands on the bacteria, injects its genome, and replicates rapidly, ending in lysis of the bacteria cell and the release of hundreds of progeny viruses.

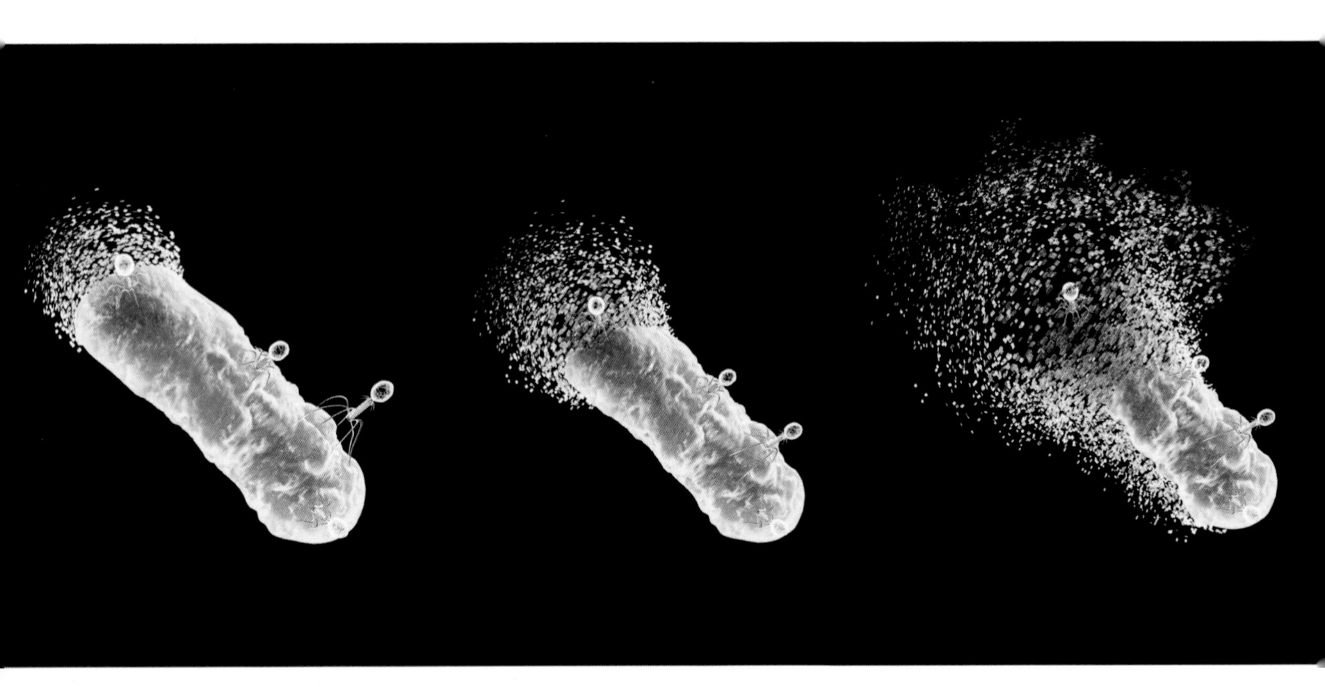

VIRUSES AND CANCER TREATMENT

Using viruses as vaccines was discussed on page 152, but they are also being explored in the treatment of various types of cancer. Benign viruses can act as delivery tools for genes that can help combat cancer, or they can be engineered to kill only cancer cells but not healthy cells. In recent years viruses have also been used to deliver other genes, such as the gene to make a retinal pigment in people who have a genetic disorder that leads to blindness. The hope is that they may be able to deliver genes to combat other genetic diseases too, including sickle-cell anemia and cystic fibrosis.

Knowledge about beneficial viruses lags far behind what we know about the pathogens. While it isn't surprising that pathogens have been studied in more detail because of their impacts on human health and the health of domestic plants and animals, there may also be an element of fascination on the part of humans for bad news that has resulted in the neglect of the "good" viruses.

PCV-1

Pepper cryptic virus 1

Cryptic virus with a visible impact

GROUP	III
FAMILY	Partitiviridae
GENUS	Deltapartitivirus
GENOME	Linear, two-component, double-stranded RNA comprising about 3,000 nucleotides, encoding two proteins
VIRUS PARTICLE	Non-enveloped, icosahedral
HOSTS	Jalapeño and related peppers (*Capsicum annuum*)
ASSOCIATED DISEASES	None
TRANSMISSION	Strictly vertical

Pepper cryptic virus (PCV-1) is a vertically transmitted plant virus that infects all jalapeño peppers. These viruses are often called cryptic, from the Latin word for "hidden," because they don't cause any symptoms in infected plants and are usually found at very low levels. They infect their hosts for many generations by being passed on to all the offspring.

Despite being a cryptic virus, PCV-1 does have an important and very visible impact on its plant hosts: it deters aphids. These insects are a menace for plants, carrying many viral diseases and damaging the plants by feeding on the sugars they make for their own growth. Many plant viruses have a close relationship with the insects that transmit them, inducing their plant hosts to make compounds that attract these insects. PCV-1's effect of deterring aphids hasn't been found in any other plant virus. Another unique feature of this virus is that aphids feeding on pepper plants infected with PCV-1 do not reproduce as well as those feeding on uninfected plants. Hence the virus has a double effect, decreasing harmful aphids.

Jalapeño peppers are a variety of the species *Capsicum annuum*, which includes many domestic peppers. It is estimated that *C. annuum* was domesticated about 10,000 years ago from the wild chiltepín pepper. Chiltepín is found throughout Mexico, and wild isolates are also infected with PCV-1. Since the virus is transmitted only vertically through the pollen or ovum, the virus has probably been in peppers for at least 10,000 years. PCV-1 has the slowest mutation rate of any known virus, and there is almost no variation between the jalapeño and chiltepín strains.

→ Cryogenic electron microscopy image of pepper cryptic virus. Because the virus has such a low copy number in infected plants, it took more than 1 kg of leaves to obtain enough virus to generate this image.

CThTV

Curvularia orthocurvulavirus 1

Virus that confers heat tolerance to its host and its host's hosts

GROUP	III
FAMILY	Curvulariviridae
GENUS	Orthocurvulavirus
GENOME	Linear, two-segmented, double-stranded RNA comprising about 4,100 nucleotides, encoding five proteins
VIRUS PARTICLE	Non-enveloped, small icosahedral
HOSTS	*Curvularia protuberata*
ASSOCIATED DISEASES	None—beneficial
TRANSMISSION	Vertical

Curvularia orthocurvulavirus 1 is more commonly known as Curvularia thermotolerance virus (CThTV) and was responsible for a renewed interest in beneficial viruses. It's an important example of multiple layers of interactions among viruses and their hosts.

CThTV was discovered in Yellowstone National Park in Wyoming, in a fungus that was colonizing Hot Springs Panic Grass (*Dicanthelium lanuginosum*) growing in geothermal soils. Most plants can't tolerate very high soil temperatures, but the panic grass has adapted to heat because it is colonized by the fungal endophyte *Curvularia protuberata*. In turn, the fungus is infected with CThTV. All three partners are required for this thermal tolerance to work. The fungus can be grown in culture, but not at high temperatures, and the plant can be colonized by a virus-free isolate of the fungus, but it then loses its thermal tolerance. In experiments in a laboratory, this heat tolerance also worked when the virus-infected fungus was transferred to Tomato plants (*Solanum lycopersicum*), showing that the virus has a very broad effect.

Since the discovery of CThTV in 2007, several other related viruses have been found in fungi, although none have its unique property of thermal tolerance. Plants in other parts of Yellowstone are colonized by similar fungi, but unless they are growing in geothermal soils the virus is not present. This mutualistic relationship is an example of how these kinds of interactions evolve. The property of heat tolerance was a happenstance: the virus somehow (accidentally) affected the expression of genes in the fungus and plant to increase heat tolerance. Once the relationship was established, there was a strong selection pressure for the plants to maintain the virus-infected fungus.

→ Hot Springs Panic Grass (*Dicanthelium lanuginosum*) grows in geothermal soils at temperatures well above what plants can normally tolerate.

MHV

Murid gammaherpesvirus 4

Latent herpesvirus that prevents infection from bacterial pathogens

GROUP	I
FAMILY	Herpesviridae
GENUS	Rhadinovirus
GENOME	Linear, single-component, double-stranded DNA comprising about 180,000 nucleotides, encoding more than 75 proteins
VIRUS PARTICLE	Enveloped, with a large icosahedral core
HOSTS	Mice (*Mus* species)
ASSOCIATED DISEASES	None in latent infections
TRANSMISSION	Direct contact

The mouse gammaherpesvirus 68 (MHV-68) strain of murid gammaherpesvirus 4 (MHV) is often used as a model for human pathogenic gammaherpesviruses. Mice make excellent models for studies that cannot be carried out in human hosts.

MHV-68 is a model for several human herpesviruses, including the Epstein–Barr virus, which causes mononucleosis, and Kaposi's sarcoma-associated herpesvirus, which causes cancer in immunosuppressed people. It is also closely related to human cytomegalovirus (HCMV). Herpesviruses often cause latent infections; the virus is found in the neural tissue of the host and replicates only slowly. Most herpesviruses are not pathogens in this latent state.

Mice with a latent infection of MHV-68 are resistant to infection by the bubonic plague bacteria, *Yersinia pestis*, and *Listeria monocytogenes*, a foodborne bacterial pathogen of humans. We don't hear much about the bubonic plague these days, but it is still a problem in some parts of the world. The Black Death was a type of bubonic plague that killed as much as 60 percent of Europe's population in the fourteenth century. There are usually a few cases of bubonic plague in North America every year, but most people are descendants of survivors of the plague, and are resistant to the bacterial disease.

MHV-68 induces an innate immune response (see page 163) in mice that is probably important in preventing infection by these bacterial pathogens. Levels of interferon (proteins that induce innate immunity) and macrophages (blood cells that devour invaders) are elevated in infected mice. In humans, HCMV prevents infection with HIV, which causes AIDS.

→ Ribbon diagrams of the ORF 52 (upper) and the capsid protein (lower) of mouse gammaherpesvirus 68. ORF 52 is involved with developing the structure of the virus inside infected cells.

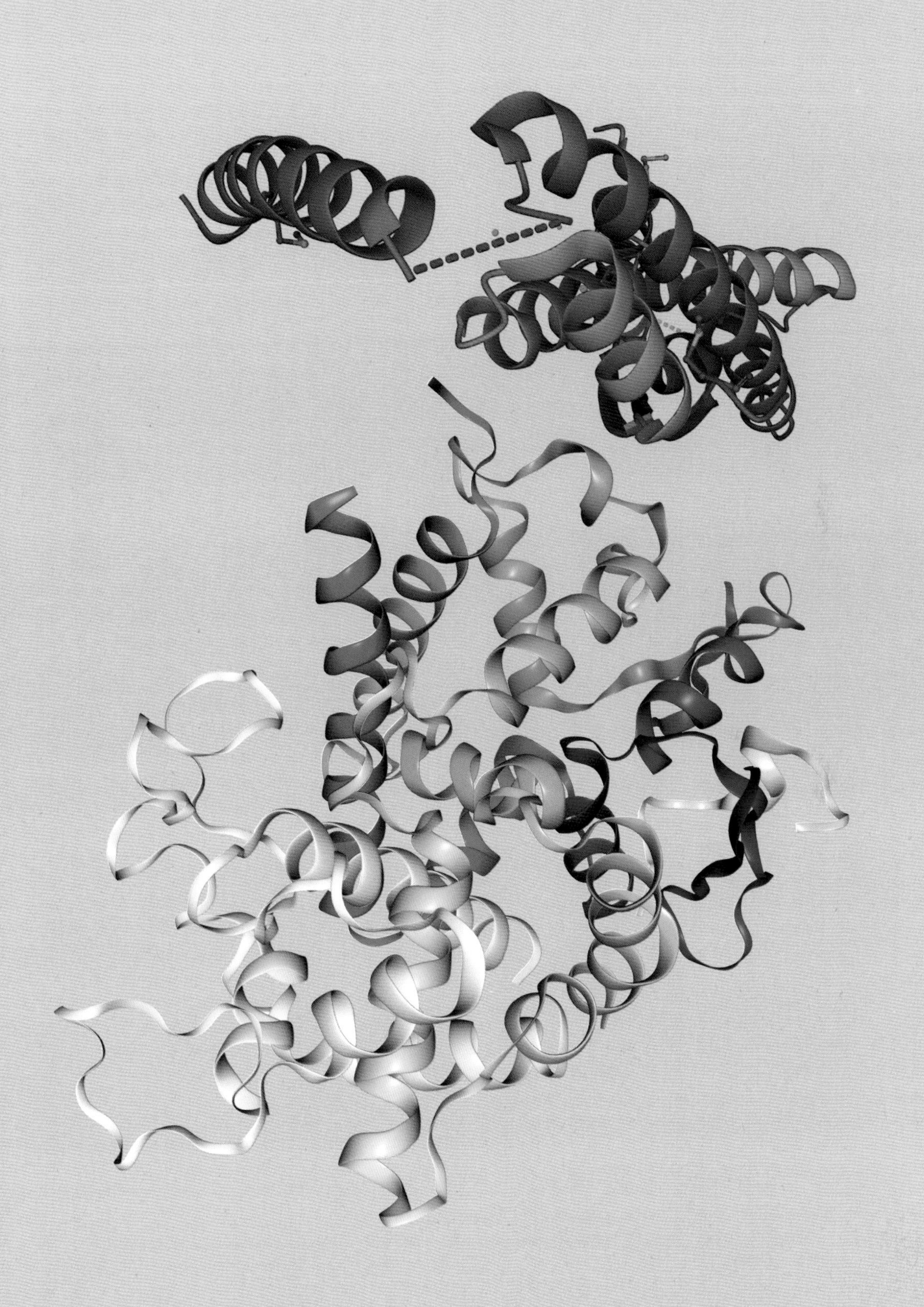

CTXφ

Vibrio virus CTXphi

Virus that triggers the cholera bacterium to cause disease

GROUP	II
FAMILY	Inoviridae
GENUS	Affertcholeramvirus
GENOME	Circular, single-component, single-stranded DNA comprising about 6,700 nucleotides, encoding nine proteins
VIRUS PARTICLE	Non-enveloped, flexuous rod
HOSTS	*Vibrio cholerae*
ASSOCIATED DISEASES	Cholera
TRANSMISSION	Waterborne

Cholera is a bacterial disease that has been a serious scourge on humans for centuries. It causes a watery diarrhea that leaves infected people severely dehydrated. Cholera was the first disease widely recognized to be transmitted in water, and cleaning up drinking water supplies has largely eliminated the disease from many parts of the world. However, it can also be found in raw shellfish and still occurs in areas where clean drinking water is not available.

Vibrio cholerae, the bacterium responsible for cholera, doesn't work alone and causes disease only when it is infected by the vibrio virus CTXphi (CTXφ). The virus encodes the toxin that is required for the bacterium to invade the human gut and cause disease. This relationship is clearly beneficial for the bacterium, as it allows it to invade the new niche of the human gut. Isolates of *Vibrio cholerae* that do not cause disease are found in many waterbodies, including Chesapeake Bay in the United States, but these don't contain the virus and so can't infect the human gut.

CTXφ isn't often found as a free virus. Instead, it is usually integrated into the bacterial genome in a state called lysogeny. Many bacterial viruses remain in this integrated state, where they replicate with the host genome and express relatively low levels of their proteins. When they are triggered to leave the genome and begin a lytic infection, they replicate to high levels and usually cause the host cell to burst open and release the infectious viruses into the environment—in this case water.

→ Computer-assisted model of an interaction between a CTXφ viral protein and a *Vibrio cholera* protein.

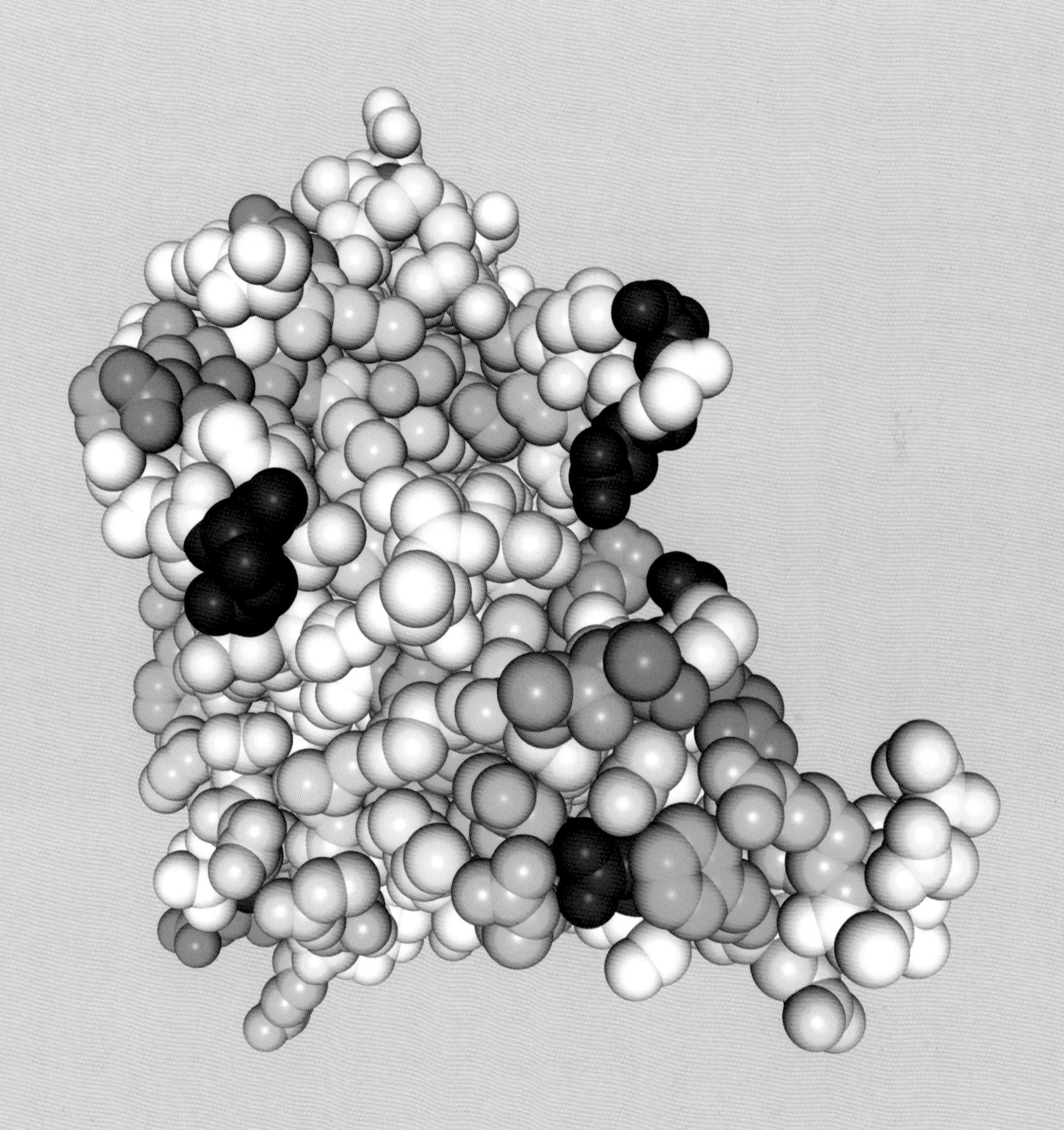

ScV-L-A

Saccharomyces cerevisiae virus L-A

Virus that allows its host to kill off the competition

GROUP	III
FAMILY	Totiviridae
GENUS	Totivirus
GENOME	Linear, single-component, double-stranded RNA comprising about 4,600 nucleotides, encoding two proteins
VIRUS PARTICLE	Non-enveloped, icosahedral
HOSTS	*Saccharomyces cerevisiae* yeast
ASSOCIATED DISEASES	None
TRANSMISSION	Vertical, yeast mating

In nature, yeasts are often found in an environment of competitors. However, if the yeast *Saccharomyces cerevisiae* is infected with Saccharomyces cerevisiae virus L-A (ScV-L-A) and one of its satellite RNAs, it can kill off its competitors with a potent toxin yet remains immune itself.

ScV-L-A is a helper virus for a satellite RNA (see page 49), encoding all the genes for replicating and encapsidating the satellite RNA. The toxin is encoded on the satellite RNA as a single polyprotein with five components. The first component tells the toxin where to locate itself in the cell and then is cleaved off. The remaining four components fold, so that two of them are brought together with a chemical structure called a disulfide bond, and then the other components are cleaved off. The remaining two bonded components exit the cell as a toxin that kills yeast cells. When the toxin gets into a sensitive yeast cell it goes into the nucleus, where it stops the cell cycle, killing the cell by preventing it from replicating.

The toxin can enter any yeast cell, but in cells that are infected with the virus it binds to the polyprotein that is being made by the satellite RNA. This deactivates the toxin, so that the infected yeast cell is immune to it.

Viruses like ScV-L-A, with double-stranded RNA genomes, keep their genome hidden inside host cells. This is true for all known double-stranded RNA viruses, probably because double-stranded RNA is often a trigger for many antiviral activities. Instead of being released from the virus particle, these viral genomes never leave their safe hiding place but instead simply exude single-stranded RNA that acts as messenger RNA and a pregenome inside the cells. The second strand of the genome is made inside the virus particle.

→ Structural model of Saccharomyces cerevisiae virus L-A deduced by x-ray crystallography data.

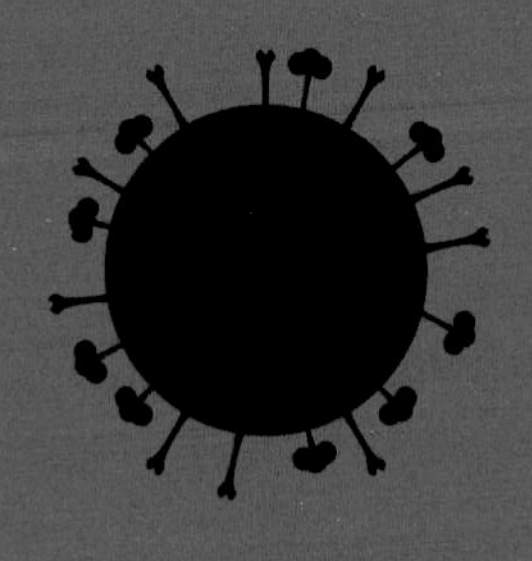

THE PATHOGENS

Introduction

Pandemic! In 2020 this word became one of the most common in the English language, and Merriam-Webster declared it the word of the year. A pandemic is a widespread infectious disease and differs from an epidemic in that it spreads around the world. Humans had not seen a lethal pandemic in 100 years—not since the 1918 influenza pandemic. Changes in human behavior, especially the huge increase in international travel, make pandemics more likely now than ever before. As we move around, we take our viruses with us, and we often also transport viruses of other animals or plants. This chapter describes three major viral human pandemics, and profiles viruses that have jumped species and spread worldwide.

Jumping species
Viruses may be endemic in their native host and are often asymptomatic. This diagram illustrates how viruses in wild plants may occasionally spill over into a domestic host when agriculture is close to wild areas. In most cases this is a "dead end" infection, but on rare occasions the virus may spread in the domestic host or jump species.

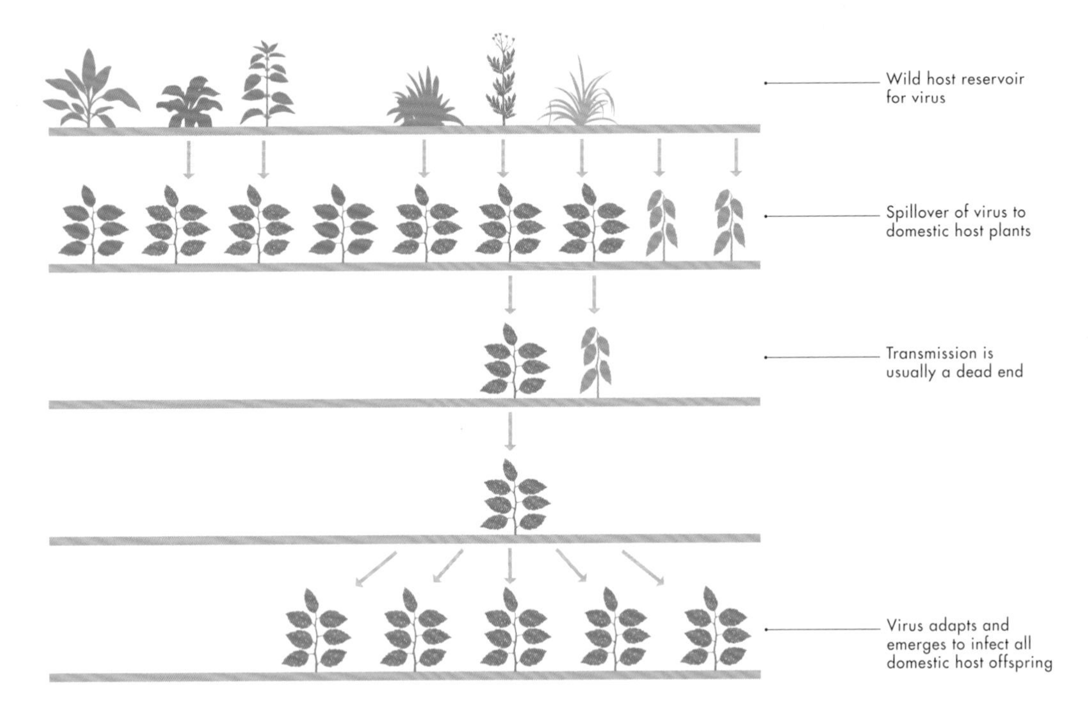

VIRUS	GEOGRAPHICAL EMERGENCE	DATE FIRST REPORTED	ORIGIN OF CROP PLANT	SOURCE OF VIRUS
Cassava mosaic geminiviruses	East Africa	1894	South America	Unknown
Maize streak virus	Africa	1928	Central America	Wild native grasses
Tomato yellow leaf curl virus	Israel	1930s	South America	Infects many wild hosts; origin unclear
Sugarcane yellow leaf virus	Southern United States, Central and South America	1994	Southern Asia	Host unknown, but originated in Colombia
Pepino mosaic virus	Peru, but has emerged in Solanaceae around the world	1980	South America	Native Pepino (*Solanum muricatum*)
Tomato torrado virus	Spain	1996	South America	Unknown; infects many Solanaceae
Iris yellow spot virus	Brazil	1981	Worldwide	Unknown, but common in weeds
Plum pox virus	United States[a]	1999	China	Unknown; may have arrived from Europe on nursery stock
Wheat mosaic virus[b]	United States	1993	Turkey	Unknown but also found in native maize (*Zea mays*) crops

EXAMPLES OF PLANT VIRUSES IN CROPS GROWN IN NON-NATIVE REGIONS

[a] Widespread in Europe; emerged in the eastern United States in 1999.

[b] Also called High Plains virus.

Most viruses adapt to their hosts, balancing their own survival with the immune response of the host. However, a virus will occasionally infect an organism that isn't its normal host. This usually involves only one individual, which may become sick but doesn't pass the virus on to any other hosts. To jump to a completely different new species, a virus has to evolve to overcome numerous barriers to infect the new host, and then an additional set of barriers to leave the host and be passed on to others (see page 148). Viruses may be jumping species more frequently because humans are now moving into areas that have previously been mainly inhabited by wildlife. In plant viruses, many species-jumping events have occurred when plants have been moved from their native environment to a new part of the world, where they encounter novel viruses. Pandemics can result after a species jump, and they have occurred in plant, animal, and human viruses.

Influenza

While the 1918 influenza pandemic is best known, there were a number of earlier influenza epidemics and pandemics. The first use of the word to describe a disease was in fourteenth-century Italy, and it derives from the Latin *influentia*, meaning "influence." The first documented influenza pandemic began in 1580 in Asia, spreading to Europe and, eventually, the Americas. There were two influenza pandemics in the seventeenth century and two in the eighteenth century, but it was the 1918 event that gained particular notoriety—until it was displaced by coronavirus disease 2019 (COVID-19).

1918 INFLUENZA PANDEMIC

Although viruses had been recognized for about 20 years before the 1918 pandemic, the viral nature of influenza was still unknown. Contrary to the popular name Spanish flu, the outbreak didn't begin in Spain. It simply seems likely that Spain had more complete press coverage of cases at the time because it wasn't involved in the First World War, and this may have led to the misconception. The first reported cases were in early March 1918 in Fort Riley, Kansas, among army recruits. From there it spread throughout military camps in the Midwest and southeastern United States. The soldiers headed to Europe to fight, carrying the virus to France in April 1918, and from there it spread throughout Europe. The initial wave of the virus wasn't as lethal as later waves. The second wave was first reported in the port city of Brest, France, in August 1918, and from there it spread to many parts of the globe. It reached dockworkers in Sierra Leone on a British ship in the same month, from where it spread throughout Africa and to Asia, and eventually reached Australia in 1919. The second wave returned to North America from Europe and spread there throughout the fall and winter of 1918. This variant was much more lethal. The virus in South America may have arrived independently from Europe or Africa.

The spread of the virus was greatly enhanced by the movement of troops fighting in the First World War, by the recent construction of railroads, and by the widespread use of steamships throughout the world. The Trans-Siberian Railway was responsible for a lot of the movement from Europe to Asia, while ocean shipping carried the virus across much of the rest of the world. By January 1919 almost nowhere in the world was free of influenza.

↗ The Trans-Siberian Railway aided the spread of influenza across eastern Europe in the 1918 pandemic.

→ During the 1918 influenza pandemic, hospitals were filled to capacity with patients suffering from the disease and makeshift units had to be set up to accommodate the sick.

Yaroslavl
Moscow
Kirov
RUSSIA
Perm
Ekaterinburg
Tyumen
Skovrodino
Khabarovsk
Omsk
Krasnoyarsk
Tayshet
Belogorsk
Novosibirsk
Ulan-Ude
Chita
Irkutsk
Vladivostok

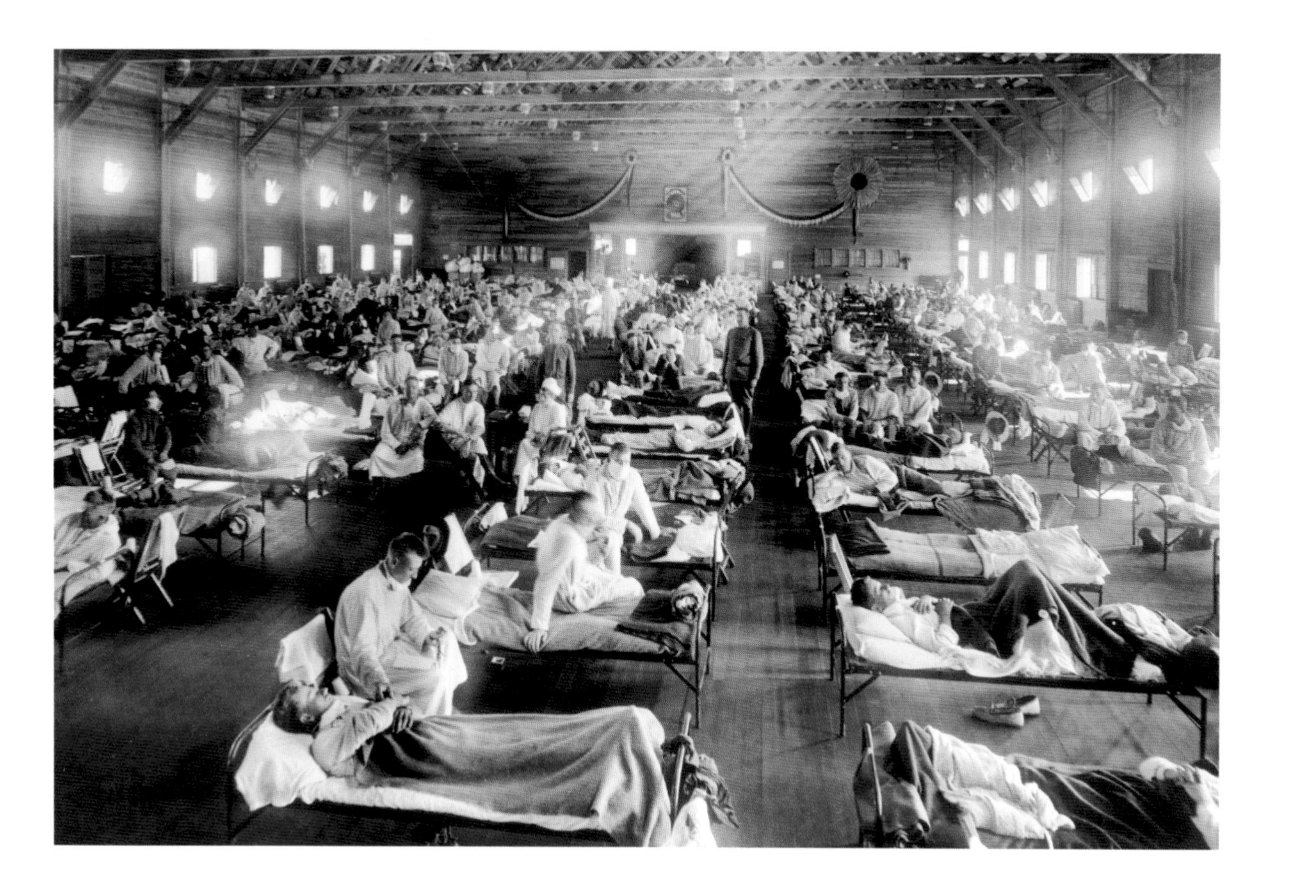

It is estimated that at least 500 million people were infected worldwide, or about one in three people. More than 20 million people are reported to have died, although these numbers are almost certainly lower than the reality. One estimate from India is that 15 million died in that country alone, and some estimates put the worldwide death rate as high as 50 million. In the United States more people died of influenza than in the First and Second World Wars combined. In many families there was no one to care for the sick because everyone was infected, although younger children generally had milder symptoms.

↖↑ Although the viral nature of influenza wasn't known in 1918, the respiratory nature of the infection was understood. Posters and newspapers carried notices advising against spitting and many people donned masks.

↗ The Red Cross Emergency Station in Washington, DC, during the 1918 influenza pandemic.

The author's mother was five years old when the second wave hit the United States, and she recalled that only one of her adult relatives was spared. He spent all of his time going from home to home, delivering food and providing care, and then in the end, when the others were finally recovering, he succumbed too. Many people in their 20s to 40s were affected, and there were many more deaths than in previous pandemics. Why was this outbreak so lethal? One reason may be that it was a new strain for humans, one to which the population at the time had no immunity. Many people contracted secondary bacterial infections such as pneumonia and died from these because antibiotics had yet to be developed. The two subsequent influenza pandemics of the twentieth century, in 1957 and 1968, were much milder.

There was no dramatic end to the 1918 pandemic. There was a spike in cases in the winter of 1920, but there weren't many after that. Most pandemics end when sufficient numbers of people are immune to the virus such that it can't find enough hosts to continue to spread. This is sometimes called herd immunity, but the details and nuances of this phenomenon are not well understood.

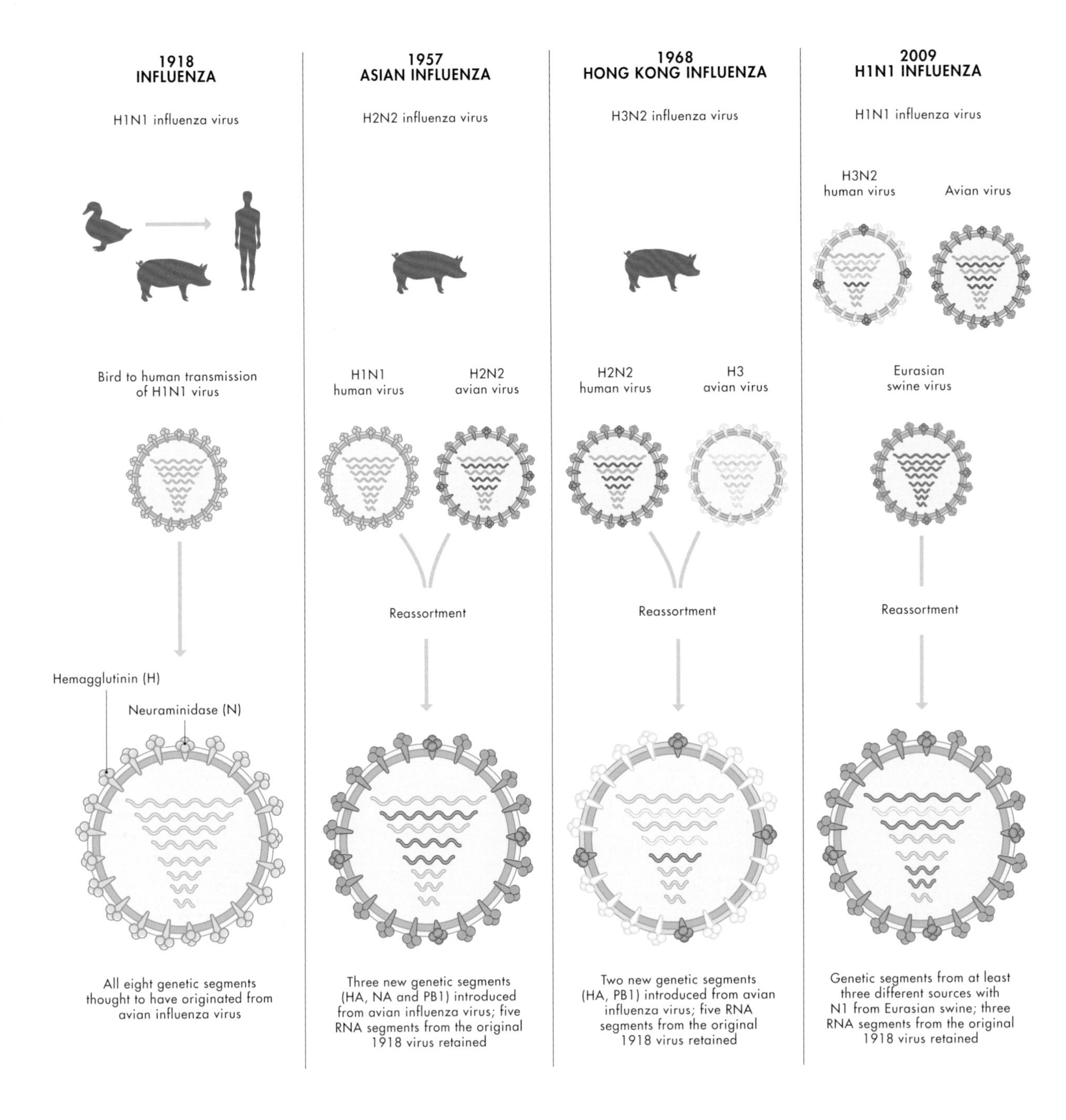

Influenza virus reassortment to produce pandemic strains

The genetics of influenza virus are complex. It has eight different segments, each encoding a different protein. Like all RNA viruses, influenza gradually evolves over time, giving rise to strain variation every year. This is called genetic drift. When two different strains of influenza infect a single host, this can lead to genetic shift and a reshuffling of the eight segments. These new strains have the potential to cause a pandemic because the human population may lack immunity to them. Influenza strains are usually named for just two segments, H and N. This is because the H and N proteins are on the surface of the virus and the majority of the host's immune response is against them.

UNDERSTANDING THE FLU VIRUS

In 2005 the complete nucleotide sequence of the 1918 virus was determined. This provided a lot of information about the early history of the pandemic strain. All of the segments of the virus originated in birds, but the virus was infecting a mammalian host for several years before it emerged in humans. The most likely candidate is pigs. It was also infecting humans for some time, perhaps a few years, before it emerged as a pandemic strain.

The most important proteins for the immune response to flu are hemagglutinin (H) and neuraminidase (N). These proteins are found on the surface of the virus, and are important for enabling it to enter cells. Most flu strains are designated by these two proteins; the 1918 flu was H1N1, and the two pandemic strains in the mid-twentieth century were H3N2 and H2N3, which is still circulating. In 2009 a new pandemic strain emerged that was also H1N1. Most of the segments were from the circulating strains, but the H and N came from a swine flu virus. There was initially a lot of concern because the strain had the same H and N serotypes as the 1918 flu, but it turned out to be milder than the other strains circulating at the time and didn't displace them.

↓ A Northern Gannet (*Morus bassanus*) suffering from bird flu in the United Kingdom. During 2022 tens of thousands of seabirds in the North Atlantic died from the disease, particularly in dense colonies in the north of the UK.

INFLUENZA IN BIRDS

All influenza originates in birds. The virus is endemic in migrating waterfowl, but these strains can't usually infect humans and need to go through another host. However, in a few outbreaks the virus has infected humans directly from domestic birds, when it is often called bird flu. Bird flu is very severe, with extremely high mortality rates, but none of these strains have acquired the ability to be transmitted from human to human, and it is unlikely they will. They are particularly severe because they infect cells very deep in the lungs. This deep locus of infection is also the reason why they can't be transmitted—for transmission, a respiratory virus needs to be present in high levels in the upper respiratory track.

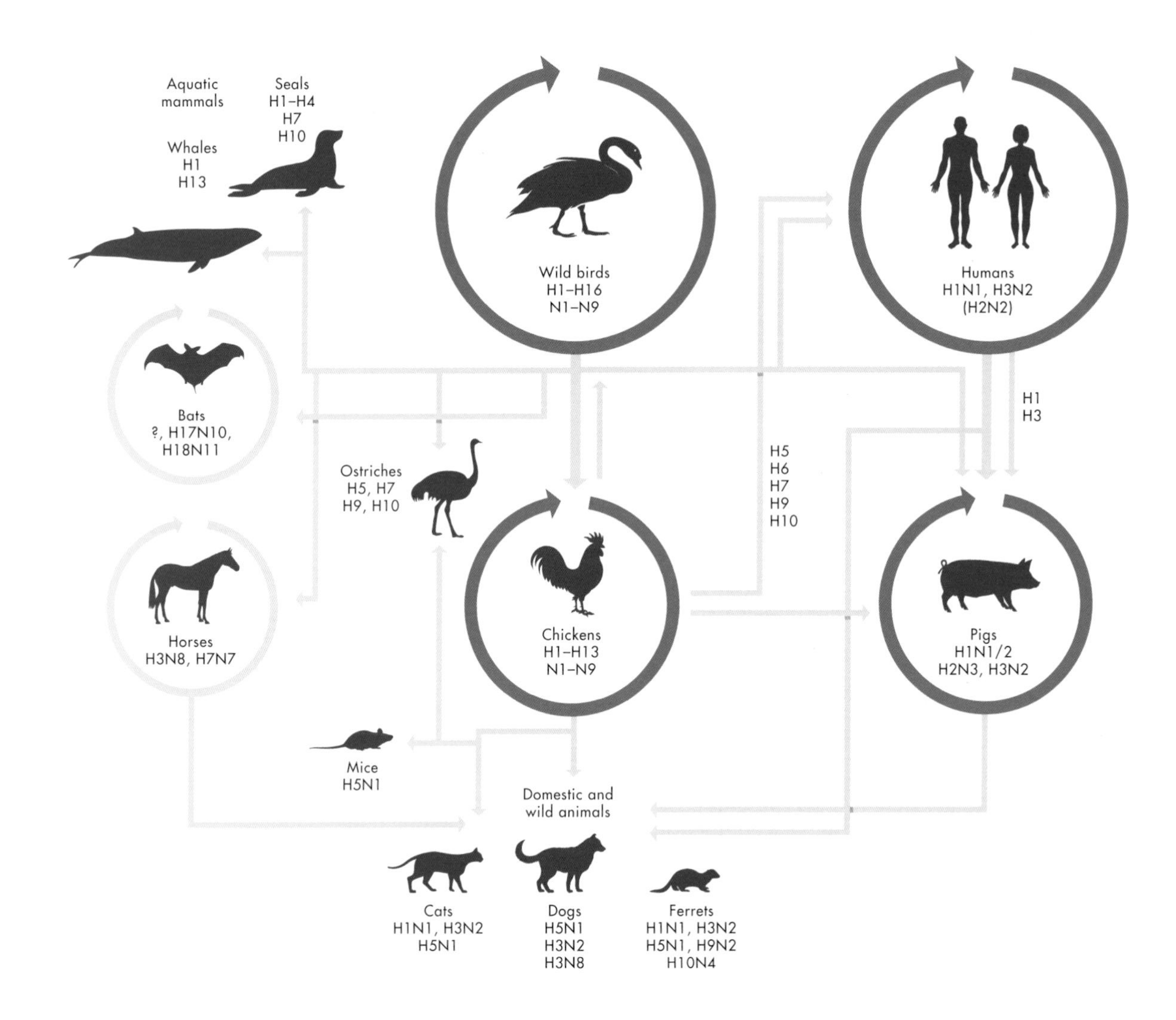

Cycles of influenza virus transmission

All influenza viruses originate in wild waterfowl, which lack symptoms when infected. The virus can infect domestic birds and many mammals, with different serotypes found in different hosts. Human hosts can't get influenza directly from birds, and instead the virus passes to them through other animals. Pigs are the most common intermediate host, and have been so for all the influenza pandemic strains over the past 100 years.

Severe acute respiratory syndrome (SARS)

In late 2002 a novel type of pneumonia was reported in Guangdong, China. Another case was reported in early 2003 in Hanoi, Vietnam, and the World Health Organization (WHO) officer who examined the patient died within six weeks. The disease was severe, and the mortality rate was high. In early March 2003 the virus emerged in a hotel in Hong Kong, and from there it spread around the world. The WHO issued a global alert. By April, a Canadian research group published the sequence of the virus, named severe acute respiratory syndrome coronavirus (SARS-CoV).

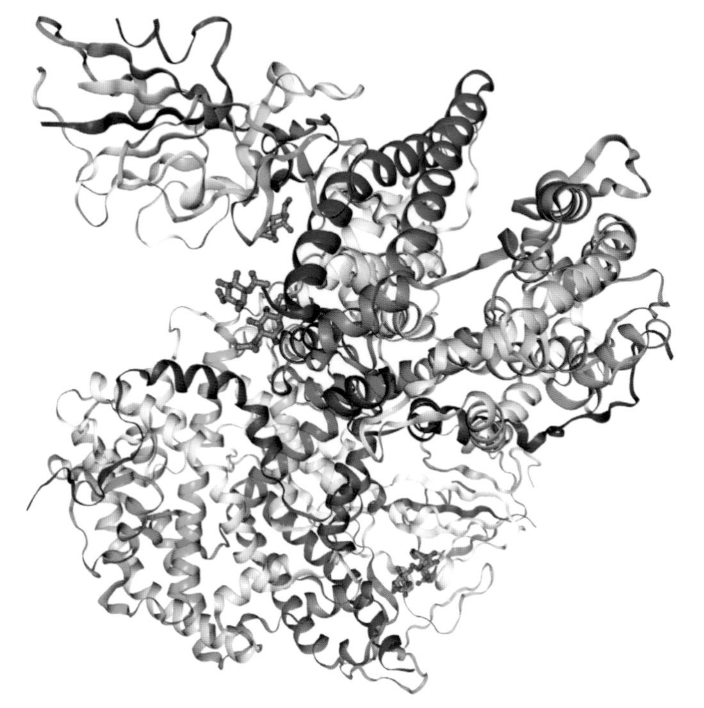

Genetic analyses of the virus indicated that it had originated in bats and had passed through an intermediate host, civets, before infecting humans. By June the pandemic was essentially over. In the end, just over 8,000 people had been infected and nearly 700 died, a mortality rate of about 9 percent.

← Crystal structure of spike protein receptor-binding domain from the 2002–2003 SARS coronavirus civet strain complexed with human-civet chimeric receptor NGL.

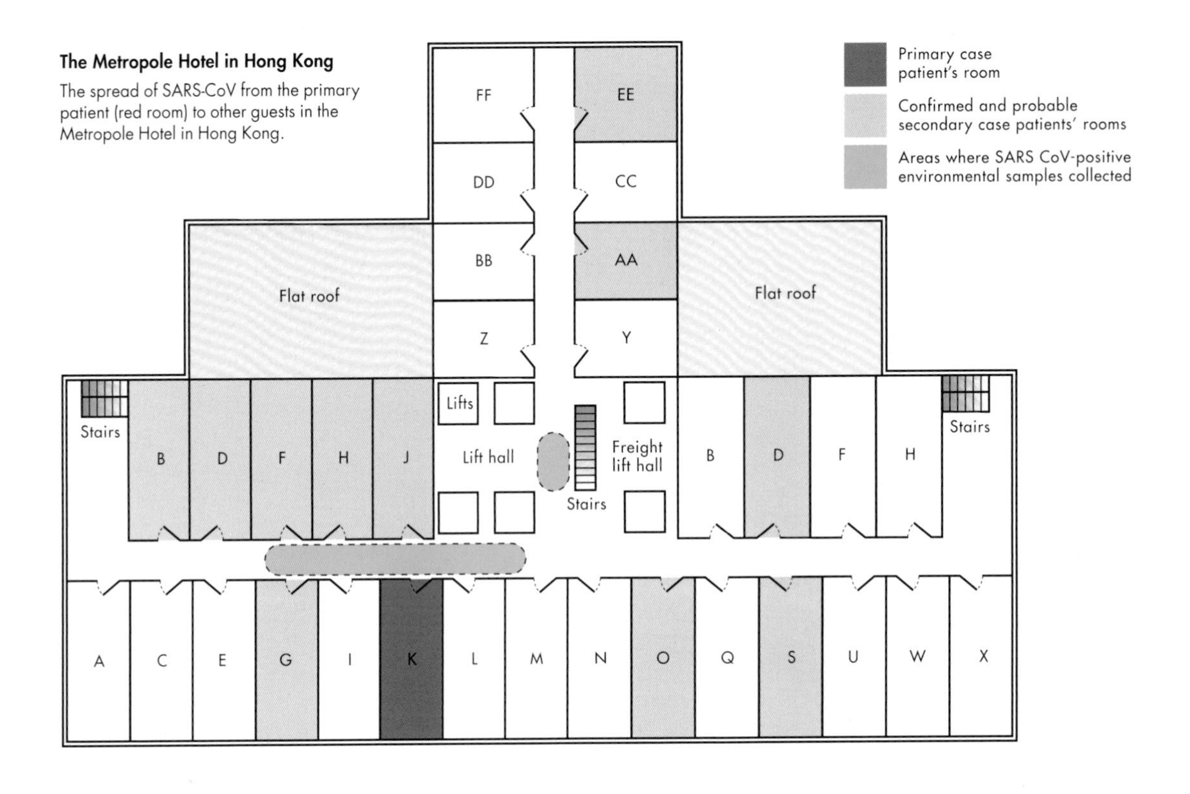

The Metropole Hotel in Hong Kong
The spread of SARS-CoV from the primary patient (red room) to other guests in the Metropole Hotel in Hong Kong.
Primary case patient's room
Confirmed and probable secondary case patients' rooms
Areas where SARS CoV-positive environmental samples collected
FF
EE
DD
CC
BB
AA
Z
Y
Flat roof
Flat roof
Stairs
Lifts
Lift hall
Freight lift hall
Stairs
Stairs
B
D
F
H
J
B
D
F
H
A
C
E
G
I
K
L
M
N
O
Q
S
U
W
X

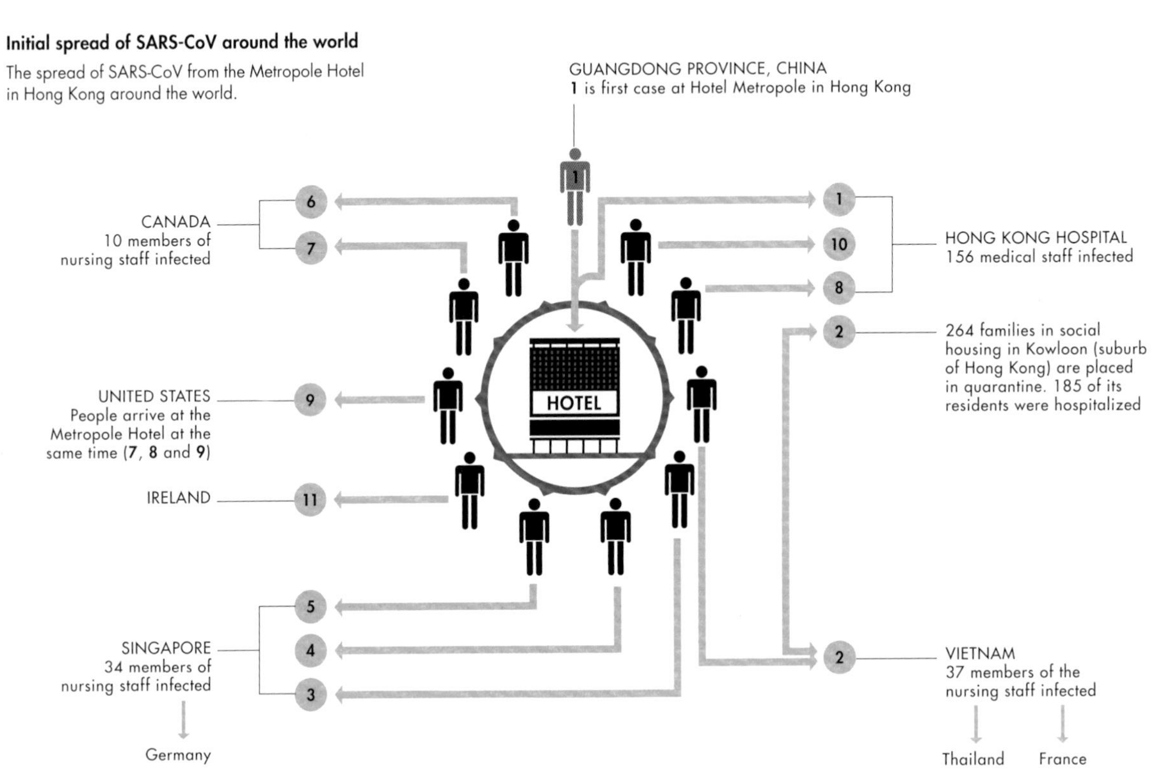

Initial spread of SARS-CoV around the world
The spread of SARS-CoV from the Metropole Hotel in Hong Kong around the world.
GUANGDONG PROVINCE, CHINA
1 is first case at Hotel Metropole in Hong Kong
HOTEL
CANADA
10 members of nursing staff infected
HONG KONG HOSPITAL
156 medical staff infected
264 families in social housing in Kowloon (suburb of Hong Kong) are placed in quarantine. 185 of its residents were hospitalized
UNITED STATES
People arrive at the Metropole Hotel at the same time (7, 8 and 9)
IRELAND
SINGAPORE
34 members of nursing staff infected
Germany
VIETNAM
37 members of the nursing staff infected
Thailand
France

In 2012 another coronavirus disease emerged, called Middle Eastern respiratory syndrome (MERS). This spread slowly in the Middle East and a few sporadic cases have been found in travelers, but it hasn't become widespread. However, it has a mortality rate above 25 percent, and so has remained a cause for concern. This virus also originated in bats, with camels being the most common intermediate host.

COVID-2019

In late 2019 another new coronavirus disease emerged, named COVID-19 and caused by the SARS-CoV-2 virus. This is related to SARS-CoV, but almost certainly didn't evolve directly from that virus. It also has an apparent origin in bats, but intermediate hosts, although likely, are not known. SARS-CoV-2 can infect a large spectrum of animals, both wild and domestic, which may act as reservoirs. The overall mortality rate for this virus had dropped to about 1.5 percent by early 2022, although it was higher at the start of the pandemic, especially before vaccines became available. Even so, this figure is lower than for many previous human pandemics of the past century, and lower than the 1918 influenza pandemic, which had a mortality rate between 4 percent and 10 percent.

In a short time, SARS-CoV-2 has become the most thoroughly studied virus in the world. It has a number of unique features. On the outside of the virus is the spike protein that allows it to attach to host cells, but there is also an unusual amount of sugar. Viruses sometimes use sugar molecules to hide from the host immune system. Once the spike has bound to its

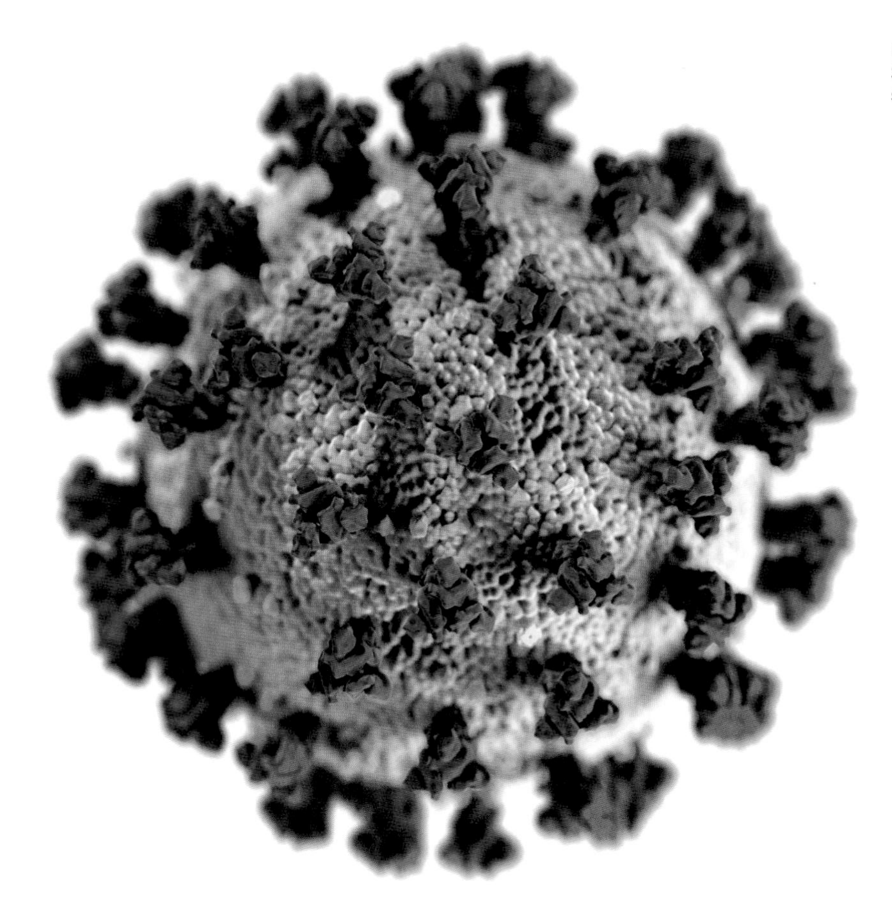

↙ Structural model of SARS-CoV-2, showing the spike proteins in red.

receptor on the host cell, it uses host proteins to fuse with the host plasma membrane and enter the cell. Once inside, it then suppresses the cell's ability to make RNA and takes over the cell's machinery to copy its own RNA. It shuts down about 70 percent of the cell's ability to make proteins, turning this machinery over to making viral proteins. The infected cells lose their ability to alert the immune system.

Once the SARS-CoV-2 virus has taken over the host cell, it induces it to form a fatty coating that makes cells fuse. This kind of cell fusion is normal in some cell types such as muscles, but it is a problem in lung cells, where large structures form from many fused cells. It is likely that these structures allow the virus to replicate even more efficiently. Most of these events occur with other viruses, but SARS-CoV-2 seems to combine all of them in one infection, and they seem to occur at a much faster rate. On its way out of the cell, the virus uses a different mechanism from other coronaviruses, leaving via cell structures that usually export waste. This isn't a very efficient mechanism and it isn't clear why the virus uses it.

There have been several waves of SARS-CoV-2, with major strains designated by Greek letters. These variants seem to have arisen independently. In other words, one might expect that Delta evolved from Beta, and Omicron from Delta, but that doesn't appear to be the case. Evolution favors the spread of a virus rather than its ability to cause disease, and it is unlikely that future strains of SARS-Cov-2 will be more pathogenic.

↑ A public transit train in Bangkok, Thailand, in March 2020, showing people wearing medical face masks to protect themselves from COVID-19.

Infection and disease cycle of SARS CoV-2

The virus enters the cell when the spike protein binds to the ACE2 receptor on the cell surface. The viral envelope and the host cell fuse, and the viral RNA is released into the cell. The viral proteins are rapidly made from its RNA genome, and work to suppress the translation of host RNAs into proteins. The virus remodels the cell's membrane network and more viral proteins are made. The virus is replicated in association with the cell membranes. The viral spike proteins are passed through the Golgi apparatus, an organelle that is important in moving proteins to the cell's plasma membrane. The virus leaves the cell through the Golgi or through the lysosomes that normally provide an exit for cell waste.

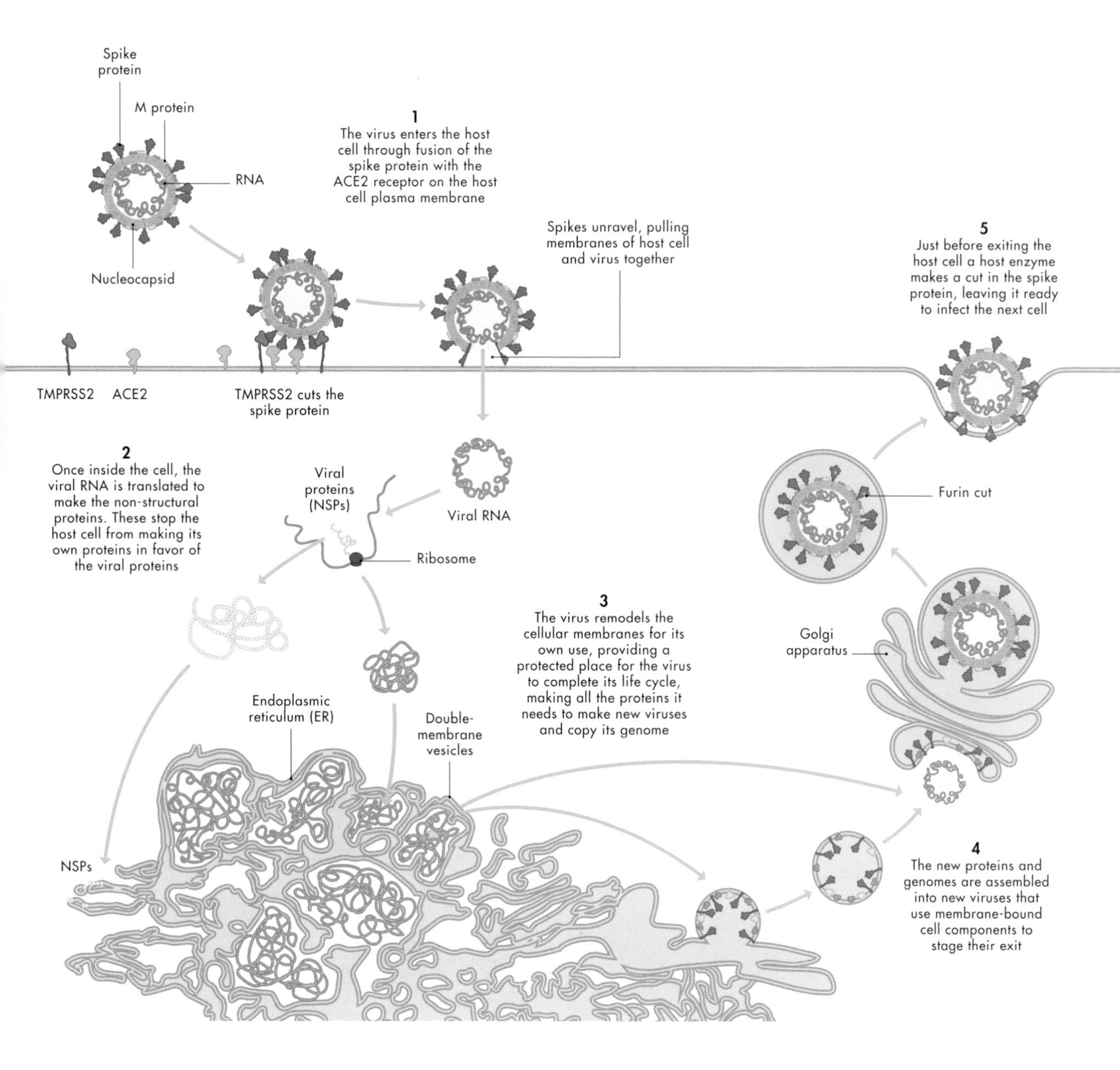

Citrus tristeza virus

There have been many plant virus pandemics over the past century, but these receive much less press coverage than human pandemics. This isn't surprising, but plant viruses are important for two reasons: directly or indirectly, human food is almost entirely derived from plants; and plant viruses can provide experimental systems and pandemic modeling that isn't possible with human or other animal viruses.

↙ A citrus tree dying from citrus tristeza virus infection.

↓ Meyer lemons are usually infected with citrus tristeza virus, but do not show any symptoms. This tolerance was part of the reason that the citrus disease spread through other citrus in the Americas.

↘ Citrus leaves infested with the Brown Citrus Aphid (*Toxoptera citricida*), which is a very efficient vector for the citrus tristeza virus.

Plant diseases have changed the course of human history in a number of examples, including ending the production of rice in some parts of the world and potatoes in others. In the Americas, the demise of citrus due to citrus tristeza virus began in the early twentieth century.

The *Citrus* genus originated in Asia, and most varieties grown there are resistant or tolerant of citrus tristeza virus. When citrus cultivation spread first to the Mediterranean region and then around the globe, propagation was done from seed. The virus is not transmitted via seeds, so most of the world was virus-free. However, when a root rot disease became a worldwide problem in citrus plants in the nineteenth century, a resistant Sour Orange (*Citrus* × *aurantium*) rootstock was propagated and used around much of the world. Most citrus plants outside of Asia were now being grown on the progeny of a single rootstock.

In the 1930s a serious disease was first noticed in South America, named tristeza (meaning "sadness") for the misery it caused. Millions of trees died and many more were rendered useless. The disease was identified as viral in the mid-1940s, and the Sour Orange rootstock proved to be especially susceptible. The virus spread throughout the world and is now estimated to have resulted in the demise of 100 million trees.

Where did the citrus tristeza virus come from? The "Meyer" lemon (*Citrus* × *meyeri*) was brought to California from China in 1908, and then introduced to Florida and Texas in the 1920s. Once scientists had developed the tools to detect the virus, they tested "Meyer" lemon trees, finding all to be infected but without symptoms. It seems most likely that the "Meyer" lemon was the carrier of the deadly disease and aided its spread to pandemic status.

Citrus tristeza virus is transmitted by aphids. Researchers have tested transmission in many different aphid species experimentally, finding that the Brown Citrus Aphid (*Toxoptera citricida*), which was introduced to South America in the early twentieth century, is the most efficient and likely contributed to the widespread infections there. The aphid moved northward through Central America in the early 1990s and arrived in Florida in 1995, where it soon became established. This invasive species led to a new surge of citrus tristeza. The aphid isn't yet found in California, and the state is working hard to keep it that way, prohibiting the import of citrus plants from outside the state.

Citrus tristeza virus is still a serious problem in citrus plants in many parts of the world. Diseases in trees are especially difficult to deal with because of the longevity and long generation times of trees. In contrast, diseases of annual crops can often be avoided by planting different crops in alternate years, or using different cultivars that are resistant to the virus.

↓ Symptoms of citrus tristeza virus on lime leaves.

THE POTENTIAL FOR FUTURE PANDEMICS

Most viral pandemics have occurred in humans and domestic plants and animals. They are fueled by monoculture, where large numbers of a single host species live together, often in crowded conditions. In agriculture this usually includes a single cultivar or breed, so there is very little genetic variation in the host.

There will certainly be more pandemics in the future, in humans and in their crops or domestic animals. Human movement around the globe is an important factor—not only are human diseases translocated in the process, but also domestic plants and animals and the vectors that carry diseases. Climate change is another factor that will very likely increase pandemics. It will lead to more human migration and different host ranges for many insect vectors of disease.

The learning curve for scientists is steep as they try to understand how the COVID-19 pandemic started and spread. Hopefully new tools will be developed from this catastrophic event that will allow similar events in the future to be predicted and prevented.

→ Symptoms of various plant virus infections. Upper left, cassava mosaic virus on cassava; upper right, maize streak virus on corn leaves; lower left, plum pox virus symptoms on an infected peach fruit; and lower right, wheat streak mosaic virus on wheat leaves.

CPPV-1

Carnivore protoparvovirus 1

Species-jumping virus of cats and dogs

GROUP	II
FAMILY	Parvoviridae
GENUS	Protoparvovirus
GENOME	Linear, single-component, single-stranded DNA comprising about 4,600 nucleotides, encoding four proteins
VIRUS PARTICLE	Non-enveloped, icosahedral
HOSTS	Wild and domestic cats, other carnivores
ASSOCIATED DISEASES	Gastrointestinal, neurological, and immune diseases
TRANSMISSION	Contact, respiratory, fecal–oral
VACCINE	Killed virus, live attenuated virus

Carnivore protoparvovirus 1 (CPPV-1) is known by several other names, most commonly feline panleukopenia virus or feline parvovirus. The disease has been recognized since the 1920s and is a very serious infection in cats and especially kittens, where it is often fatal.

Symptoms of CPPV-1 infection include lethargy, followed by diarrhea, fever, and vomiting. Kittens often do not survive without veterinary intervention. Infected animals shed huge amounts of virus, which is extremely stable, remaining on surfaces for as long as a year. Feral cats are thought to get infected in the first year of life, and if they survive they have a robust lifelong immunity.

Fortunately, there is an excellent vaccine against CPPV-1, and most domestic cats are vaccinated. Kittens have maternal antibodies until they are weaned, but there is a waiting period after weaning and before the vaccination can be given, because any remaining maternal antibodies would destroy the vaccine. Hence, there is a small window in which kittens are very vulnerable to infection.

CPPV-1 is closely related to CPPV-2, also known as canine parvovirus. CPPV-2 appeared in dogs in the 1970s, almost certainly by jumping across from cats. It causes a very similar disease in dogs, and likewise the vaccination can't be administered before maternal antibodies are cleared. It is possible for cats to be infected with CPPV-2, with a similar course of disease.

Related viruses are found in many wild species, including mink, Raccoons (*Procyon lotor*), foxes, and Wolves (*Canis lupus*).

→ Structure of carnivore protoparvovirus 1, from cryogenic electron microscopy studies, showing the capsid structure at high resolution.

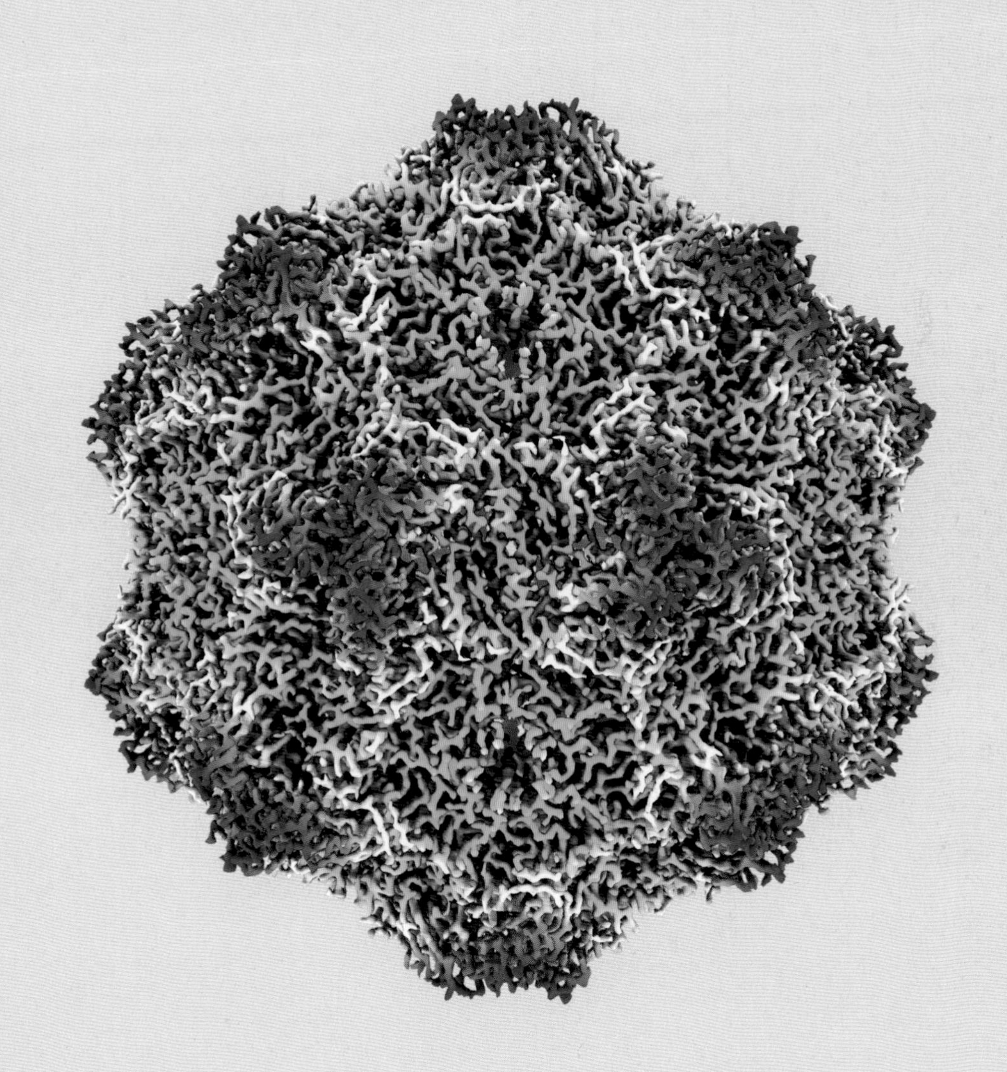

SIV

Simian immunodeficiency virus

Benign virus in its natural host, but a lethal pathogen after jumping species

GROUP	VI
FAMILY	Retroviridae
GENUS	Lentivirus
GENOME	Single-component, single-stranded RNA comprising 9,600 nucleotides, encoding seven proteins
VIRUS PARTICLE	Enveloped, spherical core containing two copies of the genome
HOSTS	Numerous primate species
ASSOCIATED DISEASES	Usually asymptomatic; causes immunodeficiency in macaques, Chimpanzees (*Pan troglodytes*), and gorillas
TRANSMISSION	Vertical, horizontal by intimate contact
VACCINE	None available

Simian immunodeficiency virus (SIV) is a typical retrovirus, converting its RNA genome into DNA and then integrating into the genome of the cell it infects. The genome remains in the DNA of the infected cell for as long as it survives, and is passed on to its daughter cells.

SIV is a common virus in wild African primates and has been infecting its hosts for millennia—it is found in primates on Bioko island, which has been separated from the African mainland for about 11,000 years. It has been studied most intensively in African Green Monkeys (*Chlorocebus sabaeus*) and Sooty Mangabeys (*Cercocebus atys*), where infection is high but there is little evidence of any disease. SIV can also infect Chimpanzees (*Pan troglodytes*), and in some cases causes an immunosuppressive disease similar to acquired immunodeficiency disease (AIDS) in humans. In captive primates it is found in Rhesus Macaques (*Macaca mulatta*), where it also causes disease. It is thought that the macaques acquired the virus from Sooty Mangabeys that were housed in the same facility.

SIV is the progenitor of human immunodeficiency virus (HIV). The evolution of HIV from SIV is a classic example of a virus jumping species and becoming a serious pathogen in the new species. By comparing the genomes of HIV strains and SIV strains, scientists can tell that the virus has jumped into humans more than once. HIV-1, the most common strain worldwide, is most closely related to a Chimpanzee strain, while HIV-2 came from Sooty Mangabeys.

→ Computer-generated cut-away structure of simian immunodeficiency virus based on cryo-EM data.

ACMV

African cassava mosaic virus

Devastating virus for an important food crop

GROUP	II
FAMILY	Geminiviridae
GENUS	Begomovirus
GENOME	Circular, two-component, single-stranded DNA comprising about 5,200 nucleotides, encoding eight proteins
VIRUS PARTICLE	Non-enveloped, twin-icosahedral particles
HOSTS	Cassava (*Manihot esculenta*)
ASSOCIATED DISEASES	Cassava mosaic disease, cassava chlorosis
TRANSMISSION	Silverleaf Whitefly (*Bemesia tabaci*)

African cassava mosaic virus (ACMV) is one of about 10 related viruses that cause severe disease in Cassava (*Manihot esculenta*), a staple crop in much of the tropical world. The disease was first reported in Africa at the end of the nineteenth century, but its viral nature wasn't known until the 1930s.

In recent years ACMV and related viruses have also been found in Asia. The viruses are transmitted by the Silverleaf Whitefly (*Bemesia tabaci*), which has spread around the world, leaving the potential for more widespread disease in other places where Cassava is grown.

ACMV is a geminivirus, so named because the particle looks like twin icosahedrons. Geminiviruses are a major plant health threat, infecting many different crops, including beans, tomatoes, beets, corn, turnips, and spinach, and many ornamental plants. They often induce brightly colored mosaic patterns on the leaves, some of which are considered desirable traits. Geminiviruses are all transmitted by insects and in some cases provide benefits to their insect vectors.

While Cassava is native to South America and the crop is still widely grown there, ACMV has not been seen in the Americas. The virus presumably originated in Africa in some other unidentified plant, and spread to Cassava after the species was introduced to Africa in the sixteenth century. Cassava has become an important source of carbohydrates in much of the developing world, owing to its ability to adapt well to poor soils and its tolerance of drought conditions. In a processed form it is known as tapioca, which is used to make puddings.

→ Model of the twin particle structure of African cassava mosaic virus. These twin particles gave rise to the family name Geminiviridae.

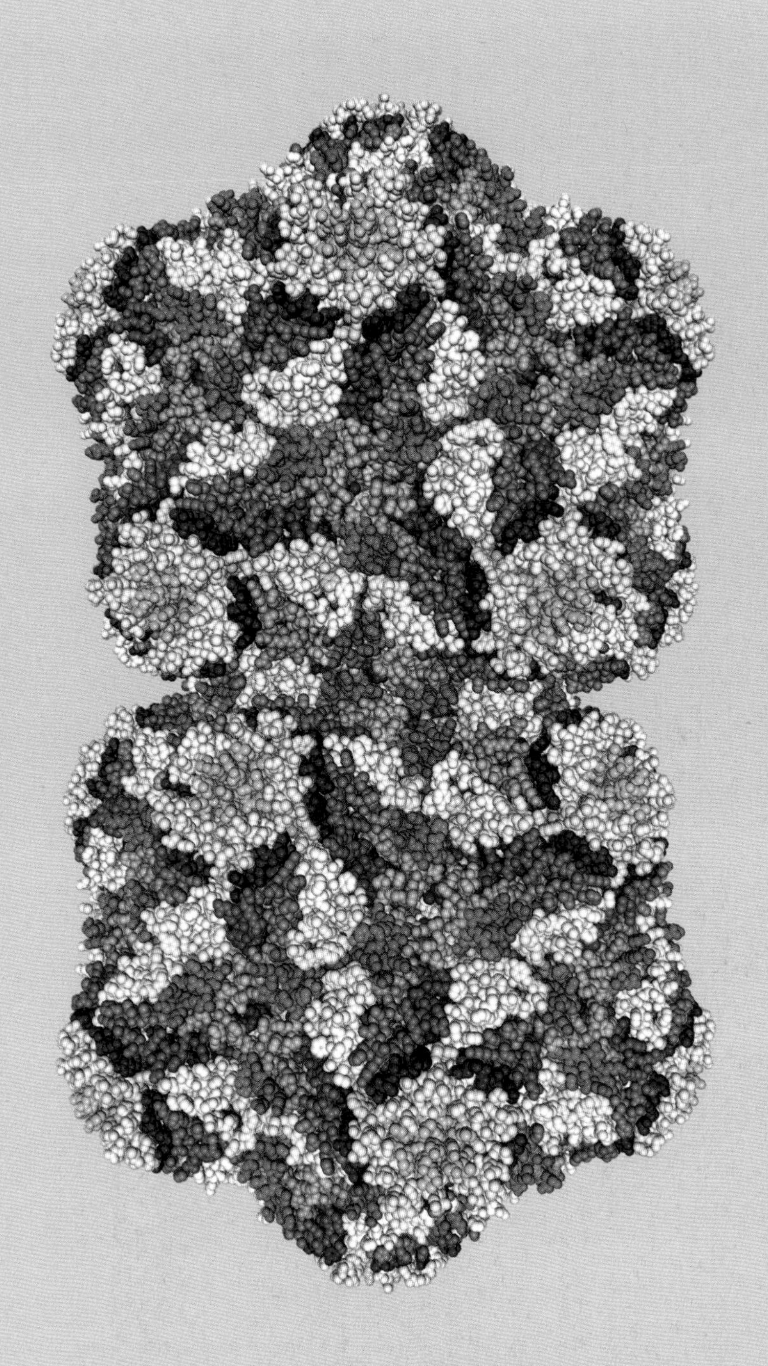

BBTV

Banana bunchy top virus

Major threat to banana production

GROUP	II
FAMILY	Nanoviridae
GENUS	Babuvirus
GENOME	Circular, six-component, single-stranded DNA comprising about 7,000 nucleotides, encoding six proteins
VIRUS PARTICLE	Non-enveloped, small icosahedral particles
HOSTS	Bananas and plantains (*Musa* species)
ASSOCIATED DISEASES	Bunchy top disease
TRANSMISSION	Banana Aphid (*Pentalonia nigronervosa*), vertically through explants

Banana bunchy top virus (BBTV) is a threat to banana and plantain production in much of the world, with the exception of continental America, although it is found in Hawaii. The vector for the virus, the Banana Aphid (*Pentalonia nigronervosa*), is not found in the virus-free areas of the world.

Because banana plants are propagated through suckers that arise from a mother plant rather than by seeds, BBTV is very hard to eradicate. Inherited in this way, the disease is usually severe and renders the fruits twisted and stunted, and these plants are not used for propagation. The virus is also transmitted horizontally by the aphid vector, but plants infected this way have milder symptoms that may go unnoticed, continuing the cycle of infection between vertical and horizontal transmission. Clean stocks of mother plants are important for continued successful propagation of bananas.

BBTV and other nanoviruses have very small particles and package each segment of their genomes separately. A heavily divided genome has some advantages for a plant virus, because the small size of the individual components makes it easier for them to move between plant cells. However, this is also a disadvantage because to fully infect a single cell all the components have to be present at once. Other nanoviruses seem to have overcome this by exchanging the protein products of their individual components among cells in a plant, so that one cell may produce one subset of the proteins while another produces a different subset. However, it isn't clear whether BBTV also uses this strategy. The virus is limited to the phloem of the plant, tissue with tube-like structures that transport nutrients. The cells of the phloem are highly interconnected, so the virus's strategy is likely highly beneficial in aiding its spread.

→ Banana plants infected with banana bunchy top virus, showing stunting and with all the leaves emerging from one point, forming a "bunchy top."

ASFV

African swine fever virus

Animal virus pandemic that may change traditional diets

GROUP	I
FAMILY	Asfariviridae
GENUS	Asfivirus
GENOME	Linear, double-stranded DNA genome comprising about 170,000 nucleotides, encoding about 160 proteins
VIRUS PARTICLE	Double envelope with an icosahedral core
HOSTS	Domestic Pig (*Sus domesticus*), wild boar, warthogs, Bushpig (*Potamochoerus larvatus*), and ticks (*Orthinodorus* species)
ASSOCIATED DISEASES	Hemorrhagic fever
TRANSMISSION	Vector borne (ticks), direct contact
VACCINE	None available

African swine fever is a very severe disease of pigs that has spread to many parts of the world in recent decades. Its horizontal transmission through animal products used as feed has increased the problems with this disease.

African swine fever is endemic in sub-Saharan Africa, with a cycle between domestic and wild pigs, but in 2007 African swine fever virus (ASFV) was accidentally introduced to the country of Georgia. From there it spread across the Caucasus and to Russia, from where it was introduced to Europe in 2014. In 2018 the virus spread to China and other parts of Asia. It is continuing to spread, and it will likely reach other parts of the globe as surveillance to contain it has been impacted by the human COVID-19 pandemic.

There are a number of different strains of ASFV, some of which are more lethal than others. The most severe strains have a 100 percent fatality rate, while the mildest strains induce only minimal symptoms. The virus is transmitted from wild pigs to domestic pigs via a tick vector, but once in domestic populations it spreads rapidly by direct pig-to-pig contact.

The impact of ASFV in China has been devastating, where pork forms a large component of the diet and is especially important during holidays and celebrations. However, the impact isn't just on the human diet. The supply of heparin, an important pharmaceutical used in the treatment of clotting disorders, is also impacted because most of the drug is produced in pigs. There have been efforts to create a vaccine against the virus, but so far these have not been successful.

→ High-resolution structural model of African swine fever virus from cryo-EM data.

PICTURE CREDITS

Illustrations by: Martin Brown 145, 149, 150; Lindsey Johns 13, 75 (bottom); Caitlin Monney (Monney Medical Media) 16, 65, 66, 70 (top), 77, 79, 80, 82, 83, 85 (top and bottom), 86, 89, 138, 163, 167, 171, 173, 198, 203, 223, 224, 256, 263; Tejeswini Padma 10, 11, 14, 31, 33, 34–35, 38–39, 70 (bottom), 72, 75 (top), 109 (top and bottom), 110, 112, 115, 117, 142; John Woodcock 58, 60, 132, 144, 175, 197, 208, 229, 233, 250, 258, 260 (top and bottom).

The publisher would like to thank the following for permission to reproduce copyright material:

Adobe Stock: molekuul.be: 25 • Alamy Photo Library Pictorial Press Ltd: 22T; Science History Images: 22B; Photo12, Ann Ronan Picture Library: 23; Larry Downing, Reuters: 28; dpa picture alliance: 29T; Ivan Kuzmin: 29B, 225; Nic Hamilton Photographic: 45R; All Canada Photos: 46; Nigel Cattlin: 61, 87, 103, 119J, 267TR & BR; Juan Gaertner Science Photo Library: 67; Rosanne Tackaberry: 69T; Scott Camazine: 71T; Cavallini James BSIP: 71B; inga spence: 76; Antonio Guillem: 78T; Kateryna Kon Science Photo Library: 93, 236; Science Picture Co: 97; Steve Gschmeissner Science Photo Library: 107T; Maxim Cristalov: 114R; Vintage_Space: 116; FineArt: 137; Holmes Garden Photos: 146; The Granger Collection: 164; Science History Images: 165; IanDagnall Computing: 176L, 253L; RBM Vintage Images: 177R; Jagadeesh N.V Reuters: 207B; History and Art Collection: 209; Nanoclustering Science Photo Library: 219; Niday Picture Library: 226; North Wind Picture Archives: 232; Granger – Historical Picture Archive: 235B; Shawshots: 254L; World of Triss: 254R; J Marshall – Tribaleye Images: 261; Biosphoto: 264L; Tim Gainey: 264R • Stéphane Blanc: 150 • Centres for Disease Control and Prevention, James Gathany: 113 • Andrew Charnesky, Hafenstein Lab, The Pennsylvania State University: 211 • Churchill Archives Centre, The Rosalind Franklin Papers, FRKN 2/31: 22BL • CNRS © AMU/IGS/CNRS Photothèque: 53 • Delft School of Microbiology Archives, Courtesy of the Curator: 20 • Dreamstime Kanokphoto: 196; Nflane: 227 • John Finch, MRC Laboratory of Molecular Biology: 19 • Flickr: Harry Rose: 21; Oregon State University: 40L; International Institute of Tropical Agriculture, Nigeria: 55; Chattahoochee Oconee National Forest: 235T; James St. John: 241; H.Holmes, RTB – The CGIAR Research Program on Roots, Tubers and Bananas: 267TL; U.S. Department of Agriculture, European and Mediterranean Plant Protection Organization Archive, France 267BL; Scot Nelson: 275 • iNaturalist: James Bailey: 205; Gilles San Martin: 230 • Invasive. Org: Rupert Anand Yumlembam, Central Agricultural University, Imphal, Manipur, India, Bugwood.org 42 • iStock: Tomasz Klejdysz: 202; Gerald Corsi: 222 • Journal of Biological Chemistry Open Access, Fig. 2 in 'Andrés et al. The cryo-EM structure of African swinefever virus unravels a unique architecture comprising two icosahedral protein capsids and two lipoprotein membranes. Volume 295, Issue 1, P1–12, (2020) https://doi.org/10.1074/jbc. AC119.011196': 277 • Russell C. J. Kightley: 69B • Heui-Soo Kim: 57 • Caroline Langley, Hafenstein Lab, The Pennsylvania State University: 5, 125 • Hyunwook Lee, Hafenstein Lab, The Pennsylvania State University: 215, 239, 269 • Library of Congress, National Photo Company Collection: 255 • Pedro Moreno: 168 • National Cancer Institute: 271 • National Institute of Allergy and Infectious Diseases, Courtesy of: 47 • National Plant Protection Organization, the Netherlands, Annelien Roenhorst: 49 • Nature Communications Open Access, Fig. 4 in 'Hesketh, E.L., Saunders, K., Fisher, C. et al. The 3.3 Å structure of a plant geminivirus using cryo-EM. Nat Commun 9, 2369 (2018). https://doi. org/10.1038/s41467-018-04793-6: 213 • PDB-101 (PDB101.rcsb.org), RCSB PDB, David S. Goodsell 157 • David Price-Goodfellow: 257 • RCSB PDB created using Mol* (D. Sehnal, S. Bittrich, M. Deshpande, R. Svobodová, K. Berka, V. Bazgier, S. Velankar, S.K. Burley, J. Koča, A.S. Rose (2021) Mol* Viewer: modern web app for 3D visualization and analysis of large biomolecular structures. Nucleic Acids Research. doi: 10.1093/nar/gkab314), and RCSB PDB, Image 2X8Q Image 2X8Q Hyun, J.K., Radjainia, M., Kingston, R.L., Mitra, A.K. (2010) *J Biol Chem* 285: 15056 Proton-Driven Assembly of the Rous Sarcoma Virus Capsid Protein Results in the Formation of Icosahedral Particles: 4T, 101; Image 2CH8 Tarbouriech, N., Ruggiero, F., Deturenne-Tessier, M., Ooka, T., Burmeister, W.P. (2006) *J Mol Biol* 359: 667 Structure of the Epstein-Barr Virus Oncogene Barf1: 5BL, 88; Image 6JHQ Cao, L., Liu, P., Yang, P., Gao, Q., Li, H., Sun, Y., Zhu, L., Lin, J., Su, D., Rao, Z., Wang, X. (2019) *PLoS Biol* 17: e3000229- e3000229 Structural basis for neutralization of hepatitis A virus informs a rational design of highly potent inhibitors: 5BC, 123; Image 5IRE Sirohi, D., Chen, Z., Sun, L., Klose, T., Pierson, T.C., Rossmann, M.G., Kuhn, R.J. (2016) *Science* 352: 467–70 The 3.8 angstrom resolution cryo-EM structure of Zika virus: 9; Image 3J9X DiMaio, F., Yu, X., Rensen, E., Krupovic, M., Prangishvili, D., Egelman, E.H. (2015) *Science* 348: 914–17 A virus that infects a hyperthermophile encapsidates A-form DNA: 34–35; Image 6P7B Li, N., Shi, K., Rao, T., Banerjee, S., Aihara, H. (2020) *Sci Rep* 10: 393 Structural insights into the promiscuous DNA binding and broad substrate selectivity of fowlpox virus resolvase: 91; Image 4V99 Makino, D.L., Larson, S.B., McPherson, A.(2013) *J Struct Biol* 181: 37–52 The crystallographic structure of Panicum Mosaic Virus (PMV): 127; Image 2H3R Benach, J., Chen, Y., Seetharaman, J., Janjua, H., Xiao, R., Cunningham, K., Ma, L.-C., Ho, C.K., Acton, T.B., Montelione, G.T., Hunt, J.F., Tong, L., Northeast Structural Genomics Consortium (NESG) Crystal structure of ORF52 from Murid herpesvirus 4 (MuHV-4) (Murine gammaherpesvirus 68). Northeast Structural Genomics Consortium target MhR28B: 243T; Image 4BML Gipson, P., Baker, M.L., Raytcheva, D., Haase-Pettingell, C., Piret, J., King, J.A., Chiu, W. (2014) *Nat Commun* 5: 4278 Protruding Knob-Like Proteins Violate Local Symmetries in an Icosahedral Marine Virus Protruding Knob-Like Proteins Violate Local Symmetries in an Icosahedral Marine Virus: 217 • RCSB PDB created using NGL (A.S. Rose, A.R. Bradley, Y. Valasatava, J.D. Duarte, A. Prlić, P.W. Rose (2018) NGL viewer: web-based molecular graphics for large complexes. *Bioinformatics* 34: 3755–58) Image 4G7X Ford, C.G., Kolappan, S., Phan, H.T., Waldor, M.K., Winther-Larsen, H.C., Craig, L.. (2012) Crystal Structures of a CTX{varphi} pIII Domain Unbound and in Complex with a Vibrio cholerae TolA Domain Reveal Novel Interaction Interfaces: 4B, 5T, 245; Image 7DWT Fibriansah, G., Lim, E.X.Y., Marzinek, J.K., Ng, T.S., Tan, J.L., Huber, R.G., Lim, X.N., Chew, V.S.Y., Kostyuchenko, V.A., Shi, J., Anand, G.S., Bond, P.J., Crowe Jr., J.E., Lok, S.M. (2021) *PLoS Pathog* 17: e1009331- e1009331. Antibody affinity versus dengue morphology influences neutralization: 5BR, 191; Image 7XDI Han, Z., Yuan, W., Xiao, H., Wang, L., Zhang, J., Peng, Y., Cheng, L., Liu, H., Huang, L. (2022) *Proc Natl Acad Sci USA* 119: e2119439119–e2119439119 Structural insights into a spindle-shaped archaeal virus with a sevenfold symmetrical tail: 34TL; Image 3J31 Veesler, D., Ng, T.S., Sendamarai, A.K., Eilers, B.J., Lawrence, C.M., Lok, S.M., Young, M.J., Johnson, J.E., Fu, C.Y. (2013) *Proc Natl Acad Sci USA* 110: 5504-5509 Atomic structure of the 75 MDa extremophile Sulfolobus turreted icosahedral virus determined by CryoEM and X-ray crystallography: 34TR; Image 6CGR Dai, X.H., Zhou, Z.H. (2018) Structure of the herpes simplex virus 1 capsid with associated tegument protein complexes *Science* 360: 155; Image 7LGE Chang, J.Y., Gorzelnik, K.V., Thongchol, J., Zhang, J.(2022) *Viruses* 14 Structural Assembly of Q beta Virion and Its Diverse Forms of Virus-like Particles: 159; Image 6HXX Kezar, A., Kavcic, L., Polak, M., Novacek, J., Gutierrez-Aguirre, I., Znidaric, M.T., Coll, A., Stare, K., Gruden, K., Ravnikar, M., Pahovnik, D., Zagar, E., Merzel, F., Anderluh, G., Podobnik, M.(2019) *Sci Adv* 5: eaaw3808– eaaw3808 Structural basis for the multitasking nature of the potato virus Y coat protein: 189; Image 7NXR Naniima, P., Naimo, E., Koch, S., Curth, U., Alkharsah, K.R., Stroh, L.J., Binz, A., Beneke, J.M., Vollmer, B., Boning, H., Borst, E.M., Desai, P., Bohne, J., Messerle, M., Bauerfeind, R., Legrand, P., Sodeik, B., Schulz, T.F., Krey, T. (2021) *PLoS Biol* 19: e3001423–e3001423 Assembly of infectious Kaposi's sarcoma-associated herpesvirus progeny requires formation of a pORF19 pentamer: 243B; Image 1M1C Naitow, H., Tang, J., Canady, M., Wickner, R.B., Johnson, J.E. (2002) *Nat Struct Biol* 9: 725–28 L-A virus at 3.4 A resolution reveals particle architecture and mRNA decapping mechanism: 247; Image 6EK5 Hipp, K., Grimm, C., Jeske, H., Bottcher, B. (2017) *Structure* 25: 1303–09.e3 Near-Atomic Resolution Structure of a Plant Geminivirus Determined by Electron Cryomicroscopy; 273; Image 3D0H (2008) Li,F. *J Virol* 82: 6984–91 Structural analysis of major species barriers between humans and palm civets for severe acute respiratory syndrome coronavirus infections: 259 • RCSB PDB, Jmol: an open-source Java viewer for chemical structures in 3D. http://www.jmol.org/: 51 • Rusty Rodriguez: 228 • Science Photo Library Laguna Design: 3, 59; National Library of Medicine: 24; Henning Dalhoff: 64; Science Source: 99; Tim Vernon: 139; Dr. Klaus Boller: 153, 185; Roger Harris: 187; Dr. Victor Padilla-Sanchez Phd, Washington Metropolitan University: 193; AMI Images: 120–21; Ramon Andrade 3DCIENCIA: 131; PR J.L. Kemeny, ISM: 180 • Jean-Yves Sgro, Protein Data Bank: 1DNV; Rasmol image by Dr Sgro,(UW-Madison, Dept of Biochemistry): 44; Protein Data Bank: 5K0U; UCSF Chimera image by Dr Sgro, (UW-Madison, Dept of Biochemistry): 129 • Shutterstock: Juan Gaertner: 2–3; Kateryna Kon: 4C, 37 (all), 73, 88T, 169, 170, 172T&B, 181; Sashkin: 10L; Bussakan Punlerdmatee: 15; Catherine Avilez: 17; Lifestyle Graphic: 32; Martin Prochazkacz: 40R; walkerone: 43; Ihor Hvozdetskyi: 44–45T; Jezper: 48; DodoDripp: 74; podsy: 78L; Kostiantyn Kravchenko: 81; homi: 95; schankz: 107L; Choksawatdikorn: 108; Tatiana Shepeleva: 111; LightField Studios: 114L; JennLShoots: 118; Evgeniyqw: 119A; Thammanoon Khamchalee: 119B; Mi St: 119C; Jamierpc: 119D; Vera Larina: 119E; Tomasz Klejdysz: 119F, 119L, 231; Wut_Moppie: 119G; EVGEIIA: 119H; F.Neidl: 119I; chinahbzyg: 119K; frank60: 121TR; Suti Stock Photo: 133; Sandra Mori: 140–41; FJAH: 148T; Creativa Images: 148BL; Laborant: 148BR; Jose Luis Calvo: 166; Everett Collection: 174, 207T; Yekatseryna Netuk: 176–77; Igor Petrushenko: 183T; Showtime.photo: 183B; AJCespedes: 199; Rejdan: 200; massimofusaro: 201; Lam Van Linh: 237L; Grandpa: 237R; The Escape of Malee: 262; LifeCollectionPhotography: 265; Theeraya Nanta: 266 • Mark J.A. Vermeij: 41 • Wikimedia Commons: EEIM: 121BR; Spencerbdavis 178; NASA/USGS image courtesy of Steve Groom: 204B; Stefan Ertmann & Lokal Profil: 253T • Willie Wilson, Marine Biological Association, Plymouth: 204T • Professor Ju-Yeon Yoon: 143 • Heiko Ziebell: 179.

All reasonable efforts have been made to trace copyright holders and to obtain their permission for the use of copyright material. The publisher apologizes for any errors or omissions in the list above and will gratefully incorporate any corrections in future reprints if notified.